Springer Advanced Texts in Life Sciences

David E. Reichle, Editor

Springer Advanced Texts in Life Sciences
Series Editor: David E. Reichle

Environmental Instrumentation
By Leo J. Fritschen and Lloyd W. Gay

Biophysical Ecology
By David M. Gates

Forest Succession: Concepts and Application
Edited by Darrel C. West, Herman H. Shugart, and Daniel B. Botkin

Marine Ecological Processes
By Ivan Valiela

Analysis of Biogeochemical Cycling Processes in Walker Branch Watershed
Edited by Dale W. Johnson and Robert I. Van Hook

Ecotoxicology: Problems and Approaches
Edited by Simon A. Levin, Mark A. Harwell, John R. Kelly, and Kenneth D. Kimball

Ecological Microcosms
By Robert J. Beyers and Howard T. Odum

Robert J. Beyers   Howard T. Odum

Ecological Microcosms

With 292 Illustrations

Springer-Verlag
New York   Berlin   Heidelberg   London   Paris
Tokyo   Hong Kong   Barcelona   Budapest

Robert J. Beyers
Department of Biological Sciences
University of South Alabama
Mobile, AL 36688 USA

Howard T. Odum
Department of Environmental
Engineering Sciences
University of Florida
Gainesville, FL 32611-2013 USA

Series Editor:
David E. Reichle
Environmental Sciences Division
Oak Ridge National Laboratory
Oak Ridge, TN 37830 USA

QH
541
.15
.C55
B48
1993

Library of Congress Cataloging-in-Publication Data
Beyers, Robert J.
  Ecological microcosms / Robert J. Beyers and Howard T. Odum.
    p. cm. — (Springer advanced texts in life sciences)
  Includes bibliographical references and index.
  ISBN 0-387-97980-8 (New York). — ISBN 3-540-97980-8 (Berlin)
  1. Closed ecological systems. 2. Human ecology. I. Odum, Howard
T., 1924– . II. Title. III. Series.
QH541.15.C55B48 1993
574.5 — dc20   93-290

Printed on acid-free paper.

© 1993 Springer-Verlag New York Inc.
All rights reserved. This work may not be translated or copied in whole or in part without the written permission of the publisher (Springer-Verlag New York, Inc., 175 Fifth Avenue, New York, NY 10010, USA), except for brief excerpts in connection with reviews or scholarly analysis. Use in connection with any form of information storage and retrieval, electronic adaptation, computer software, or by similar or dissimilar methodology now known or hereafter developed is forbidden.
The use of general descriptive names, trade names, trademarks, etc., in this publication, even if the former are not especially identified, is not to be taken as a sign that such names, as understood by the Trade Marks and Merchandise Marks Act, may accordingly be used freely by anyone.

Production managed by Ellen Seham, manufacturing supervised by Jacqui Ashri.
Typeset by Best-set Typesetter Ltd., Hong Kong.
Printed and bound by Edwards Brothers, Ann Arbor, MI.
Printed in the United States of America.

9 8 7 6 5 4 3 2 1

ISBN 0-387-97980-8 Springer-Verlag New York Berlin Heidelberg
ISBN 3-540-97980-8 Springer-Verlag Berlin Heidelberg New York

*To our long supporting wives, Barbara J. Beyers and
Elisabeth C. Odum.*

Preface

The experimental study of ecological systems in containers has been important to the development of theoretical ecology and to the very pragmatic purposes of finding the way ecosystems are affected and adapt to such stresses as toxic wastes. Partially isolated ecological systems are often called ECOLOGICAL MICROCOSMS. Even when experiments were done for other purposes, the pervasive ubiquity of ecological self organization was found wherever enclosures have been studied. This is a review of the state of knowledge of ecological microcosms, their self organization, interdependencies, chemical cycles, and controls.

The book reviews types of microcosms and methodologies. It considers ecosystem theory, many aspects of which have been developed through the study of microcosms. Included are methods for testing ecosystem responses to environmental treatments and impacts.

Many important laboratory studies have been conducted on populations of one or two species in closed containers including continuous cultures and chemostats. Although these are ecological microcosms of great interest and importance, they are not emphasized in this book which focuses on multispecies microcosms.

The variety of intricate, small, experimental worlds constructed by various investigators rivals that of nature developed without human hands. In each study there may be original containers and controls, development of interesting ecosystem types, and perhaps some principle demonstrated as a byproduct of the original study purpose.

Ecological microcosms played a major role in the careers of the authors, showing us, in our student days, how highly organized ecosystems are, the way populations are controlled, the universality of common patterns, the importance of the maximum EMPOWER concept, the way chemical cycles are tuned for maximum metabolism, the way life controls media, the priority to be given to holistic concepts, and the power of modeling from the top down.

Both authors have used microcosms as a starting point in the teaching of ecology, limnology, marine science, and general biology. A classroom full of microcosms, one developed by each student, is a very exciting place for showing process, biogeochemistry, forcing functions, energetics, succession, competition, information functions, metabolism, systems principles, and models. This book may be useful as a text for an ecology course organized around students' microcosms.

Most importantly, we hope our readers can share our love of little systems, their mystery, their creativity, their domesticity, their immortality, and the guidance they provide for the larger realms. As living models, microcosms help bridge the details of reality with the abstractions of general systems, revealing the principles of the way all systems work.

Divisions of the Book

Part 1 contains principles, relationships, models, and their examples. Chapters consider metabolism, self-organization and succession, chemical cycles, diversity, and stress and toxicity. Examples concern microcosms in extreme conditions, under stress, and experimental impact treatments. These include high temperature, high salinity, low oxygen, radiation, heavy metals, and organic toxins.

Part 2 considers various kinds of microcosms, presenting some of the great variety of data on each. Chapters consider aquaria, stream microcosms, terraria and soils, wetland microcosms, ponds and pools, reefs and sediments, plankton columns, and thermal, anaerobic, and brine microcosms.

Part 3 contains microcosms in human interface and use, for food, for wastes processing, and for possible support in space. As an important technique of ecological engineering, possible roles for microcosms are suggested for the future. Finally, appendices contain instructions for setting up and measuring the metabolism of a microcosm and a paper by staff of Biosphere 2 on gases in the "Test Module" during its support of humans in 1988–89.

Other Summaries

The reader is also referred to the following reviews of microcosms, mesocosms, and other similar entities.

Cooke, in 1971, reviewed the scientific literature on laboratory systems up to that time including community metabolism, ecological succession, community regulation, and experimental investigations on the ecology of one or more small groups of species. He provided considerable evidence that microcosms are similar to natural systems.

Pilson and Nixon (1980) reviewed marine microcosms with a historical table of 21 different initiatives. They listed three purposes: to understand more complex systems and ecological theory, to study impacts and toxicities, and to study the biogeochemical behavior of substances. They noted a lack of theory for scaling, and little agreement about how closely a microcosm should represent the larger system for which it was designed to be similar. These problems raised questions concerning size, predation, and stochastic sampling.

Boyd (1981), in reviewing studies of planktonic food chains in bags and tanks, considered mainly their similarity to open waters and how useful they might be for understanding those waters. Included were pictures of the devices and discussions of turbulence, controls, surface–volume ratio, and carnivore exclusion.

Grice and Reeve (1982) edited a collection of chapters on marine mesocosms. Although Banse, in the second chapter, called for more integration with models, the book mainly considers isolated parts and transients after enclosure. Some understanding was developed of chemical, organismal, and population responses after waters were contained, but the ecosystem was considered in only two chapters.

Gearing (1989) reviewed the massive literature on the subject of ecotoxicological research in microcosms, particularly large marine systems, and microcosms' relationships to natural systems. Representative results of research were given.

Wimpenney (1988) edited a two-volume set of review papers on laboratory models for microbial ecosystems, which considered combinations of cultures, compound chemostats, gradient cultures, and systems ecology of the properties which result from combining species of known functions.

With *Dynamic Aquaria, The Construction and Operation of Microcosms, Mesocosms, and Aquaria*, Adey and Loveland (1991) summarized their successful work in culturing complex ecosystems, such as coral reefs and marshes, for public exhibition and research. As part of their review of ecosystem cultures, they considered the species which have been cultured in this way and which are most suitable for sustainable mesocosms.

Lalli (1990) edited a volume reviewing the status of enclosed experimental marine ecosystems, covering pelagic waters, hard bottoms, soft bottoms, a Baltic example, comparison of microcosms and mesocosms, uses for pollution evaluation, a discussion of statistical evaluation, and discussion of three microcosm models.

Other summary books and papers include: Warren and Davis (1971), laboratory streams; Balch et al. (1978), Scripps Deep Tank and Dalhousie Aquatron; Hood (1978), upwelling impoundments; Zeitzschel and Davies (1978), Kelly (1984), benthic chambers; Muschenheim et al. (1986), laboratory flumes; and Draggan (1976), Kinne (1978), Giesy (1978), Parsons (1978), miscellaneous microcosm papers.

Simulation Programs for Models

The simple simulation programs for the models in this book are in BASIC and are available on disk for PC and Macintosh, or as printout listings of one or two pages each. These programs can be run as is to gain insights, conduct "what if" experiments, and calibrate for particular situations, or they can be modified to include additional factors. Using simulations along with this book is one way of teaching systems ecology and simulation. Also available are elementary introductory texts and workbooks on microcomputer simulations of ecological minimodels, and a disk with 44 additional minimodels (Odum et al. 1988a, 1988b; Odum and Odum 1991).*

Robert J. Beyers and Howard T. Odum
March 1992

---

*Items available from
Center for Wetlands,
Phelps Lab, University of Florida,
Gainesville, Fl., 32611:

Microcomputer Disks
1. Programs in BASIC for "Ecological Microcosms" (PC or Macintosh)
2. Minimodel programs in BASIC for instruction books (PC or Macintosh)
3. Macintosh disk with energy symbol blocks for EXTEND

Instruction books using microcomputer disks
1. Computer Minimodels and Simulation Exercises for Science and Social Science — H.T. Odum and E.C. Odum
2. Environmental Systems and Public Policy — H.T. Odum, E.C. Odum, M.T. Brown, D. LaHart, C. Bersok, and J. Senzimir
3. Environment and Society in Florida — H.T. Odum, E.C. Odum, and M.T. Brown

Acknowledgments

We acknowledge the encouragement and support of colleagues, administrators, and students of the University of South Alabama at Mobile and the University of Florida at Gainesville. Work was aided by sabbatical periods, which allowed for collaboration. Research and workshops on microcosms were supported over the years by the National Science Foundation, the Atomic Energy Commission, and the Environmental Protection Agency. Joy Earp developed the bibliography, Barbara Lemont provided new illustrations, and Joan Breeze served as secretary and arranged permissions.

Publication was aided by funds from the Crafoord Prize of the Royal Swedish Academy of Sciences and from the Rhode Island Sea Grant Information and Education Office.

Contents

Preface   vii
Acknowledgments   xi

Part 1   Microcosms: Concepts and Uses   1

Chapter 1
Introduction to Microcosmology   3

Chapter 2
Metabolism and Homeostasis   11

Chapter 3
Succession and Self-organization   41

Chapter 4
Chemical Cycles and Limiting Factors   62

Chapter 5
Diversity and Information   88

Chapter 6
Hierarchy, Control, and Oscillation   108

Chapter 7
Stress, Toxicity, and Adaptation   148

Chapter 8
History   178

Part 2   Kinds of Microcosms and Mesocosms   189

Chapter 9
Aquaria   191

Chapter 10
Stream Microcosms   209

Chapter 11
Terraria and Soil Microcosms   234

Chapter 12
Wetland Microcosms   260

Chapter 13
Ponds and Pools   272

Chapter 14
Reefs and Benthic Microcosms   294

Chapter 15
Plankton Columns   322

Chapter 16
Thermal and Brine Microcosms   341

Part 3   Microcosms for Society   355

Chapter 17
Food Microcosms and Mesocosms   357

Chapter 18
Waste Microcosms and Mesocosms   376

Chapter 19
Human Microcosms and Space   397

Chapter 20
Microcosm Perspectives   433

Appendices 445

A. Directions for Classroom Microcosms 445

B. Directions for Monitoring Microcosms 452

C. Experiments on the Closed Ecological System in the Biosphere 2 Test Module 463
   A. Alling, M. Nelson, L. Leigh, T. MacCallum, N. Alvarez-Romo, J. Allen, and R. Frye

Bibliography 481

Index 547

Part 1
Microcosms: Concepts and Uses

In Part 1, concepts involved in understanding, developing, and using microcosms are presented with examples. These are the principles and ideas of ecology, a frontier of science to which microcosmology research contributes. In each chapter, simulation models help to clarify concepts and relate mechanisms, patterns of organization, and trends with time.

Chapter 1 defines microcosms, their variety, and some of their theoretical and practical uses. Chapter 2 covers the basic metabolic processes of ecosystems and the way enclosures change them. Chapter 3 describes processes of self organization and succession. Chapter 4 considers biogeochemical cycles in microcosms, the conservation of matter, and limiting factors. Chapter 5 explores the complexity, diversity, and information found in microcosms. Chapter 6 covers food chain hierarchies and oscillation. Chapter 7 considers stress, toxicity, and adaptation. Chapter 8 briefly reviews the early history of microcosms.

Chapter 1
Introduction to Microcosmology

Ecological microcosms are small ecosystems held in containers. Starting originally as a way to bring the beauty and complexity of nature into schoolrooms and living rooms the world over, these small "worlds" have become a major research tool. They are useful for studying the way ecosystems work and for the very practical purpose of determining what happens to toxic substances in ecosystems. Microcosms are important because they are a way to study whole, simplified ecosystems, and because they can be replicated for experimental studies at reasonable cost. This book is a review of the extensive but scattered papers and reports on microcosms, with emphasis on the concepts of systems ecology that emerge from the hypotheses and experimental tests.

It is sometimes difficult to define exactly what a microcosm is, since microcosms have so many different shapes, sizes, and compositions. Wimpenny (1988) states that there is a consensus among scientists that a microcosm possesses some or all of the following properties:

1. Origin: Microcosms derive from natural ecosystems.
2. Isolation: The microcosm, whatever its origin, is physically enclosed and no longer in contact with natural ecosystems.
3. Size: Though variable in size, microcosms tend to be compact subsets of the natural systems from which they came.
4. Genotypic heterogeneity: With some exceptions, most microcosm research makes use of natural, mixed populations of microorganisms.
5. Spatial heterogeneity: Although spatial heterogeneity is not in any way implied by the term microcosm, virtually all retain this property to some degree.

6. Temporal heterogeneity: Most microcosms are closed or partially closed to exchanges of species and materials. Time dependent changes in the physical, chemical, and biological properties of the system are seen.

Ecosystem Boundaries

In approaching the subject of ecological microcosms, we were faced with the old ecological problem of boundary definitions. In nature, where does one ecological system begin and the adjacent ecosystem end? Almost all ecosystems are connected to adjacent ones, and all are influenced by large scale phenomena. By adapting to various exchanges with its surroundings, the ecosystem of one area gains more production and diversity than if it were isolated. Thus, isolated ecosystems tend to be displaced by those that incorporate useful exchanges. In other words, all ecosystems are part of the great interconnected ecosystem of the biosphere as a whole. For experimental studies, however, ecosystems with free exchanges are hard to study and control. Starting in the mid 1950s, several ecologists attempted to simplify ecosystem studies by creating artificial boundaries around "pieces of nature" in the laboratory. These were usually contained in flasks, carboys, or aquaria, and were called by several names: balanced aquaria, artificial ecosystems, microecosystems, etc. However, these laboratory ecosystems were most commonly called microcosms.

In the early years, laboratory constructs were small and usually held under rather rigidly controlled conditions of light and temperature, while simply being observed or undergoing various experimental manipulations. As the use of the microcosm technique expanded, the size of the systems became both larger and smaller, special containers were fabricated, and the conditions under which they were held became almost infinitely variable. Typical microcosms are shown in Figure 1.1. There are aquaria, terraria, circulating streams, long tubes to represent deep waters, and microcosms large enough to include people. Ecological microcosms can be as small as a few milliliters (Taub 1969a, 1969b) or as large as an armory (Chapter 19). For each of the examples in Figure 1.1 there is a chapter later in the book, covering some of the interesting results the many researchers have found in the last four decades.

Scope

Due to the advantages of controllability, replicability, and low cost, microcosm experiments have been used in practically every area of modern ecology, terrestrial and aquatic. Although any assemblage of living organisms in a container could be called a microcosm, we consider here those that have a variety of species which self-organize to form the common characteristics of ecosystems, such as food chains, hierarchies, production–consumption coupling, mineral recycling, diversity, and animal

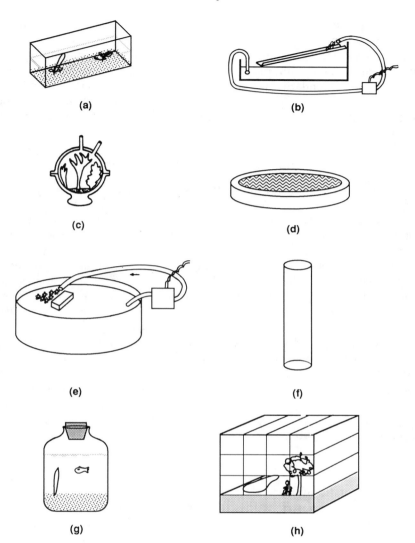

Figure 1.1. A variety of microcosms. (a) Aquaria; (b) stream microcosm; (c) terrarium; (d) artificial pond; (e) reef microcosm; (f) plankton column; (g) sealed microcosm; (h) human-containing microcosm.

control of plants and microorganisms. Thus, most laboratory population studies involving only one or two species were not included.

Many ecosystems in nature have sharp boundaries, such as the banks of a stream and the borders of a pond. Small ecosystems in nature have been called microcosms, as in the often quoted essays by Forbes (1887) and Hutchinson (1964). This book, however, considers only those isolated in artificial containers created by humans. Larger ones are sometimes called mesocosms.

To summarize, the following are the criteria for consideration here: (1) the system contains multiple species, thus excluding uniorganismal cultures, (2) the boundaries are artificial, with containers created through human, not natural, causes (3) there is at least a partial isolation of the experimental system from the external world in terms of matter, energy, and/or species dispersal.

Boundary Conditions and Exchange

After the container defines the boundary of an ecological microcosm, the boundary influences (sources) must be defined if meaningful studies are to be made. Inflows (sources) may include light energy, stirring energy, regular additions of nutrients, regular additions of organisms, or any combination thereof. For example, an important principle of water exchange rate was studied by Ketchum (1954) in determining the presence of zooplankton and benthos with larvae. Estuaries with long flushing times could develop faunas uniquely adapted to local conditions during self-organization. However, those estuaries which flushed rapidly, developed the faunas of the larger system with which they were exchanging. For an indigenous population, the turnover time of life history stages in the water has to be faster than the flushing time of the estuary. Water exchange rate is critical to those microcosms where there is regular water input and output. Sanders and Cibik (1985) studied arsenic effects in tanks flushed with bay waters on a two-day turnover time. In spite of the short time, there were substitutions of phytoplankton species, especially flagellates and *Chloromonas*.

Initial Conditions

The initial conditions need to be defined as thoroughly as possible. This means defining major categories of chemical composition and introduced species. For many purposes, an excess of potentially suitable species is added to the system. As a result of self-organization, only a few are sustained.

Variability

Because of the self-organization that takes place, each microcosm develops unique properties. When several are started at the same time and kept isolated, the ecosystems which emerge are variable in composition. Many scientists trained to study one mechanism at a time, find the complexity and uncertainty unscientific. Yet the real essence of ecosystems is their self organization, adaptation, and maintenance. Excluding the complexity excludes the reality of ecosystem functions. Out of necessity, each level of science tends to deemphasize study of larger and smaller scales. Microcosmology is at the systems ecology level, dealing more with systems mechanisms and less with component mechanisms. Almost all microcosm studies are concerned to some degree with variation among replicate microcosms, and similarities between microcosms and the parent community.

Scaling Microcosms for Larger Inferences

The hierarchy of small and large systems may be represented by the generalized spectrum of territory and turnover time as in Figure 6.1. Any particular system includes a window of several orders of magnitude from among the range of many orders of magnitude. On an abstract and energy basis, each level is similar except with reference to turnover time and territorial size. A system at one level may be a model for a higher level, if both the size and time are properly scaled. Therefore, one may use a computer simulation model for one realm as a model for another realm, by changing the definitions for what a second of computer time and signal amplitude represent. A simulation scaled for phytoplankton can be applied to another time–space scale, changing the axis labeling on the output graphs. Parsons (1982) provides a graph of the increases in time and space which correspond to the increasing scale of microcosms. In other words, by scaling up, the growth of zooplankton in a tiny microcosm may be compared to the niche of fishes in a larger realm. A means for changing scales using energy units is given by the transformity, which is the energy flux at one level which is required to create an energy flux at a higher level in the energy hierarchy. As Figure 6.1 suggests, terrestrial ecosystems tend to be at a higher position in the hierarchy than aquatic ones.

Multiple Seeding, Self-organization, and Maximum Power

Some of the most successful investigators have emulated nature by fostering waves of immigration into their microcosms. This is usually termed multiple seeding. Organisms from similar, but geographically different sources are placed in the microcosm, and the system is allowed to establish its own web of interactions. This principle is called self-organization. Self-organization allows the developing microecosystem to establish feedback pathways which reinforce those processes that contribute most, according to the maximum power principle (Lotka 1922a, 1922b; Odum 1967, 1982).

Synthetic Versus Natural Assemblages

Organisms in microcosms interact in order to develop microecosystems, but usually not all of the species are known. However, with special care to sterilize all media and isolate each species as a pure culture before introduction, the species present can be fully determined, including microorganisms. Ecosystems with completely-known species complements, have been termed gnotobiotic (Taub 1969a,b,c; Nixon 1969; Ingham and Coleman 1983). In some cases, investigators isolate the organisms from an environment where the species occur together as a natural assemblage. In a different approach, species are selected that have been well studied physiologically, and are available from various culture banks. In this case, an artificially synthesized microecosystem develops with species which may not normally be found together. Advantages have been claimed for the synthetic type of system when the component species are already well studied, because this allows the various ecological functions of the various organisms to be selectively included or omitted. On the other hand, transplanted natural assemblages may have had thousands to millions of years to evolve mutualistic behavior and biochemical and ecological interactions. Therefore, the system made of previously associated species may more closely resemble nature and may have greater efficiency.

Natural Microcosms

In wild landscapes and in cities, small, partly isolated ecosystems occur which have fascinated many scientists who have sometimes called these systems microcosms. The name "phytotelmata" was used for holes in trees, the cups of water in bromeliads, standing hollow bamboo canes, the pools in rocks, tide pools, water in discarded bottles, small ponds, springs, etc. Information on what developed in these natural containers indicates what can develop over long periods of self organization in nature.

Dissemules, Propagules, and Succession

Microcosms are used in theoretical studies of succession, water–sediment interactions, hot springs, temporary ponds, and hypersaline lagoons. By their very nature, microcosms are isolated to some degree from the other ecosystems of the biosphere. Often, by aseptic techniques, experimenters attempt to make this isolation complete, as far as living organisms are concerned. However, they are often surprised by the sudden appearance of some "foreign" plant, animal, or bacterium, particularly during the initial, successional stages in the life of the microcosm. This appearance is usually attributed to faulty technique or accidental inoculation of the interloper into the system. Often the presence of cysts, spores, eggs, or other resistant resting stages is not considered. Such propagules may have come in with the organisms originally seeded into the system, unnoticed by the investigator. When conditions become suitable, they may "hatch," reproduce, and make their presence known. This is the aquatic equivalent of the well known phenomenon of soil, taken from a climax forest and placed in flats in the green house, producing early successional herbs and forbes, even though such plants may not have been present in that particular area for fifty or one hundred years.

Cross Seeding and Replication

Even when they are started similarly, microcosms often develop differently. Where duplicates are desired, replication can be enhanced by regularly transferring some of the contents of one container to another. This mixing replaces lost species, replicates population densities, and makes chemical contents similar. Even synthetic gnotobiotic microcosms may need cross seeding (Taub 1969a, 1969b). Random events within microcosms lead to differential mortality of species, and species must be replaced, especially during the initial, successional, setup period.

Experiments on Mechanisms which Retain Ecosystem Complexity

Although many sciences concentrate on eliminating the complexities of a system in order to study a relationship, microcosms provide another way to study mechanisms. Instead of removing a part of the system and isolating it in the laboratory, the ecosystem complexity is retained, and only one factor is changed. All factors are kept the same except for the treatment variable.

Metabolism and Biogeochemistry

Much microcosm literature concerns metabolic measurements. Since most microecosystems contain autotrophic and heterotrophic elements, metabolic studies must include production (photosynthesis) as well as respiration. A generalized, simplistic metabolic equation can be written for overview purposes. Reading from left to right the equation is driven by light. It defines photosynthetic production designated P. In the reverse direction energy stored as organic matter is consumed. This defines community respiration R.

$$CO_2 + H_2O + Minerals \rightleftharpoons Organic\ Matter + O_2$$

Since the rate of any chemical process can be determined by measuring the appearance or disappearance of any product or reactant, theoretically, any of the items in the above equation could be used for metabolic measurements. In actuality, water is usually too abundant, and most of the minerals in too low a concentration for easy measurements. Thus oxygen, carbon dioxide, and organic matter have usually been measured to determine microecosystem metabolism. Microcosms are often the subject of thermodynamics, a field that deals with overviews and total processes. Energy processing was usually inferred from metabolic measurements, but the development of microcalorimeter apparatus now allows the energy transformations to be sensed directly. See the review by Lasserre (1990).

Making available additional resources allows self-organization to increase energy processing, biogeochemical cycles of inorganic materials, especially phosphorus and nitrogen, have been frequently studied in microcosms. Studies of biologically active elements, such as calcium, iron, sulfur, and the trace elements, often include radioactive or heavy stable isotopes as tracers.

Practical Uses

Some early microcosm studies concerned themselves with sewage and industrial wastes in an effort to improve water treatment. This was followed by using microcosms to study the fate and action of pesticides, heavy metals, and other toxins in aquatic and terrestrial systems. An added advantage of the microcosm technique was the safety factor of not having to add toxins to any portion of the environment. Some investigators dealt with the whole microecosystem, others with only one or two trophic levels, and still others with only one indicator species within the system. In a similar vein, microcosms have been used to study the effect of elevated temperatures on whole ecosystems and some of their components. These studies were devised to examine the effects of the massive amounts of cooling water generated by the power generating and other industries. Since thermally elevated water is discharged to both streams and lakes, these studies have involved both flowing and still water microcosms.

Finally, as an adjunct to the national space program, microcosms have been developed for the investigation of biological life support systems. Initial efforts involved attempting to balance humans with aquatic plants. More recently, artificial biospheres have been designed to contain humans (Chapter 19).

Microcosms for Teaching Ecology

Over the years, the authors have used student-designed microcosms as the center of laboratory work in ecology, limnology, and marine science courses. By watching the process of self-organization, by measuring metabolism, diversity, and nutrients, and by simulating the results with models, a student learns to think of the ecosystem first, before getting distracted by interesting details. A classroom full of original microcosms is a fascinating place after two months of self-organization. Each is different and ideas are generated rapidly as a result of trying to account for what is happening. Students learn from all the examples. See suggestions given in Appendix A.

Chapter 2
Metabolism and Homeostasis

The processes occurring in microcosms are the same as those found in ecosystems, but they are simplified since the system is enclosed and isolated. We consider in Chapter 2 the metabolic basis of microecosystems. The main processes are photosynthetic production, consumption, and the recycling of nutrients. Oxygen and carbon dioxide are produced, consumed, and exchanged with the atmosphere. Organic biomass is produced, accumulated, or used up, depending on conditions. There is a self-regulating homeostasis in the coupling of production and consumption, which makes microcosms remarkably stable. In this chapter, using models and simulations, we examine the metabolism of various types of microcosms and the patterns of diurnal variation, succession, and temperature change. Special emphasis is placed on the internal responses to external sources.

Models and Concepts

Simplified concepts about how systems work are called models. Making models explicit to explain old data and predict what will happen with new experiments is one of the main scientific methods for advancing knowledge. When the models are expressed in a mathematical equation form, quantitative inferences can be made and experimentally tested. Microcomputer simulation of a model shows the consequences of the model in time or its response to a treatment. Simulation models are like controlled experiments. They show what would happen given certain premises, if other factors are held constant. Models that work, summarize in compact form the knowledge of microecosystems, simplifying for the human mind extensive facts and experimental data. Simulations help establish in our minds some basic patterns which can be expected in the microcosms we examine.

When the simulation performance of a model is first compared with data, the fit is not usually very good. The model has to be changed repeatedly in form or calibration, until data and concepts are consistent. Many simulation models satisfactorily represent observed time series of microcosms for a short time. Simulations of models don't predict longer periods very well. This tells us that our concepts of ecosystem principles, which qualitatively seem to explain events, are not yet adequate.

Because microcosms are simpler than outdoor ecosystems, they are themselves models — living models. Therefore, simulation models of microcosms may be easier to develop, which may be why microcosms have helped develop ecosystem theory. This chapter includes microcosm simulation models.

Diagrams and Equations for Representing Models

To see more of the parts and relationships of a system simultaneously, diagrams of parts and connecting pathways are usually drawn. By using symbols that have mathematical equivalents, the diagram becomes a way to connect the mind's eye view of components, flows, storages, actions, structural relationships, etc. with mathematical equations. The ecosystem diagrams are a principal means for representing one's models to another person, for comparing models, and for generating new models easily.

In this book, models are each represented with the energy diagram symbols (Odum 1967, 1972, 1983a) and with mathematical equations. Symbols are given in Figure 2.1. A typical diagram has outside sources (cross boundary influences) in circles around the box that represents the boundary selected for the system. For microcosms, this boundary is the container. After one is familiar with the network language and which functions the symbols represent, he or she can overview systems much more rapidly than one can the equivalent mathematical equations. Sources and components are arranged from left to right in order of solar transformity. Solar transformity of a flow or storage is equal to the solar energy joules (called solar emjoules) required to generate one joule of that flow or storage. For example, the solar transformity of a zooplankton population may be 40,000 solar emjoules per joule. The higher the solar transformity, the higher an item is in the food web. By arranging symbols in order of solar transformity, one is arranging them in order of natural energy hierarchy. Since every symbol has an appropriate place on the diagram, two people modeling the same system will produce similar diagrams. Each person can readily read another person's diagrams and easily compare them. The arrangement by transformity also eliminates unnecessary crossing of pathway lines, providing one draws all feedback control actions as counterclockwise loops to the tops of symbols.

Writing equations from energy systems diagrams is rigorous. However, there are many different system configurations with the same mathematics.

Diagrams and Equations for Representing Models

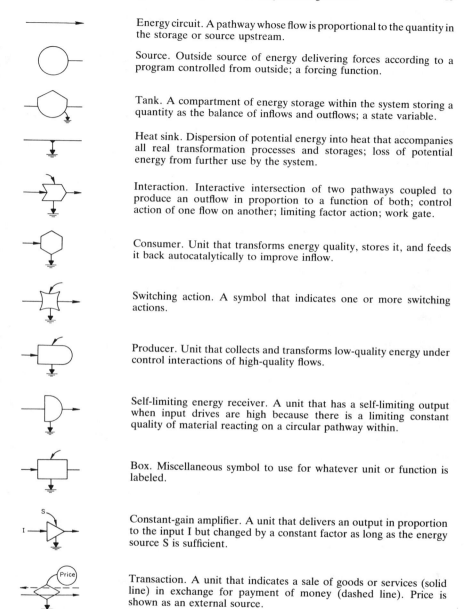

Energy circuit. A pathway whose flow is proportional to the quantity in the storage or source upstream.

Source. Outside source of energy delivering forces according to a program controlled from outside; a forcing function.

Tank. A compartment of energy storage within the system storing a quantity as the balance of inflows and outflows; a state variable.

Heat sink. Dispersion of potential energy into heat that accompanies all real transformation processes and storages; loss of potential energy from further use by the system.

Interaction. Interactive intersection of two pathways coupled to produce an outflow in proportion to a function of both; control action of one flow on another; limiting factor action; work gate.

Consumer. Unit that transforms energy quality, stores it, and feeds it back autocatalytically to improve inflow.

Switching action. A symbol that indicates one or more switching actions.

Producer. Unit that collects and transforms low-quality energy under control interactions of high-quality flows.

Self-limiting energy receiver. A unit that has a self-limiting output when input drives are high because there is a limiting constant quality of material reacting on a circular pathway within.

Box. Miscellaneous symbol to use for whatever unit or function is labeled.

Constant-gain amplifier. A unit that delivers an output in proportion to the input I but changed by a constant factor as long as the energy source S is sufficient.

Transaction. A unit that indicates a sale of goods or services (solid line) in exchange for payment of money (dashed line). Price is shown as an external source.

Figure 2.1. Symbols used for energy network diagrams (Odum 1983). Q usually refers to a quantity stored. K usually refers to a pathway coefficient in equations and computer programs. Within the interaction symbol, × means to multiply.

If one is given equations for a model, one may make more than one configuration diagram. Thus, it is possible to misinterpret an author's model if only equations are given, unless they are explained. For example, the algebra for an internally recycling limiting factor and an input from an outside limiting factor is the same. Theoretical ecology has been unnecessarily restrained because authors and their editors have not published rigorous model diagrams of their ideas. Often, simulation results have been printed, but diagrams, equations, and computer programs have been omitted, so the models remain as mysterious black boxes that cannot be understood or compared. In this chapter, both diagrams and equations are given for each model and simulation. Programs are available on disk.

Limitations on Outside Sources

In any ecosystem, phenomena are driven by outside sources that include solar energy, inflows of materials and organic matter, and/or controlling actions and information. Because of the enclosing container around a microcosm, many outside influences are eliminated. Microcosms are thus simplified, but some normal characteristics are also eliminated. Natural ecosystems are usually connected to larger realms that control them. For example, there are animals with large territories which have large, although infrequent actions on a small-sized system. However, when a small ecosystem is enclosed in a container, the action of the larger realm may be excluded. Thus, influences acting over a longer period tend to be excluded. The microecosystem developing in the microcosm may be more stable without the long-period surges of outside influence. Of course, long-period pulses can be included as part of the boundary conditions of the system.

Relaxation Microcosms

If all outside inputs are eliminated, the inside processes run down as the residual storages of resources are used up. Sometimes called "relaxation", this kind of microcosm has been useful in isolating consumer processes for study. Examples are the BOD (biochemical oxygen demand) bottles which are filled and placed in constant temperature baths for study of rates of decomposition. Another example is the decomposition sequence in marine waters put in storage for several months by Von Brand and Rakestraw (1941). Nitrogen was observed to go from the organic state, to nitrite, and finally to nitrate like water below the thermocline in the sea. The decay curves of the storage and of the rate of consumption are often exponential. The simplest model of relaxation is a single "tank" storage with outflow drains, but no input (Figure 2.2). Over a short period, such as a few days and nights, the consumer populations do not change greatly.

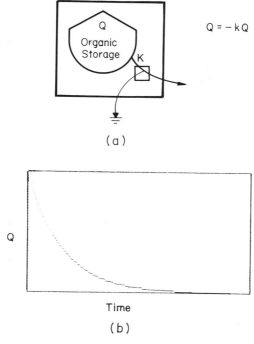

Figure 2.2. Single storage model for a relaxation microcosm (Program: DRAIN). (a) Energy diagram and equation; (b) simulation results.

Therefore, the fraction consumed per unit time can be adequately represented as the constant K.

Production Models with Linear Consumption

Many microcosms are closed except for inflows of light energy from daylight or from pulsed artificial light. Input regimes which alternate between on and off, produce characteristic patterns of flows and storages in systems according to the turnover times. In considering microcosms, it may be helpful to show such basic patterns first. For microcosms without larger consumers, consumption and recycling are done by many small units, such as microzoa and microbes. Small components have fast turnover times and develop rapid, quasi-steady-state responses to their available food supplies. Though individual units and populations have quick periods of growth, collectively they may consume in proportion to the stock of available food. The several production–consumption models, introduced in Figures 2.2–2.4 and 2.10–2.12, include linear consumption. They have been successful in simulating short-term patterns such as diurnal variations.

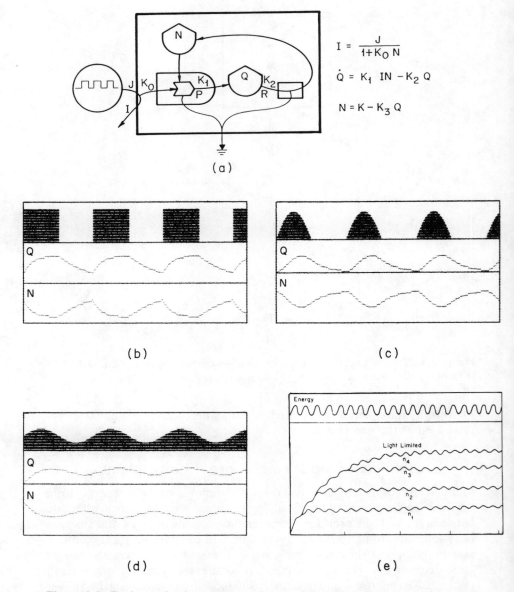

Figure 2.3. Basic production–consumption model of diurnal processes in sealed aquaria (Program: PR). (a) Simulation model and equations; (b) response of organic storage and available nutrients to an on and off light regime; (c) response of organic storage and available nutrients to a half sine wave representing a clear-sky solar regime; (d) response of organic storage and available nutrients to a sine wave representing a simplified temperate latitude seasonal solar regime (Program: YEARPR); (e) successive growth runs with increasing content of nutrients (n_1, n_2, n_3, etc.).

Basic Production–Consumption Model

One simple model which successfully represented the main metabolic processes in sealed aquaria receiving alternating daylight and night darkness is shown in Figure 2.3. This model was simulated with a special purpose electric analog device (Odum et al. 1963) and used for the whole biosphere (see also Odum 1971). Mathematically, the model is identical with the Michaelis-Menten model for enzymes (Michaelis and Menten 1913). Here, the 1963 model by Odum et al. is given a limited, renewable energy source appropriate for available light. Production is a product of recycled materials (nutrients) and available light. Production is stored until consumed in proportion to the amount stored. The consumption (respiration) releases the materials needed for more production. As with the relaxation model, the simplest consumer function is the linear one in Figure 2.2.

Shown in Figure 2.3 are the responses of this simple model to several light regimes. Light is turned on and off, creating a "square wave" input in Figure 2.3a. The response is an alternating rise and fall of organic product storage, Q, and the accompanying mirror image decline and rise of raw materials, N (Figure 2.3b). The response to diurnal types of light input (half sine waves) is similar but more rounded (Figure 2.3c). The response of the model to a seasonal type of undulating light input is even more rounded (Figure 2.3d).

Nutrient Cycles and Limits

Any production process (indicated by the pointed block interaction symbol in Figures 2.1 and 2.3) has two or more required inputs, usually including light and nutrient materials such as phosphorus, carbon dioxide, and nitrogen. Any of these may be locally limiting, although the self-organization processes develop pathways that reduce shortages. In a microcosm closed to matter, the available nutrients may be limiting because they do not circulate through the inside cycle fast enough. In Figure 2.3e are the results of successive simulation runs of the simplest production–consumption model, starting with a very low concentration of nutrients. Nutrients are added with each additional simulation run, resulting in higher predicted rates of production and organic storage. As the nutrients become less limiting than the available light, the simulation reaches its maximum curve, responding no further to nutrient additions. More on chemical cycles follows in Chapter 4.

Rate Graphs for Representing Microcosm Performance

Given in Figure 2.4b are the rates of gross production and respiration Figure 2.4a shows the storages (as already given in Figure 2.3). Since raw materials are most available after a dark period, gross production starts

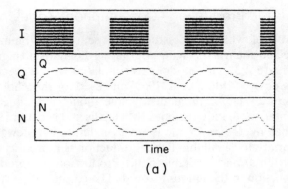

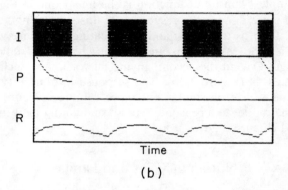

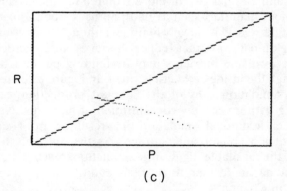

Figure 2.4. Graphs of rates of production and composite respiration from the simulation of the simplest production–consumption model in Figure 2.2 with square wave light (Program: PRRATES). (a) Nutrients, N and biomass, B; (b) production, P and respiration, R; (c) phase plane graph of production, P and respiration, R.

high and declines even when the light intensity is constant. After a light period, organic storage is high and therefore respiration is high. In the dark, the storage declines, and with it the dependent respiration. Net production is the algebraic sum of production and respiration(consumption) as shown in Figure 2.6b. The net production is positive during the day and negative at night.

Phase Plane Graph of Production and Consumption Rates

Figure 2.4c has the phase plane plot (one state variable as a function of another) of production rate and respiration rate, as the model responds to the turning on and off of available light. For square wave light, the points plot a line oscillating back and forth. The production rate is higher but production operates only in the daytime, whereas consumption operates 24 hours a day at a lower rate. At steady-state, as in Figure 2.3b–3d, production and consumption balance during a diurnal cycle. With a seasonal undulation of available light, the phase plane plot circles above and below the diagonal line of equal production and consumption rates.

Generalized Production–Consumption Pattern

That this pattern of production and respiration rates in alternating light regimes was a general one was pointed out by Beyers (1963a) and it was simulated with a special purpose electric analog system (Odum et al. 1963a). However, by providing more components within the production–consumption model, we can simulate particular microcosms more accurately.

Diurnal Rates of Photosynthesis and Respiration

Beyers (1963a, 1965) used aquarium microcosms derived from different aquatic ecosystems to illustrate a common pattern of diurnal metabolism. The systems he used were: (1) benthic fresh water systems whose primary producers were *Vallisneria* and *Oedogonium*, and whose major animal component was an oligochaete (*Sutroa*) and the snail, *Physa*, taken from the San Marcos River in Texas (Beyers 1963b), (2) a hypersaline algal mat microcosm from the Laguna Madre in Texas, containing principal community members *Lyngbya*, *Desulfovibrio*, *Oscillatoria*, and purple sulfur bacteria (Armstrong and Odum 1964), (3) another algal mat system from Mimbres Hot Springs, New Mexico which was isolated and held at 51°C and contained *Phormidium*, *Oscillatoria*, *Anabena*, and thermophillic bacteria, (4) a brine system derived from the evaporating pans at La Parguera Salt Works, Puerto Rico, the principal components of which were the alga, *Dunaliella* and the brine shrimp, *Artemia*, and (5) a temporary pond-type system from the pools on the top of Enchanted

Rock, Texas, consisting of cladocera, ostracods, and small green flagellates. All of these systems were held under conditions of constant temperature and artificial light.

Nighttime respiration in all systems was at a maximum in the early evening after the lights were extinguished. The respiratory rate then dropped until the lights came on again. The net photosynthesis showed a similar pattern, with production at a maximum immediately after the lights were turned on, and with decreasing photosynthesis during the rest of the day (Figure 2.5). This pattern can be generated by a computer model such as that shown in Figure 2.3, and is frequently encountered in other still water microcosms (Kuhl and Mann 1962). Although the biota of all five types of systems varied greatly, the net photosynthesis (0.17 to 0.80 mM CO_2/liter per 12 h) and nighttime respiration (0.11 to 0.69 mM CO_2/liter per 12 h) were similar under the similar light conditions of the experiments (467 to 1,000 foot-candles). The P/R ratios varied between 1.00 and 1.47.

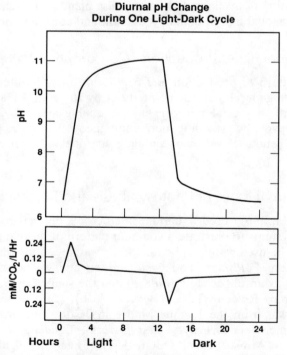

(a)

Figure 2.5. Diurnal curves showing net photosynthesis and night respiration. (a) Typical graphs of pH and carbon dioxide metabolism; (b) rates of carbon dioxide uptake (−) and release (+) in a marine algal mat microcosm, a hot spring microcosm, a temporary pond microcosm, and a brine microcosm. The pattern of maximum metabolism in the first half of the light or dark period is displayed in all (Beyers 1965).

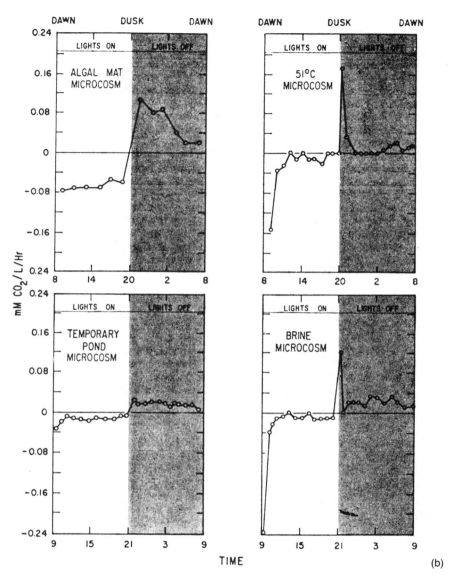

Figure 2.5. *Continued*

Basic Aquatic Model with Oxygen

Whereas the simplest production–consumption model in Figure 2.3a has oxygen aggregated as part of the product storage, Figure 2.6a shows oxygen as stored separately, reacting with organic matter so that respiration is proportional to the product of their concentrations. Dissolved oxygen

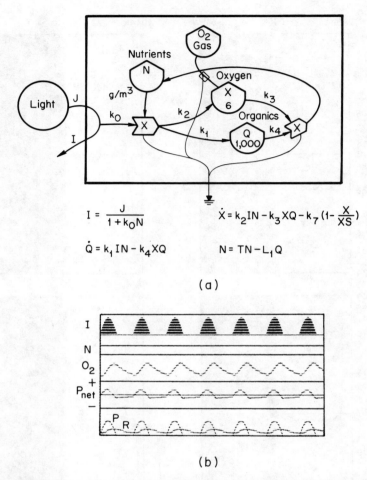

Figure 2.6. Model of production and respiration for sealed aquatic microcosms with dissolved oxygen in the water and in a gaseous space above the water (Program: CLOSEAQ). (a) Energy diagram and equations; (b) simulation results.

concentrations in waters are small (compared to air) and much affected by the metabolism. Oxygen is often measured in aquatic microcosms as a sensitive indicator of total metabolism. When much more oxygen is produced than saturates the water at that temperature, oxygen diffuses into the air above the water. If the container is sealed without an air space, bubbles form and a gas phase forms, as included in the simulation model in Figure 2.6a. The results of the simulation are given in Figure 2.6b and show the diurnal rise and fall of dissolved oxygen due to the rise and fall of net production (P_{net} = production, P minus respiratory consumption, R).

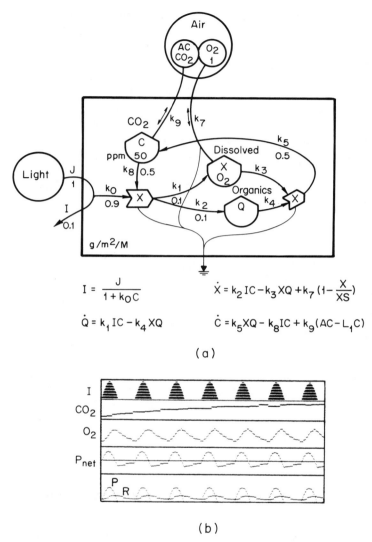

Figure 2.7. Model of production and respiration for sealed aquatic microcosms with gaseous exchange of carbon-dioxide and oxygen with the atmosphere (Program: OPENAQ). L_1 is the fraction of biomass which is bound nutrient. (a) Energy diagram and equations; (b) simulation results.

Microcosms Open to Air

Most aquatic ecosystems are open to the air and stirred by the winds. Since oxygen and carbon dioxide in the water can exchange with the almost constant concentrations of these gases in the air, the atmosphere acts as a buffer preventing either oxygen or carbon dioxide from becoming extremely high or low. The "ocean of air" is a worldwide buffer of life.

In microcosms, atmospheric buffering is imitated by supplying a stream of air bubbles by means of an air stone, with care that the source of air is from outside or regulated at known partial pressures of oxygen and carbon dioxide. In Beyers' (1962a, 1963b) studies, aquaria were allowed to self-organize under these conditions. The model in Figure 2.7 is appropriate. The amplitude of the diurnal rise and fall of oxygen depends on the vigor of the mechanical energy applied to the atmospheric exchange. Most aquatic ecosystems in nature have similar patterns.

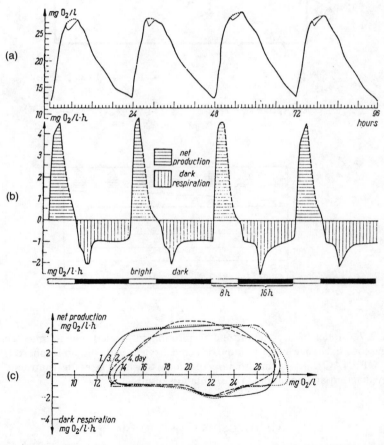

Figure 2.8. Simulation of diurnal patterns in a microcosm by Uhlmann and Cramer (1975). (a) Diurnal pattern of dissolved oxygen; (b) diurnal pattern of metabolism; (c) phase plane of net production as a function of oxygen storage.

Example of Diurnal Metabolism

An example of a closed aquatic microcosm with a diurnal pattern like the simulation result in Figure 2.7b is given by Uhlmann and Cramer (1975). These investigators also use another kind of phase plane graph to represent their results: a graph of net metabolism and oxygen concentration (Figure 2.8).

Labile and Dead Organic Storage

The models in Figures 2.3–2.7 have all the organic matter storage driving the respiration process. However, the total organic matter may become large and be located in less active storage sites such as sediments or

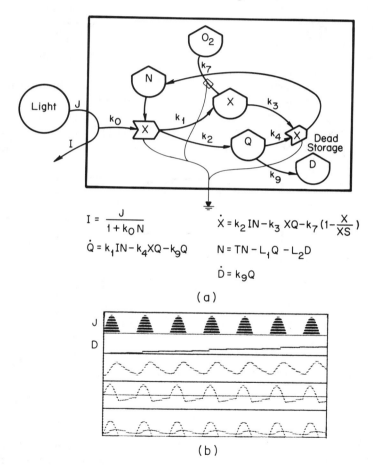

$$I = \frac{J}{1 + k_0 N}$$

$$\dot{Q} = k_1 IN - k_4 XQ - k_9 Q$$

$$\dot{X} = k_2 IN - k_3 XQ - k_7 \left(1 - \frac{X}{XS}\right)$$

$$N = TN - L_1 Q - L_2 D$$

$$\dot{D} = k_9 Q$$

(a)

(b)

Figure 2.9. Model of production and consumption with oxygen exchange and dead organic storage (Program: DEADPR). (a) Diagram and equations; (b) simulation results.

skeletons, little affected by short-term periods of light and dark. The model in Figure 2.9a has two types of storage for organic matter, one labile, driving respiration, and another inert, slowly receiving organic matter from the first one. Ecosystems with a great excess of nutrients, causing an excess of net production, expel oxygen to the air and organic matter to dead storage.

Production–Consumption Model for a Terrestrial Microcosm Using CO_2

Production and consumption in terrestrial microcosms are represented in simplest form in Figure 2.10a. The recycling material is carbon, which is available to plants as gaseous carbon dioxide. In terrestrial microcosms, oxygen content is high and short-term variations involve a tiny, negligible percent of the gas available. It is the carbon dioxide that is in short supply and responds to system metabolism, thus, carbon dioxide is usually measured during studies of terrestrial ecosystems.

The metabolism of terrestrial microcosms adapted to low light intensities on the floor of a rain forest were studied by Odum and Lugo (1970). The pattern of the response to alternating light and dark (Figure 2.10b) was that of the production–consumption model. Burns (1970) simulated the diurnal metabolism, but because a simple constant was used for light

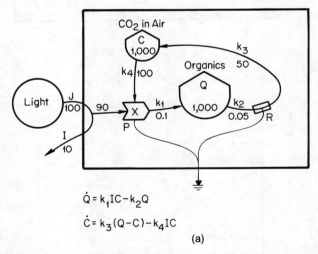

$\dot{Q} = k_1 IC - k_2 Q$

$\dot{C} = k_3(Q - C) - k_4 IC$

(a)

Figure 2.10. Production and respiration in closed terrestrial microcosms which involve measurements of gaseous carbon dioxide (Program: TERRCO2). (a) Energy diagram and equations; (b) observed record in microcosms adapted to the shade of a rainforest; (c) simulation results showing square wave light J, carbon dioxide concentration in air space, and the rates of production, P and respiratory consumption, R.

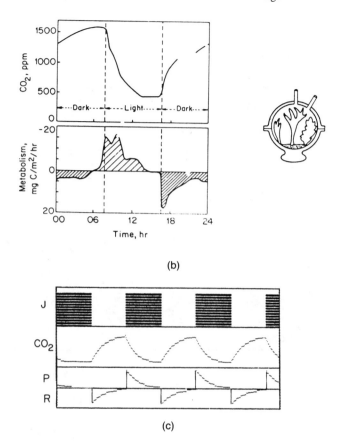

Figure 2.10. *Continued*

energy available regardless of load, the only limit was nutrient recycle. See Figure 4.7. The model in Figure 2.10a has source–limited light. The simulation results are given in Figure 2.10c and can be matched with the pattern observed in Figure 2.10b.

Terrestrial Microcosms with Gaseous Exchange

Like most terrestrial ecosystems, many laboratory microcosms are open to the air in varying degrees. Such terraria may tend to dry out as water is evapotranspired. Carbon dioxide levels may equilibrate with those of the air which is around the microcosm. This level may be a low value when microcosms are outdoors or a very high value when microcosms are located in buildings where human respiration contributes large amounts of carbon dioxide to the air. The model in Figure 2.11 simulates the response of terraria when water is added to replace that lost through

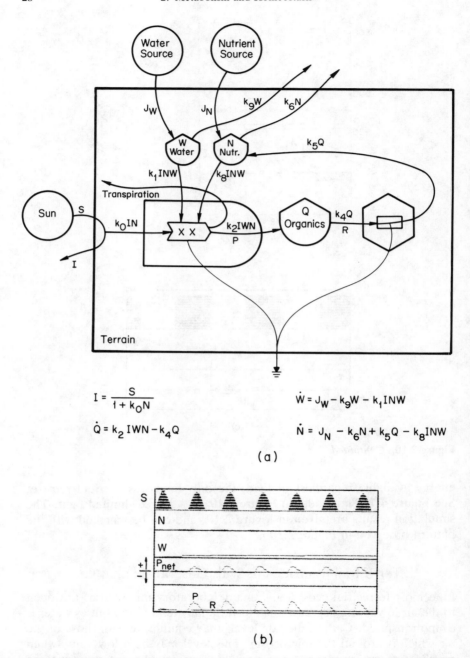

Figure 2.11. Simulation model of a terrestrial microcosm open to the atmosphere, with water added to replace evapotranspiration (Program: TERRAIN). (a) Energy diagram; (b) typical simulation results.

exchange, but oxygen and carbon dioxide are assumed to be constant. Since they are not variables, they become part of the constant coefficients of the model.

Effect of Large Initial Storages

When microcosms are set up, there are initial quantities of materials and energy bearing components. The effect of an initial large storage is to accelerate the user pathways temporarily until the excess is reduced and

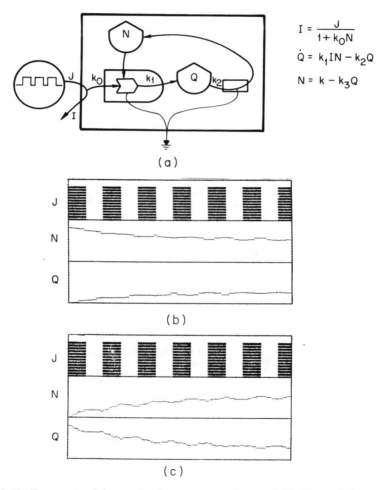

Figure 2.12. Response of the production–consumption model in Figure 2.3 started with large initial storages (Program: WEEKPR). (a) Energy diagram; (b) results of simulation with large initial nutrient storage; (c) results with large initial organic storage.

downstream parts of the system can come into balance. After a time, the effects of the initial condition disappear. In Figure 2.12, the effects of initial stored nutrients (Figure 2.12b) and organic matter (Figure 2.12c) are shown. An example is the famous early hay infusion microcosm experiments of Woodruff (1912). These experiments were started with a large organic content supporting an excess of consumer development, which is heterotrophic metabolism (Figure 8.1). Ultimately, the released nutrients stimulate photosynthetic components, if light is available. Another example is the typical schoolroom terrarium started with a thick bed of organic litter. Excess autotrophic metabolism occurs when initial

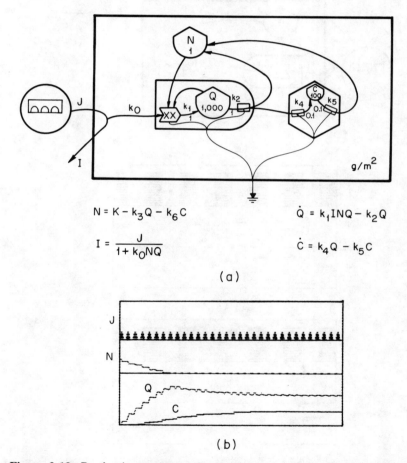

Figure 2.13. Production–consumption model with autocatalytic production and linear consumption. (Program: GROWTHPR). Abbreviations: nutrients, N; plant and detritus biomass, Q; consumers, C. (a) Energy diagram and equations; (b) results of simulation of successional growth and overshoot due to large initial nutrient storage. K is total nutrient in the system.

conditions are rich in inorganic plant nutrients without much initial organic matter. If initial storages are large enough, they may cause an overshoot during the successional period. Overshoots are observed often in nature. The model in Figure 2.13a, when started with a large initial charge of inorganic nutrients, produces an overshoot in the stock of organic matter, Q, as shown in Figure 2.13b. A model with only two storages, like those in Figures 2.3–2.12, cannot overshoot. Figure 2.13a contains an autocatalytic plant production unit and a third storage for consumers.

Microcosms with Dominant Inflows and Outflows

Microcosms which are intended to represent the phenomena of flowing streams, have strong inflows and outflows. The turnover due to outside exchanges is high in relation to the internal recycling. The model in Figure 2.14 is the basic production–consumption pattern, with oxygen gas exchange and inflows and outflows of nutrients and organic matter added. The simulation results in Figure 2.14b show the increase of nutrients, oxygen, and production which occurs when there is successional growth based on incorporation of additional nutrients from outside.

Autocatalytic Plant Production

A variation on the basic producer–consumer model includes autocatalytic production. Such a loop results in a concave growth curve for producers, but the steady-state level is not affected. The autocatalytic feedback is appropriate where the producers are small with fast turnover times. See the example in Figure 2.13. Slightly more complex is the model given by McIntire (1973) to represent the periphyton in his experimental stream microcosms. His simplest model included autocatalytic plant growth maintained by light and nutrients. There was a metabolism-accelerating interaction with temperature, and the total organic matter, as periphyton, was autocatalytic. There were rises and falls of periphyton in the simulation results, which corresponded to the observed data.

Production and Autocatalytic Consumption

When there are larger consumers, such as larger animal populations, the turnover times are longer and patterns of growth or decline may generate oscillations. These oscillations may be in addition to those due to variations of input sources such as light. An example of an autocatalytic predator configuration in a production–consumption model is given in Figure 2.15. The model includes a linear pathway as well, since the mathematical model should include the same choices available to microcosms capable of self-organization. This model gradually builds up the organic storage and the rate of consumption to that threshold at which the autocatalytic

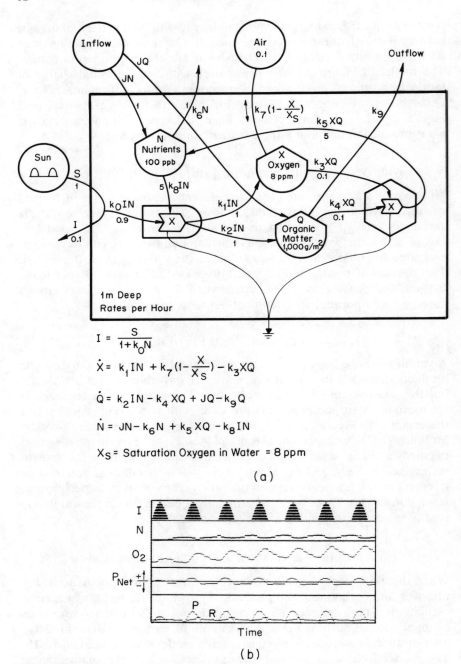

Figure 2.14. Basic production–consumption system with inflows and outflows dominant (Program: AQUATICM). (a) Energy diagram and equations; (b) results of simulation with some growth by inflow of added nutrients.

loop out-competes the linear pathway, and a sharp consumer–recycle pulse occurs. Referring again to McIntire's (1972) flowing stream model, when autocatalytic periphyton consumers were added, regular oscillations developed with total numbers of grazers increasing and periphyton mass decreasing, followed by the reverse. Addition of an autocatalytic prey–predator configuration results in a general systems model applying to

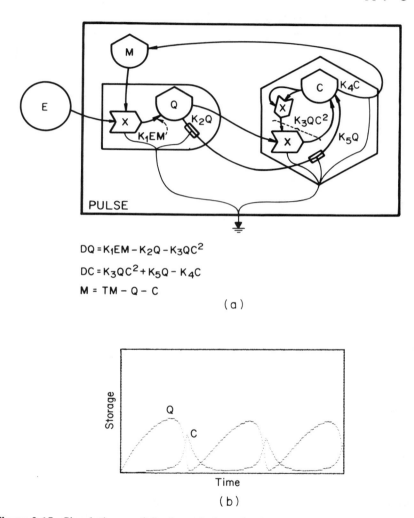

Figure 2.15. Simulation models of coupled production, consumption, and material recycle, which may develop a steady-state or a pulse regime. (a) Energy systems diagram of pulsing paradigm; (b) simulation (Program: PULSE) with alternating surges of production–consumption due to long-period autocatalytic consumers; (c) energy systems diagram of model with growth transition and steady state; (d) results of simulation without pulsing (Program: AQUARIUM).

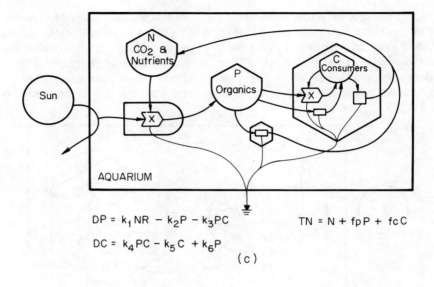

$$DP = k_1 NR - k_2 P - k_3 PC$$
$$DC = k_4 PC - k_5 C + k_6 P$$

$$TN = N + fpP + fcC$$

(c)

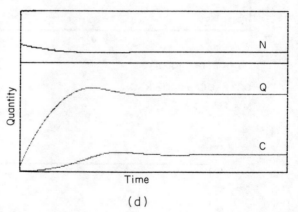

(d)

Figure 2.15. *Continued*

many kinds of systems. For example, it is almost the same as the chemical reactions studied by chemists considering the emergence of structure from chemical reactions (Nicolis and Prigogine 1977).

Extensive simulations suggest that the alternation of production and pulsed consumption in this class of models generates more performance and resource use than a non-pulsing pattern (Richardson and Odum 1981).

Consumption Accelerated by an External Energy Source

Because external energy sources often drive production, they may also be coupled to drive consumption. An important example is photorespiration, which is important in microcosms having excess organic matter and factors limiting regular consumption. Nixon (1969) and Odum et al. (1971) studied

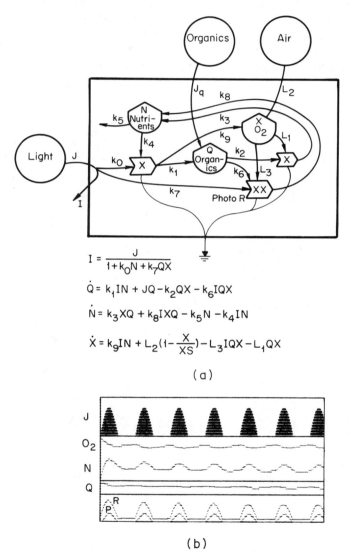

Figure 2.16. Simulation of light accelerated consumption (photorespiration) in aquarium microcosms (Program: PHOTOR). (a) Energy diagram and equations; (b) results of simulation with small inflow of organics.

the role of light in driving photorespiration of rhodopsin-containing brine bacteria, *Halobacterium*. A passive electrical analog simulator was created, making use of a photoconductor to simulate photorespiration. A general production–consumption model with a photorespiration pathway is given in Figure 2.16. The simulation which produced the results shown in Figure 2.15b, included a small, steady inflow of organic matter. The model simulates the observed down-turn of the oxygen level when light becomes available; when the light is no longer available, levels of oxygen may increase. Without an inflow of organic matter, the model predicts a rate of photorespiration which is less than the rate of photosynthesis, causing the apparent oxygen-dependent photosynthesis to seem small.

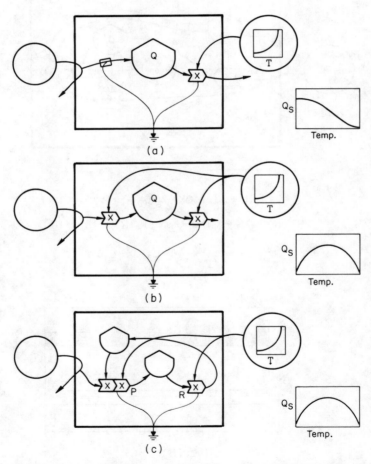

Figure 2.17. Types of temperature actions in simulation models with limited flow source. (a) Accelerated output ("pull"); (b) "push-pull" action; (c) "push and pull" in a production-consumption model.

Since most ecosystems include some photorespiration, most photosynthesis rates calculated from oxygen or carbon dioxide exchanges may be inaccurately small, especially where production and consumption are of a similar magnitude. Developing an appropriate model, such as that in Figure 2.16, is one way to separate photosynthesis and photorespiration.

The Role of Temperature in Microcosms

Since all the chemical and biochemical processes of a microcosm are a function of temperature, microcosms with varying temperatures may require temperature to be included in the simulation model. As shown in Figure 2.17, temperature action may be considered for each pathway, and then for the interplay among several pathway interactions. For a single pathway, the increase of temperature may be exponential. The effect of temperature rise increasing outflow is to pull down a storage (Figure 2.17a).

In aggregated models, many pathways represent compound chains of smaller components. Any storage with temperature action on inflow and outflow experiences a push-pull effect. If the push is from a source that cannot be substatially increased, then the pull effect becomes more important than the push. In this case, the effect of increased temperature is a parabolic curve (Figure 2.17b). The push and pull actions, within the closed loop of the basic production–consumption ecosystem, produce different responses. For example, the short-term effect of increased temperature on adapted aquarium microcosms (Beyers 1962b) was actually

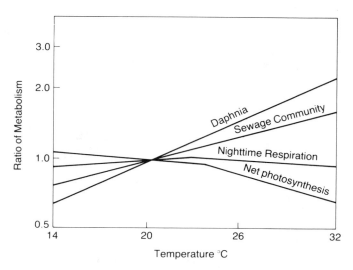

Figure 2.18. Response of aquatic microcosms to temperature change after being adapted to one temperature (20°C) in a growth chamber (Beyers 1962).

lowered metabolism, presumably because of limitations in the cycle of materials (Figure 2.18). Figure 2.19 describes a model simulating the response of bottle microcosms adapted to heating tapes (Kelly 1970). Temperature increased flow. Reasonable fit between observed results and model-generated results was obtained when increased flow was added to each of the four flows in the closed production–consumption system.

Showing More Detail in Production and Consumption

To some extent, the items included in models are the items of interest to investigators and/or those items measured by them. These items tend to be included with the main external driving sources, cycles, and dominant

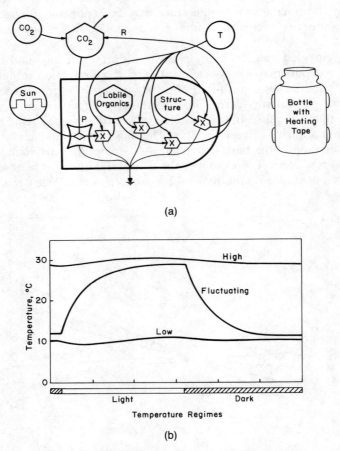

Figure 2.19. Simulation of temperature roles in thermal microcosms redrawn from Kelly (1970). (a) Energy diagram; (b) three temperature regimes tested; (c) simulated curves through a light and dark cycle and observed data points.

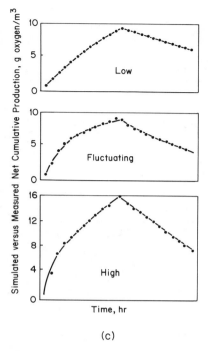

(c)

Figure 2.19. *Continued*

categories. The models given so far have aggregated production processes. Showing more detail for a specific time interval requires including more detail to the model. In other words, if the window of interest is a shorter time, then details concerning the producers are added. For example, Sciandra (1986) developed a model which predicted details of growth, including a short-term oscillation which made use of the details of phytoplankton and organic fractions. The model has the same general structure of production–consumption and recycling.

By way of introducing the variety of more detailed models referred to in other chapters, Figure 2.20 is provided. This diagram includes more of the ecological web typical of ecosystems, as viewed from the perspective of current ecological concepts. Included are several outside sources of differing energy quality, several producer categories, several hierarchical levels of consumers, feedback reinforcement pathways from larger units to populations of smaller units, recycling of materials, and export flows. In general, the more complex models for microcosms are not different from the models developed for larger systems in nature. However, the data for the simulations are often better, with less uncertainty about large influences that may have been operating undetected. In this chapter,

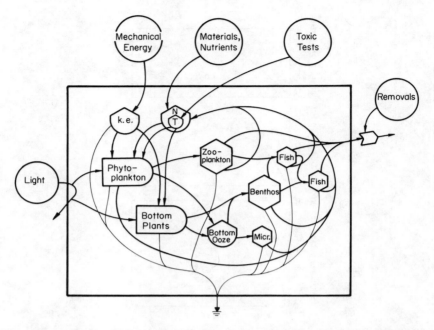

Figure 2.20. Typical microecosystem model of a microcosm. k.e. is kinetic energy of water motion.

models were used to introduce basic concepts necessary for considering microcosms. Systems diagrams are used in the following chapters to represent models of other authors, interpret observed data, and aid in comparisons.

Chapter 3
Succession and Self-organization

The self-organization process by which ecosystems develop structure, functions, and diversity from available energies and matter is called succession. These processes occur in microcosms, often rapidly, when the components are small with short turnover time. What actually occurs depends on the initial seeding of matter and available species. On a longer scale of time, the mature states which follow initial growth may oscillate, alternating production, consumption, and diversity. The simulation models in Figures 2.12 and 2.13 show how growth of biomass is related to initial conditions, available nutrients, and organic matter. However, succession is more than growth; it includes species additions and change. Figure 3.1 is a simulation minimodel that includes species seeding and diversity, and their contribution to production efficiency. In Figure 3.1b, the simulation first develops biomass, which supports more species, in turn, increasing efficiency, but additional species put additional demands on the biomass. The simulation with a small overshoot produces results which parallel what is observed in many microcosms, and suggest some consistency between model simulations and theory. Figures 3.2–3.9 show sequences of species in microcosm succession.

Classical Concept of Succession and Climax

The concept of ecological succession, like so many other concepts in science, has gone from description of observed phenomena, to qualitative explanation, to mathematical treatment. In pioneering work, Cowles (1889) and Shelford (1913) observed the orderly replacement of plant and animal species in the development of ecosystems on newly deposited sands of the Indiana Dunes. After a sequence of species replacements, a stage was

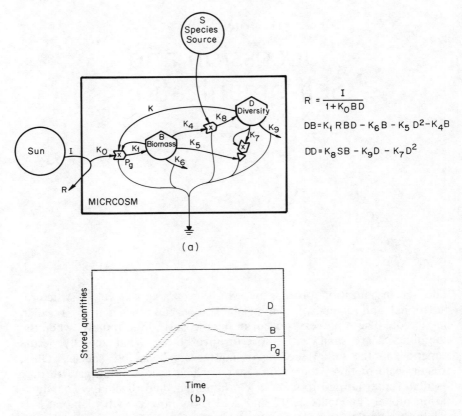

Figure 3.1. Minimodel of succession (Program: MICRCOSM) including biomass and diversity, with outside energy sources for productivity and species. (a) Energy systems diagram; (b) simulation results: D, diversity; B, biomass; P_g, gross production.

established which was self-renewing. The name climax was applied to this quasi-steady-state which followed the period of rapid growth and successional change. Each step in the process leading to the climax was named a "seral stage or sere." Microcosm succession was described by Woodruff (1912) in cultures of boiled hay. In these examples, diversity increased during succession. Summarizing a half century of successional studies, Allee et al. (1949) perceived ecological succession as a progressive sequence of replacement of communities. E.P. Odum (1971) defined succession as follows:

1. It is an orderly process of community development that involves changes in species structure and community processes with time that is reasonably directional, and therefore, predictable.
2. It results from modification of the physical environment by the community . . .

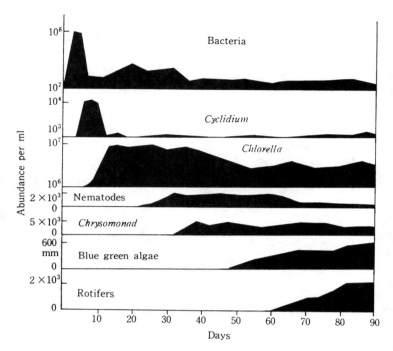

Figure 3.2. Changes in population densities in a microcosm undergoing succession. Each population represents a dominant in a sere (after Kuriharaa 1978).

3. It culminates in a stabilized ecosystem in which maximum biomass (or high information content) and symbiotic function between organisms are maintained per unit of available energy flow.

A fourth item could be added. Utilizing external catastrophes or internal mechanisms, it destroys part of its structure and starts a new cycle, apparently as a means of sustaining maximum productivity in the long run.

Also, it is known that, while the usual pattern is the achievement of a steady state or a steadily repeating pattern (climax), a reverse succession involving decreases of biomass or species diversity may occur (Figure 3.7).

Mechanisms of Self-organization

Succession is increasingly recognized as a self-organization process based on processes which select species from the choices provided by seeding. Some selection is by competition, especially in early stages, but other selections are determined by the species at higher levels of the ecosystem, reinforcing those at lower levels which make mutualistic contributions.

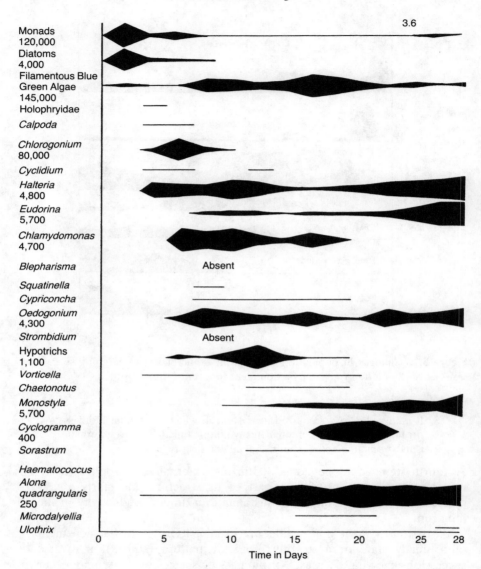

Figure 3.3. Changes in population densities in a temporary pond microcosm undergoing succession (after Tribbey 1965).

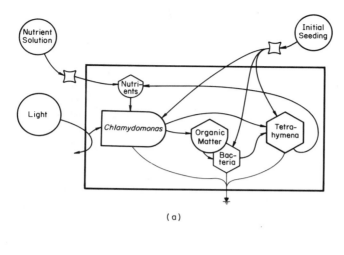

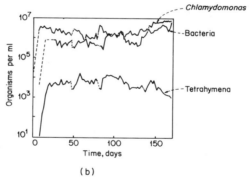

Figure 3.4. Defined (gnotobiotic) microecosystem synthesized from pure cultures, including algae (*Chlamydomonas*), protozoa (*Tetrahymena*), and bacteria (after Taub 1969). (a) Energy diagram; (b) observed biomass of component organisms.

Examples are the reinforcement achieved by nutrient recycling, the animal control of plants, and the role of keystone species in maintaining diversity. In these ways, ecosystem networks can develop patterns that tend to maximize power input, useful transformations, and efficiencies (Odum and Pinkerton 1955; Odum 1983a, 1983b). The orderly replacement of one suite of species by another is widely observed in outdoor ecosystems and microcosms. Displacement of species at one stage may occur through invasion of one seral stage by representatives from a subsequent one in the surrounding area. The conditions generated by one species, determine the conditions for the next. The relationships of the seres can be modeled as in Figure 3.6 to account for species changes such as those described in Figures 3.2 and 3.3.

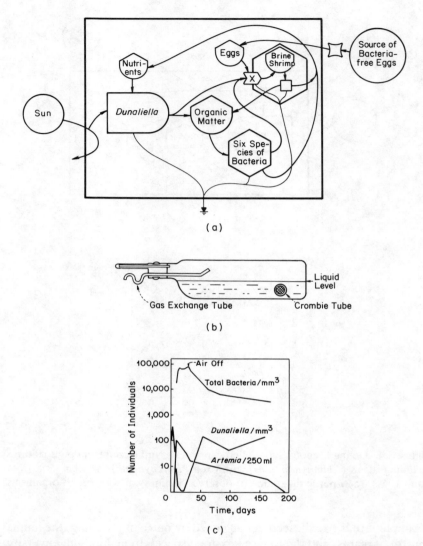

Figure 3.5. Brine microecosystem resynthesized from naturally mutually occurring components. Organisms were pure cultured and then recombined (after Nixon 1969). (a) Energy diagram; (b) microcosm; (c) numbers of organisms during succession.

Bowden (1991) studied uptake of nitrogen by *Polyhtrichum* moss from artificial rain and air flow in twelve moss-filled microcosms. An autocatalytic reinforcement was observed, in which efficient uptake of nitrogen by moss and the organic substrate, increased the biomass and its ability to catch more nitrogen.

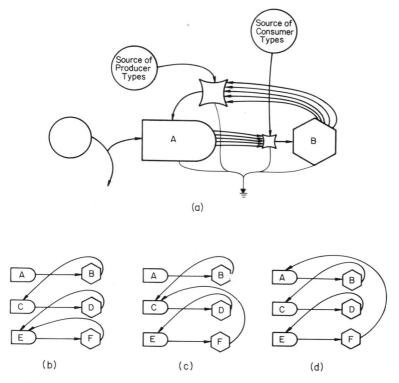

Figure 3.6. Succession of producers and consumers in microcosms and other ecosystems. Source of consumer and producer types in various resting stages within the system or immigration from without (after Odum 1983). (a) Diagram in which changing flows initiate substituted components. Substitution sequences are as follows: (b) last stage in climax; (c) cycles to an intermediate stage; (d) cycles to start.

Rodgers and De Pinto (1981) grew the bluegreen alga, *Anabaena flosaquae*, and the green alga, *Chlorella vulgaris*, in microcosms which included multi-species bacterial assemblages. They found interspecies reinforcement in which the combination of nitrogen-fixing and recycling generated soluble nitrogen for the support of all the components.

Especially in isolated areas such as microcosms, the inflow of species and genetic information may be limiting to the successional process. Robinson and Dickerson (1987) and Maguire (1980) relate the species present to the continuing inflow of invading organisms (Figures 5.1c, 5.2). Multiple seeding is the addition of many more species than are needed or sustainable. Some microcosms receive only initial seeding, whereas others are more like usual ecosystems in receiving some regular seeding on a continuing basis, Where seeding of materials and species is duplicated,

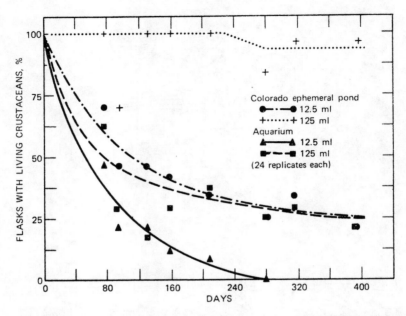

Figure 3.7. Decrease in animal diversity after start of microcosms (Maguire 1980).

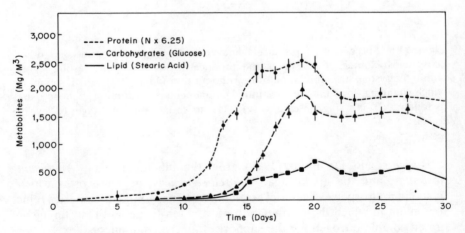

Figure 3.8. Increase in protein, carbohydrates, and lipids in a large plastic bag microcosm reflecting increase in biomass during succession in a marine microcosm (Anita et al. 1963).

similar results have been obtained (Cooke 1967). McCormick, Smith, and Cairns 1991 studied succession in 200 plastic pools for 170 days and related the results to enclosures inhibiting seeding. Dominance of mid and late successional species was reduced with more exclosure. Exclosures

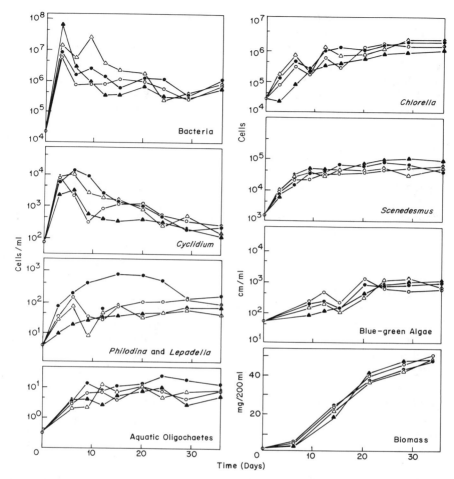

Figure 3.9. Changes in population density of various component organisms and total biomass in a microcosm. Different symbols represent treatments with various concentrations of Beta-BHC (Sugiura et al. 1976, Copyright 1976, Pergamon Press).

that held back dominants had more effect on earlier successional stages. They concluded that increased species availability increases importance of interspecies interactions.

Dormant organisms from preceding stages may also be present, accounting for population shifts in systems that have no access to organisms from the outside. In microcosms without solid input, the dissemules for succeeding stages are frequently found in the form of spores, seeds, resistant eggs, and other resting forms. The creation of a microcosm entails placing an assemblage of organisms under a set of conditions which are different from those under which they lived previously. This usually results in considerable mortality (Figure 3.2). Even when species are continually added, the microcosms often support fewer than the environments from which the microcosms are drawn. Typical successional sequences are found in such diverse microcosms as the gnotobiotic systems of Taub (1969a,b,c), shown in Figure 3.4, and Nixon (1969), in Figure 3.5, the 20-ft diameter plastic bag of Anita et al. (1963) (Figure 3.8), and the 200-ml flask microcosms of Sugiura et al. (1976b) (Figure 3.9).

Nutrient Control

When plants are isolated from an ecosystem and given a nutrient medium, those nutrients in relatively short supply may limit growth, as is often demonstrated with graphs of production as a function of the factor added. However, as Gilliland (1973, 1975) showed, an excess of one factor may depress production, because other required nutrients are depleted to limiting levels by productive use (See Chapter 4). However, when given some time for organization and genetic seeding, ecosystems reorganize and reinforce pathways that eliminate limiting factors, by storing, recycling, developing species that provide the missing nutrient required by others, and favoring species with low nutrient requirements. Addition of a nutrient factor may still stimulate growth, if the new addition represents greater total resource with which self-organization can develop a higher level of total function. Adding a nutrient in order to evaluate whether a factor is limiting, does not evaluate what was limiting before, and gives little indication of what will be limiting after reorganization occurs and the system has adapted to the new condition. On the Pamlico River estuary in North Carolina, Carpenter (1971) added phosphorus enriched water to eight 15-m^3 plastic pools which were multiply seeded with oligohaline natural water containing small species, oysters, and Rangia clams. There was an increase in the amount of nitrogen-fixing blue-green algae, one of the species which contributes to eliminating material limitations.

Eutrophication in Microcosms

The addition of either inorganic nutrients or organic matter, increases both photosynthesis and respiration. This is a situation called eutrophication, and is a condition of higher metabolic energy processing. As Nixon et al. (1986) conclude in their review, increased eutrophication in mesocosms, just as in nature, initially increases the total numbers of microheterotrophs to a greater degree than it increases the numbers of the higher members of the food chain. The larger members in higher trophic levels are more essential to recycling in oligotrophic rather than in new eutrophic systems. However, natural systems of high productivity, such as coral reefs and kelp beds, which have had long periods of self-organization and adaptation, do have well developed and complex higher food chain levels. Also, theory suggests that more abundant resources can support longer food chains. Conditions which result in better-developed eutrophic ecosystems remain controversial. When new conditions arise in nature, self-organization by trial and error and reinforcement develops an ecosystem adapted to and prevailing over patterns with lower performance and reinforcement. The species that survive constitute an information package, and if they are seeded into similar conditions, the new ecosystem can develops more rapidly with less trial and error. If the conditions are uniform, species prevail that can make use of the environmental stability by putting their energies into performance. If the conditions are variable, generalist species prevail, since they can live over a wider range of conditions. The concept is similar to that for adaptability of single species, those adapted to a narrow range of conditions being labeled stenohaline or stenothermal, whereas those adapted to a wide range of conditions are labeled eurythermal or euryhaline.

Transitions Following Enclosure

Information on the process of reorganization and readaptation comes from studies in which plastic enclosures were placed around a column of water, thus separating the inner ecosystem from the larger lake or estuary around it. The plastic enclosure eliminates exchanges of chemicals and species, but retains similar temperature regimes and some turbulence. Starting around 1960, many researchers used plastic bags to enclose planktonic waters in fresh- and salt-water environments as a way to study the ecosystem before it changed. Lund and his associates (1975, 1978) placed large plastic cylinders in English lakes (Figure 3.10) and studied the sequence of changes. The results shown in Figure 3.11 were from a

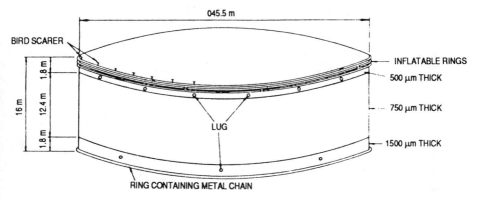

Figure 3.10. Plastic cylinder enclosing plankton ecosystem in English lakes (Lund 1978).

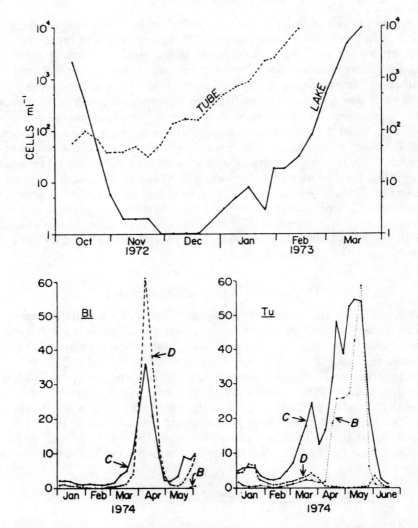

Figure 3.11. Sequence of changes after initiating and fertilizing lake enclosures (Lund 1978). Bl, Blelham Tarn; Tu, Tube; B, blue-green algae, mm/100 ul; C, chlorophyll, ug/l; D, diatoms, cells/5 ul.

cylinder which had been fertilized. Conditions inside were very different from those outside, due to both the fact of enclosure and the fertilization.

Reynolds (1988) presented data from Blelham Tarn, England which showed that plankton developing in lake enclosures exhibited patterns comparable to those in land succession, with initial net production, increase in biomass, rising respiration matching later photosynthesis, and nutrients bound in the living cycle (Figure 3.12).

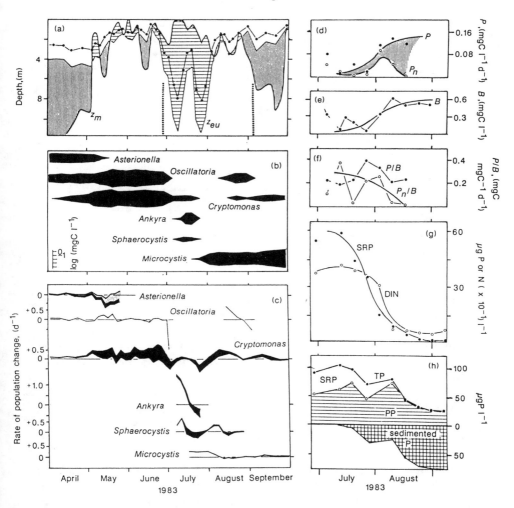

Figure 3.12. Sequence of algal populations in the succession in lake waters enclosed in a plastic chamber within a lake (Reynolds 1988).

Transition in estuaries was summarized in a multi-authored book, *Marine Mesocosms*, edited by Grice and Reeve (1982), containing 30 papers. Emphasis was placed on understanding outdoor plankton ecosystems. The enclosures were used to study parts, processes, and cycles in a state as much like the original ecosystem as possible. However, some of these enclosures were in place for two months or more, during which self-organization and succession occurred and transitions to a changed system were studied. An example of study of transitions is the CEPEX (controlled ecosystem populations experiment) conducted in Saanich Inlet, British

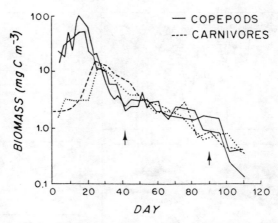

Figure 3.13. Patterns of transition observed after estuarine water columns were enclosed (Reeve et al. 1982).

Columbia. Harris et al. (1982) and Reeve et al. (1982) observed the ecosystems during transition (Figure 3.13). With enclosure, some chemical inflows were disconnected from the system, as were some larger consumers. The result was a wave of consumption, passing up the food chain, culminating in a surge of growth of the largest members present, followed by a general decline.

Ravera (1989) used replicated enclosures for short term (3 weeks) measurements of the effect of added chemical substances as compared to control enclosures. He compared cylindrical enclosures in contact with the bottom sediment with those isolated from the sediment. For example, in an experiment with copper, the sediment reduced the effects of copper on phytoplankton. The intent in this work was to discover the effects of pollutants on the lake system, not to study reorganization and adaptation, for which longer observation periods would be appropriate. In a review of previous papers on enclosures for the study of original systems, Ravera cites Goldman (1962), Lacase (1971), Shapiro (1973), Lack and Lund (1972), Gamble et al. (1977), Gachter (1979), Ravera and Anoni (1980), McQueen and Lean (1983), Kuiper (1981), Uehlinger et al. (1984), and Salki et al. (1985).

For 11 weeks in summer, Raymont and Miller (1962) studied two 20-m^3 concrete tanks, 1.3 m deep, filled with sea water and fertilized with monobasic potassium phosphate and sodium nitrate. Waters were filtered through a coarse plankton net, thus eliminating larger components. Blooms of *Nannochloris*, *Exuviella* and *Nitzschia* developed as most other phytoplankters disappeared. The initially diverse zooplankton populations disappeared, being replaced with dominant populations of *Acartia tonsa* (Figure 3.14). Longer periods are required to develop a more complex,

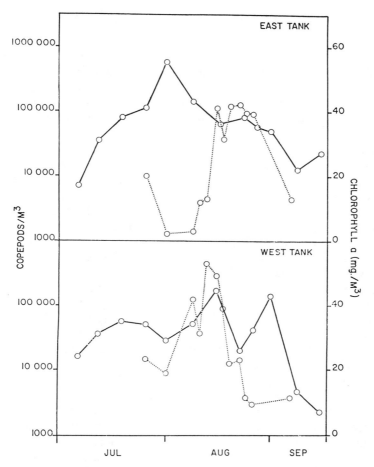

Figure 3.14. Plankton developing in tanks after filling with salt water (Raymont and Miller 1962).

adapted system. Loss of diversity is the general rule in early stages of microcosm organization, possibly because of the loss of space, species which require larger territories, and available niches. Also, intensive inorganic nutrient-type eutrophication turns on a colonization regime in which competition for maximum net growth prevails initially. In cold climates, where only part of the year is available for succession, small, quickly developing organisms can complete growth and succession in one season. These systems have some properties in common with land ecosystems which may require 100 years to mature. See examples from Reynolds (1988) (Figure 3.12).

Heterotrophic and Autotrophic Successions in Microcosms

If a microcosm receives organic matter such as amino acids, peptones, or biomass, consumers and respiration are stimulated, as the model in Figure 2.14 shows. Successional changes are said to be heterotrophic, dominated by bacteria, fungi, small animals, and other consumers. If the inflow of organic matter is continued regularly, a steady-state develops with respiration in excess of photosynthesis, although the latter is also stimulated by the large quantities of nutrients being released. If the organic matter addition is temporary, as a batch, the excess respiration is soon replaced by a surge of photosynthesis, and later both may decrease to lower levels. The relationship between production and respiration in semiclosed microcosms follows a somewhat predictable course. If a source of inorganic nutrients, such as phosphates and nitrates, is added to a microcosm, successional changes will be led by high rates of photosynthetic production (P > R) and large numbers of nutrient-opportunistic species appropriately labeled autotrophic. As the model in Figure 2.14 shows, the inorganic nutrients stimulate photosynthesis. If the nutrient inflow is continuous, a steady-state results, with continuing net production and organic matter either accumulating or being exported, if there is an outflow. If the nutrient addition is temporary, there is a batch response. The excess of photosynthesis and photosynthetic organisms soon disappear, with levels returning to balance, sometimes with a burst of respiration and an increase in the numbers of consumers dependent on the previous burst of organic production. Eventually a steady-state or an oscillating steady-state may be achieved. The relative predominance of autotrophic or heterotrophic conditions is readily indicated by the P/R Ratio: P/R > 1 for autotrophic conditions and P/R < 1 for heterotrophic conditions. Most aquatic microcosms were set up with large initial storages of inorganic nutrients (Taub and Dollar 1964) or organic nutrients (Gordon et al. 1969).

Energy Regimes

One of the well-known problems in scaling physical models of coastal embayments from microcosm-sized experiments, is keeping turbulence proportionately realistic. Special frictional devices are often added to experimental systems to retain turbulence which tends to be lost when physical dimensions are reduced. Nixon et al. (1980) studied the MERL (Microcosm Estuarine Research Laboratory) tanks (Figure 15.6) in relation to Narragansett Bay, Rhode Island, measuring turbulent energy and adjusted the plunger mechanism for the tanks. Biological energy flux was greater than the physical energy flux.

To compare energy contributions of different kinds, each type of energy can be expressed in solar EMERGY units. Calculating solar emergy-determines the amount of one kind of energy required to make each of the others. The term "Emergy", spelled with an "m", was defined as the available energy of one kind, previously used up to make a product, and its unit was defined as the emjoule (Odum 1986, Scienceman 1987). Four of the main energy sources for the MERL microcosms in Rhode Island are compared in Table 3.1. First the energy flux of each is given in Joules per m^2 per day and then the flux is expressed as solar "Emergy" in solar emjoules per m^2 per day. Expressed in Joules, received

Table 3.1. Energy and EMERGY flux contributing to Self Organization in MERL Microcosms.

Note Source	Energy Flux J/m²/day	Solar Transformity sej/Joule	Solar EMERGY Flux, million sej/m²/day
1 Sunlight	4,102,000.00	1	14.1
2 Physical stirring	669.00	25,000	16.7
3 Nitrogen flux	8.00	1.7 E6	13.8
4 Phosphorus flux	1.63	8.0 E6	13.0

* Terms and abbreviations: transformity, sej/J = solar emjoules per Joule; solar EMERGY flux, sej/m²/day = solar emjoules per square meter per day.

[1] Insolation from Nixon et al. (1980) Table 9.

[2] Stirring energy: 0.015 erg/cm³/sec from Nixon et al. (1980)
(0.015 erg/cm³/sec)(13 E6 cm³) (86,400 sec/day)/(10⁷ erg/J)/(2.52 m²) = 669 J/m2/day

[3] Potential contribution of nitrogen in inflow waters. Although inflow waters often had the same or lower nitrogen concentration than the established microcosms with stores from introduced sediments, the potential contribution of the outside influx to a developing system was about 5 ug-atom/l more than in outflow waters.

Nitrogen flux into tank, 735 millimole/m²/yr (Nixon et al. 1986)

Free energy of added nitrogen = $RT(Log_e(C2/C1))$
= (0.33 J/mole/deg)(300 deg)(Log_e (5/1))/(14 g/mole)
= 287 J/gram N

Free energy influx due to nitrogen (735 mm/m²/yr)(14 × 10⁻³ g/mm)(287 J/gN)/(365 d)
= 8.1 J/m²/day

[4] Potential phosphorus contribution of influx waters (Pilson, Oviatt, and Nixon 1980). Although influx waters often had no higher phosphorus concentration than the established microcosms with strong phosphorus recycling from introduced sediments, the potential contribution of the outside influx to a developing system was about 1 ug-atom/l more than in outflow waters.

Free energy difference = $RT(Log_e(C2/C1))$
= (0.33 J/mole/deg)(300 deg)(Log_e(2.0/1.0))/(35 g/mole)
= 49 J/gram P

Phosphorus input:
(480 l/d)(2 ug-at/l)(35 × 10⁻⁶ g/ug-at)(49 J/g) = 1.63 J/day.

per year sunlight energy was the greatest, with stirring energy second, and the chemical potential energy in the added nutrients last. Whereas the individual energies differ greatly, the solar Emergy fluxes are similar. According to the transformity effect hypothesis, self-organization reinforces arrangements in which items that require more energy to make are used to greater effect, thus justifying their role. Similar solar Emergy values may mean that the factors under consideration have similar contributions and effects.

Over long periods of time, unconfined ecosystems are normally exposed to catastrophic stresses, the result of pulses in the larger, surrounding systems. To prevail over long periods, ecosystems develop strategies for sustaining function in spite of catastrophe. The maximum power hypothesis suggests that well adapted ecosystems may have pathways for channeling catastrophic energies for useful purposes, such as accelerating recycling or dispersing propagules. Self-organization under varying conditions results in a microcosm which can maintain its generalist species and characteristics over a wide range of adaptations. Quiet fresh waters integrate large components, such as rooted plants and large animals, which can be damaged and dispersed by catastrophe. Two strategies for adapting exist: (1) some species remain small, reducing the loss caused by catastrophe, and small components can grow back quickly, (2) other species may direct their energies into building resistant structures. Thus, both tiny oceanic plankton and whales are adapted to severe storms and frequent turbulence, but in different ways. In microcosms, many catastrophic stresses and large animals are excluded, but the plankton and its rapid regrowth are a means for the microecosystem to adapt to catastrophic energy surges.

Standardized Microecosystems

Once an ecosystem has matured through self-organization, it can be reestablished or transferred more easily than a developing system, because the necessary species are already selected and collected together. Similarly, a microecosystem already developed in a microcosm can be more readily replicated and more rapidly started under similar conditions. Such standardized microecosystems are useful for research studies and testing of pollution effects.

Beyers' Standardized Successional—Eutrophic Microecosystem

We describe next a case history that was important in developing the standardized microecosystem approach. A standard experimental system was developed for general experimental use in the same way that *Drosophila*, mice, or guinea pigs have been developed for use as standard experimental animals. Thus it was called a "white mouse" microecosystem. Because the conditions were varied during species selection, a very adaptable set of organisms was obtained. The microcosms consisting of these standard assemblages are usually contained in flasks provided with chemically defined media. Aseptic techniques are usually maintained. These systems are capable of undergoing either a heterotrophic or autotrophic succession, depending on the initial composition of the medium.

The assemblage of species was collected from a sewage oxidation pond in the summer of 1959. The system kept in a window showed stability after an initial successional phase. Its gross appearance remained unchanged for several months. At that time there was much discussion as to whether an ecosystem could be reproduced in toto. An in-vitro cultivation was attempted with autoclaved tap water and rat feces, which was then inoculated with a sample from the microcosm. A succession lasting several weeks began with a bacterial bloom, followed by an algal bloom, and then a settling of the particulate matter with an increase in the numbers of grazing ostracods. The final climax stage was identical with that observed in the original vials collected from the pond. Since the succession started with excess organic matter, the succession was led by heterotrophic processes.

In the next step, the medium was a more completely defined chemical environment. Half strength Taub medium (Taub and Dollar 1964) was substituted for tap water, rat feces were replaced with soluble proteose peptone, and once again the medium was autoclaved and an inoculum introduced from the climax microcosm. A succession similar to the one with the rat feces took place. The climax community contained 11 types of bacteria, probably of the genera *Xanthomonas*, *Flavobacterium*, *Pesudomonas*, and *Bacillus*, plus the organisms mentioned above. Five of the 11 types were demonstrated to secrete thiamine (a nutrient required by *Chlorella*) into the medium. The question then arose as to whether an autotrophic succession could be established with the same species by omitting the organic matter and substituting higher levels of nitrate and phosphate and the required amount of thiamine. When this was done, the succession that followed was autotrophic in nature, with the first event being an algal bloom, followed by settling and an increase in the ostracod population. The final climax community was the same (Gorden et al. 1969; Fraleigh 1971; Ferens and Beyers 1972). This is an example of different successions leading to the same end, but following different pathways, a concept much like the original one of Cowles and Clements (Cowles 1899) when they first described the phenomenon of succession.

Total particulate biomass increased from 50 mg/l at the start to 900 mg/l on day 30, while dissolved organic material decreased rapidly from 500 mg/l to 50 mg/l in the first 8 days (Figure 3.15). The constancy of the levels of dissolved organic and inorganic material during climax suggests the operation of homeostatic mechanisms like those modeled in Chapter 2.

Microecosystem metabolism increased during heterotrophic succession. The climax systems showed a diurnal metabolism typical of other aquatic microcosms and balanced aquatic ecosystems. Maximum photosynthesis

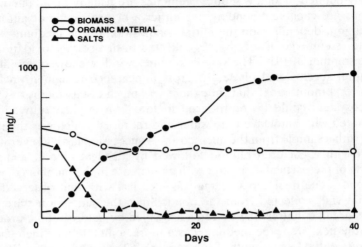

Figure 3.15. Changes in biomass, dissolved organic matter, and dissolved inorganic salts in the microcosm during succession.

and respiration occurred in the first few hours of the day and night respectively. Since the medium was not highly buffered, rather large pH swings (6.5 to 11.0) were observed diurnally.

The biota of these systems consisted mainly of microorganisms, i.e., bacteria (*Pseudomonas spp.*, *Bacillus spp.*, *Flavobacterium spp.*), algae (*Chlorella sp.*, *Schizothrix calciocola*, *Scenedesmus sp.*), protozoa (*Paramecium bursaria*), rotifers (*Lepadella sp.*, Philodina sp.), and an ostracod (*Cypridopsis vidua*). Succession took place in this system very rapidly (45 days). Many of the organisms had resting stages (spores, resistant eggs, etc.). These stages, as well as the active forms of the organisms, served as a preprogramming to accelerate succession. *Chlorella* cells increased to 1.5×10^7 cells/ml, gradually leveling off as the excess plant nutrients were used up. After that, photosynthesis was based on nutrients derived from respiratory regeneration. The final concentration was 6×10^7 cells/ml after 30 days. Thereafter, the population numbers remained constant. *Schizothrix* abundance also increased during succession. As it is a filamentous alga, it was measured in meters of filament per liter (6.3 m/l at climax). *Scenedesmus* appeared sporadically. Among the animals, numbers of the two flagellates peaked around 16 days and then diminished. The rotifers reappeared after 30 days. The ostracod numbers increased to 80 per liter on day 5, no doubt feeding on the bacterial bloom. They disappeared by day 15. However, eggs were present and the ostracods reappeared in the samples on days 17, 22, and 28, indicating that some hatching was taking place. This appearance and disappearance of species with time is quite characteristic of ecological succession in many environments.

In one experiment, a planarian and a daphnid were separately added to the climax system and both species survived. The planarian was eating the ostracods, but did not decimate the population. The *Daphnia* ate the *Chlorella* and reproduced without overgrazing. Gorden et al. (1969), Cooke (1971), and Fraleigh and Dibert (1980) used this microecosystem; Cooke (1967) and Kurihara (1978a, 1978b) used similar systems. The community assemblage was capable of existing as a stable climax over a long period of time. Several of the systems were hermetically sealed in glass flasks in 1969. Living organisms were still present in 1992 as evidenced by green color in the flasks. However, no ostracods were observable and the green pigment had faded somewhat.

Chapter 4
Chemical Cycles and Limiting Factors

The circulation of materials is a necessary part of any system, and the biogeochemical cycling of the chemical elements is a major part of ecosystems, including the macrocosm of the biosphere. When a microcosm is enclosed, the chemicals fall into a pattern of circulation within the small container. If material is neither added nor lost, the principle of conservation of matter applies. An important part of ecosystem studies is chemical cycling, evaluating the storages of material, evaluating the main flows, and determining the processes controlling each pathway.

Studies of material cycles often involve the use of radioactive tracers. Other studies compare two chemical cycles, ascertaining chemical ratios for perspectives on processes. The cycles of major chemical elements sometimes influence the flow of minor chemical elements. Major and minor elements are required for production processes. They can limit production, provide homeostasis as described in Chapter 2, and be organized in compartmental distribution sites by the feedback interactions, maximizing production and utilization. This chapter includes concepts for understanding chemical cycles and their roles in microcosms. Examples are given of studies of cycles, tracers, element ratios, and limiting factors in microcosms.

Essence of Material Cycling in Self-organizing Systems

The simplest case of material cycling in a self-organizing system which has the essence of microcosm thermodynamics, may be the Bernard Cell (Figure 4.1a), discussed by Swenson (1989a, 1989b). The flow of potential energy down a heat gradient generates hexagonal convection cells, and, as in living organisms, a complex structure is maintained which feeds back

Essence of Material Cycling in Self-organizing Systems

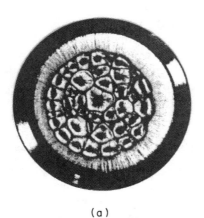

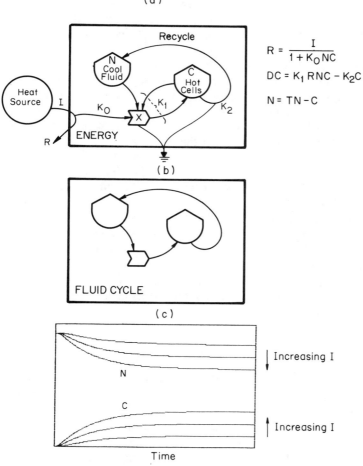

Figure 4.1. Non-living Bernard Cell microcosm with energy-driven material circulation through structure maintained away from equilibrium. (a) Sketch from Swenson (1989, Copyright 1989, Pergamon Press); (b) energy systems model; (c) overlay of model showing fluid cycle; (d) simulation results.

$$R = \frac{I}{1 + K_0 NC}$$

$$DC = K_1 RNC - K_2 C$$

$$N = TN - C$$

to circulate materials, reinforce input processes, and maximize useful power transformations. In Figure 4.1b, the essence of the heat engine microcosm is represented with two compartments (storage tank symbols) between which fluid circulates with a conservation of matter. Fluid in each compartment is at a different temperature and state of turbulence complexity. Consider the more complex ecological microcosms in which there are many materials participating and circulating.

Networks of Compartments and Flow Pathways

In the natural effort to simplify for study, papers on chemical elements often consider only one chemical substance at a time. Compartmental diagrams are drawn to help visualize storages and flows. Numerical values are often written on the diagram or presented in an adjacent table. The flow into or out of a compartment can be compared to the storage in the compartment, in order to see the turnover time (time constant) of that pathway. This process of description is aided by the principle of conservation of matter. The flows into a compartment must be accounted for either as an outflow or as an increase in storage. Each part of the ecosystem where there is a storage is sometimes called a compartment, and the quantity of material located there is a state variable. Diagrams based on this type of premise often omit much of the causal dynamics. See examples of these compartments in Figure 4.2.

Many investigators have used rectangles as the symbol for storage, following Forrester's (1961) notation (Figure 4.2a). This is unfortunate because the rectangle has even broader use in science and engineering as a miscellaneous box for subsystems (black box, white box, etc.). Often, the box is used for a complex subsystem that contains a storage, making the presentation potentially ambiguous as to the kinetics intended. The tank symbol as a representation of storage, as in Figure 4.2b, avoids ambiguity; the rectangle is saved for miscellaneous subsystems. Papers in which models are presented as networks of boxes without equations or computer codes are not rigorous. Simulation results in such papers cannot be checked, understood, or interpreted.

Closed- or Open-to-Matter Microcosms

Storages of materials in microcosms are the sum of the inflow, outflow, and rate of increase (or decrease). If flows of a material from outside the microcosm are high relative to the recycling rate, then processes are less affected by the variation generated within the resident ecosystem. When the cross-boundary exchanges are small, the internal states may vary greatly, driven by growth and oscillations of the microecosystem. Special limitations may develop if the microcosm is closed to matter.

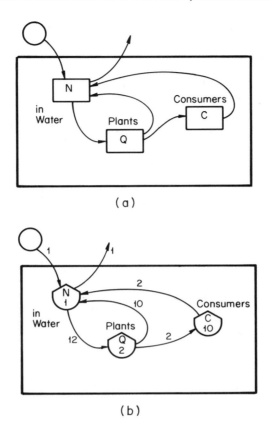

Figure 4.2. Network of compartment symbols representing storages of a chemical element in different parts of a microcosm. Pathways represent flow of the material element in, out, and recycled, without considering energy relationships. (a) Chemical cycle using rectangular boxes as the symbol for compartmental storage; (b) same as (a) but with tank symbol used for compartmental storage units.

Linear Materials Model for a Non-linear Systems Minimodel

Figure 4.3a has the chemical cycle of Figure 4.2 included within a minimodel representation of a system, including the sources, sinks, interactions, and transformations which cause the flows of material. The system is simplified in Figure 4.2b, where the causal variables Y_1, Y_2, and Y_3 are made constant, eliminating nonlinear terms. Since the coefficients refer to a material cycle that is conserved in and out of each process and storage, inflow coefficients and outflow coefficients are equal, as shown. The result of the simplification in Figure 4.2b is a linear model for the material flows of the system. Since the equations are different, simulation results are different.

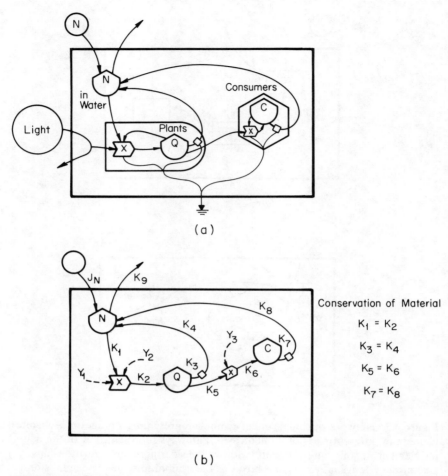

Figure 4.3. Chemical cycle considered as part of a microecosystem. (a) Model of a whole microecosystem in which the materials network in Figure 4.2 is included; (b) biogeochemical diagram as a congruent overlay to the diagram in (a), showing coefficients in and out of each process, storage, or transformation. Where coefficients refer to a conserved material cycle, inflow coefficients equal outflow coefficients, as listed.

Mineral Cycle Overlay for Causal Minimodel

The simplified mineral cycle diagram in Figure 4.3b may be useful for representing the flows of material as an overlay of the whole system diagram in Figure 4.3a. By leaving the symbols for transformations and interactions in the diagram, one can visualize what is driving these pathways. The diagrams in Figure 4.2 are even simpler, but they can be misleading because they omit the transformations and interactions. Nothing shows of the sources, pumping processes, production, etc.

Closed Cycle Homeostasis

Lotka (1925) and Kostitzin (1939) formulated equations for material cycles in systems closed to matter with all pathways behaving linearly (interacting variables constant or absent). Omitting inflow J_N and outflow K_9N from Figure 4.3, leaves a closed-to-matter system with the equations in Figure 4.4. This is a system to which the Lotka theory applies. Results of a simulation are given in Figure 4.4b. However, Lotka (1925) primarily

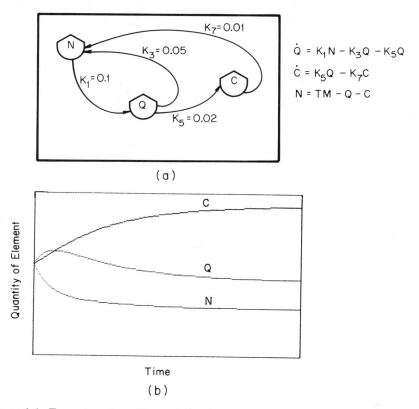

Figure 4.4. Ecosystem from Figure 4.2b with the interactive causal variables (Y's) constant and with external pathways eliminated as appropriate for a closed-to-matter microcosm. The resulting system has closed loop linear equations as modeled by Lotka (1925). (a) Energy system diagram and equations; (b) linear simulation results of the BASIC program CLOSLOOP starting with equal quantities in three compartments. T_m = Total materials; N, Q, and C are in the same mass units.

discussed the steady-state material cycle. A self-regulating homeostasis tends to form as the flow in one part is forced to adjust to that elsewhere in the cycle. The flow through one pathway of the cycle is proportional to the storage preceding it. Storages increase until each outflow is equal to its corresponding inflow. Soon the flows through each part of the cycle have become equal. The result is a distribution of materials in the system that is inversely proportional to the conductivity of the pathway. Materials accumulate ahead of bottlenecks. This Lotka circle homeostasis is one of the self-organizing principles of ecosystems. The concentration of a substance in one place in the system is determined by the rest of the system. Thus, the fate of an individual species is determined by the organization of the rest of the system which controls the materials available to it. The Lotka circle stability has been invoked to explain the homeostasis of the biosphere's chemical cycles (Odum 1950, 1951).

Compartments within a Dynamic Model

Contrast the network of material compartments in Figure 4.2 or Figure 4.4a with the more realistic Figure 4.3a, where more main energy sources and causal interactions are included. The compartments in the dynamic model, where materials are included, are still represented by the tank symbols, but the causal interactions are also included. The model in Figure 4.3 shows the chemical cycle as it is driven by the sources and causal interactions. One should never make a model for a chemical cycle diagram until the system in which it is a part has been diagrammed and modeled as a whole. For example, the chemical cycle in Figure 4.3b overlays the system diagram in Figure 4.3a, where it can be seen to be driven by other variables in several places. If the coefficients refer to a circulating material that is conserved, their in and out coefficients are equal (Figure 4.3b).

Other examples of storage tanks with inflows and outflows of a necessary circulating material (nutrients, carbon dioxide, etc.) as part of a highly aggregated but whole system were given in Chapter 2 (Figures 2.3, 2.6, 2.7, 2.8, 2.9, 2.10, etc.).

Representing Conservation of Material in Closed-to-Matter Microcosms

In closed-to-matter microcosms without inflows or exports of matter, each chemical element is "conserved" (Figure 4.4). This means that the sum of the amounts in each compartment equals a constant total (T_n). For a three-compartment system, a chemical element in each of three compartments adds up to the total in the microcosm.

$$T_n = N_1 + N_2 + N_3$$

If the chemical element is contained within biomass in two compartments, with biomass quantities Q_1 and Q_2, but free in the medium as the third compartment, the conservation equation is:

$$T_n = N_1 + F_1 Q + F_2 C$$

where F_1 and F_2 are the fractional weight of Q and C due to the materials N_2 and N_3. When simulation equations are written for the materials in the system, only two difference (or differential) equations are written. The third compartment is the difference between the total and the calculated quantities in the first two compartments.

$$N_1 = T_n - F_1 Q - F_2 C$$

In Figure 4.4 if N, Q, and C refer to the same circulating substance and are represented with the same mass units, then the conservation of matter can be represented as

$$N = T_M - Q - C$$

Materials Not Conserved

Where microcosms are closed to matter exchange, the total of each chemical element is conserved, not increasing or decreasing. Sometimes there are increases or decreases with uptake or release from the container surfaces. Some elements may be carried out with vapors. Many chemical compounds are not conserved because they are synthesized or decomposed in the system's processes. Whatever the pathway of exchange, synthesis, or decomposition, it is indicated as source, sink, or interaction in the minimodel.

Linear and Dynamic Models

It was a natural step in modeling and simulation to take a chemical cycle diagram, such as the one in Figure 4.4, and regard its pathways as linear (proportional only to the upstream donor quantity). In the process of separating out a chemical cycle, those seeking simplification dropped out all the interaction functions by which the pathways are driven by other sources and storages. Sometimes those doing chemistry were not aware of the dynamics of ecosystems, and were drawn to models without biological control pathways. When a symposium was held concerning a chemical cycle, there was a psychological pressure to see what could be done with only that one cycle. However, all the cycles are energetically coupled, as is to be expected from the maximum EMPOWER principle. Substances of different quality receive more reinforcement, achieve more performance through amplification of interactions, and thus, the interactive multiplier tends to emerge in self-organization. In other words, the system drawn in

Figure 4.4 to suggest linear pathways is really driven by the whole system shown in Figure 4.3a, with nonlinear sources and other multipliers. In other chapters we show important and realistic transients that result from nonlinear interactions. Even though a system is simplified by aggregation it may be important to retain the interactions of production and consumption.

Steady-State Chemical Cycles

It is often useful to make a steady-state diagram in which the inflows and outflows from each compartment must be equal. If data has been collected concerning some flows but not others, missing rates may be inferred from the available data, in order to make inflows equal outflows in all the compartments. Few systems are in steady-state for long, but one may calculate long-range averages as a way to define a conceptual steady-state. See Figure 4.2b for an example.

The storages of systems in steady-state are constant and thus, any process that is nonlinear because of these storages, becomes linear at steady-state. The system in Figure 4.3a could be simplified to the linear one in Figure 4.3b for steady-state conditions. Models of interactive production processes at steady-state may be simplified to linear condition. However, by such aggregation, one eliminates the ability of the model to respond realistically to large and non-steady-state perturbations, as already discussed.

Consequences of Linearizing an Ecosystem Model

To use linear mathematics, with its useful tricks, many have determined average flows and storages, considered the system as a long-range steady-state, and justified the linear simulations as what would happen if the system stayed close to steady-state. This is another way of saying that the model applies only when the perturbation energy is small. This may not be useful when the purpose of modeling is to represent important changes, not small ones. Using linear mathematics to study stability of nonlinear systems is equivalent to determining the system's stability to tiny perturbations, but not asking about its stability to important perturbations. However, linear modeling for purposes of calculating indices may be useful even though linear simulations are not realistic. For example, the relative amount of recycling can be calculated by means of an index from matrix algebra.

Linear Aggregation of the Small-Scale Components Only

The dynamic oscillations generated by the small scale of small, rapidly turning over components are absorbed into the larger effects of consumer compartments which have longer turnover times. The effect on the system

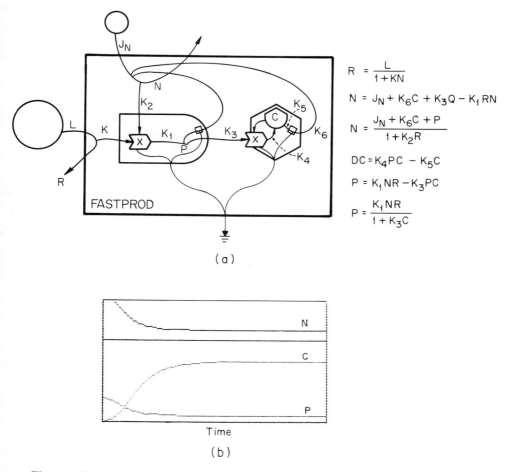

Figure 4.5. System in Figure 4.1 with small, fast components (nutrients and plant producers) aggregated into linear flow without storage, but components higher in energy-space-time hierarchy (to the right in energy systems diagrams) retaining their storage and autocatalytic feedback pathway. (a) Energy systems diagram to compare with Figure 4.3a; (b) simulation results.

of these small oscillations is that of their average state. Models can readily represent these parts of the system as linear, as if they were steady-state. Thus, the small, fast components can be aggregated into linear pathways when the window of time and space of interest is larger. See the example in Figure 4.5, where the small, autocatalytic processes are linearized but the larger ones, which determine the larger-scale periods of perturbation, are described by nonlinear, autocatalytic reinforcement pathways.

Productive Interactions

A production process, represented with an interaction symbol, combines two or more necessary but different commodities to make a product. Often the degree of interaction is proportional to the local availability and/or concentration of the materials as in chemical reactions. Thus, in Figure 4.6a, the primary production process can be isolated from the rest of the system, as shown in Figure 4.6b. The product (K_1AN) is the mathematical product of the source availability, A and the local concentration of material, N. Increasing the input, A while holding N constant, causes a straight line increase in product. Considered in this close-up way, there is no question of limiting factors, because local concentrations were stated as controlled despite the rate of use by the production process.

Limiting Factor to a Production Process

To discover what really happens as a result of chemical limitations, one must increase the scale of view in order to consider the effects of local source concentrations on the process in question. The larger frame in Figure 4.6a is shown in Figure 4.6c. If one increases the availability of input requirement A, the local concentration of N is used more quickly and its concentration is diminished so that the rate of production, P does not increase very much. A typical limiting factor curve results (Figure 4.6c).

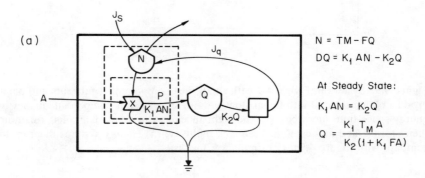

$N = TM - FQ$

$DQ = K_1 AN - K_2 Q$

At Steady State:

$K_1 AN = K_2 Q$

$Q = \dfrac{K_1 T_M A}{K_2(1 + K_1 FA)}$

Figure 4.6. Effect of concentration of material on production process. (a) Energy systems diagram of production process, P within a system with a closed material cycle; (b) close-up view [inner frame in (a)] and the effect on P of setting A at different constant values with N also set constant; (c) limiting factor curves which result when a larger view [outer frame in (a)] of the interaction and local concentrations is considered so that factor N can decrease as A is increased, but recycling, Jq is held constant; (d) limiting factor curve which results when the whole system in (a) responds to increasing A; (e) example: *Spartina* core studied as a microcosm (Gallagher and Daiber 1973).

Limiting Factor to a Production Process

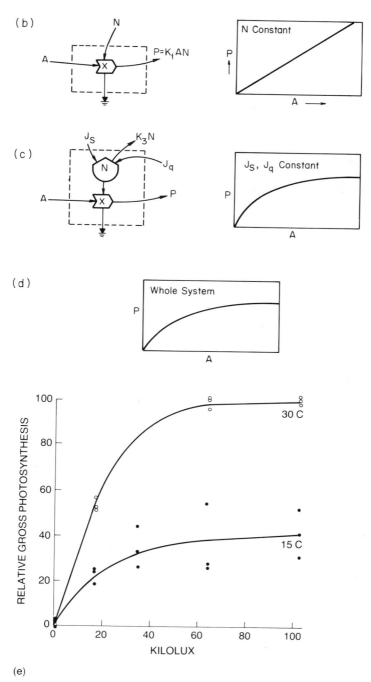

Figure 4.6. *Continued*

Limitations When Considering the Whole System

When the whole system (Figure 4.6a) is considered, and all the equations are included, a different limiting factor curve results (Figure 4.6d). This curve is the result of simulation program RECYCLMT. Within the larger part of the system, the potentially limiting material factors are part of the self-regulating closed-loop cycle. The action of the closed circle is to eliminate any local limitations through the dynamics of production and recycling. If the material, N is limiting, then the regenerative recycling process catches up with production, increasing available N. If production is faster than recycling, then products accumulate, accelerating the recycling process. Thus, the system automatically prevents either the material nutrient availability or the product availability from becoming limiting to the cycle. They become equally limiting. The system as a whole has the relationship described by the Micahelis-Menton equations, which were first developed for enzyme reactions.

Limitations of Total Available Material in a Cycle with Inflow

If the total amount of a necessary material in the whole system is small, but there is an inflow–outflow exchange of that material (Figure 4.6a), then the entire cyclic process of the system builds up the material quantity in the circulating loop until the material is no longer limiting to production.

Limitations of Total Available Material in the Cycle Without Inflow

Using the model in Figure 4.7, the effect of increasing availability of energy, A is considered with successive simulation runs, where the total available material in the microcosm has been raised with each successive run. With larger quantities of materials available, more of the energy is used and transformed.

When severely limited by material shortages, ecosystems with a gene pool of available species may respond by developing different small components. These new components can process the same amount of material faster by reusing the material more rapidly, resulting in higher production and utilization. Higher temperatures favor fast reuse of materials.

Effect of an Excess of One Nutrient

When a requirement is in excess, beyond the concentration where it no longer has a stimulative effect (to the right in Figure 4.6c), it may have an inhibitory effect. Increasing one of two requirements which are in local

Effect of an Excess of One Nutrient

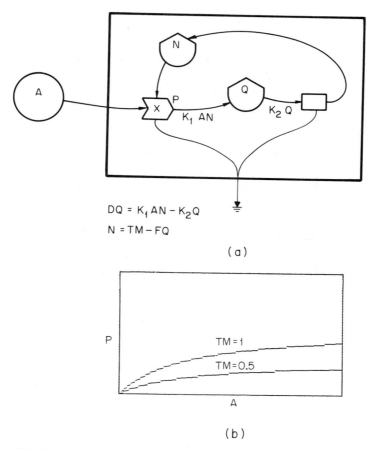

Figure 4.7. System of production, consumption, and recycling when the total material (TM) available for the cycle is small. (a) Energy systems diagram and equations; (b) simulation of production as a function of increasing availability of source, A with program RECYCLMT. More total material in the system increases gross production.

excess, causes the other to be held to such low concentration that the overall process is dampened. Also, there may be toxicity.

Questions have arisen about the excess phosphate released by mining into surrounding waters of ordinary composition. Carpenter (1971) measured productivity in estuarine microcosms, receiving a flow of enriched fresh water mixed with salt water. This was accomplished within a rectangular enclosure connected with an estuary. Photosynthetic production was not greatly stimulated, but populations of bluegreen algae developed. The ecosystem was somewhat reorganized, with nitrogen-fixing plants increasing the capacity of the system to use the excess phosphorus.

Radioactive Tracers

Radioactive tracers have been an important tool in biogeochemical studies. The need to retain radioactive substances within the laboratory has encouraged the use of microcosms. In the 1950s, until the time of maximum radioactive fallout in 1962 due to nuclear testing, studies of the elements of fallout or those with convenient isotopes were widely funded by the atomic energy authorities of several nations.

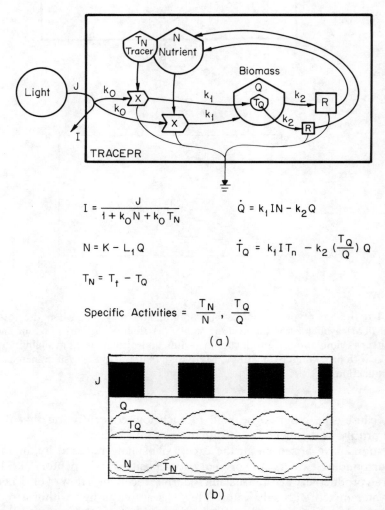

Figure 4.8. Simulation resuls of a nutrient tracer in a basic P and R microcosm (Program: TRACEPR). (a) Energy systems diagram and equations; (b) simulation results where T is the tracer (T_Q in organic biomass; T_N in nutrients in the aquatic medium).

When a radioactive tracer is added to one part of a system with closed circulation, it spreads throughout the system and will be found wherever its non-radioactive counterpart is found. The tracer element and the more abundant isotope (carrier element) are processed with almost the same coefficients as in the example in Figure 4.8. The tracer and its carrier are in separate chemical cycles, circulating similarly. The ratio of tracer to carrier is the "specific activity". When a tracer is added, it flows more rapidly into compartments without tracer until the tracer inflow–outflow relationships are similar to those of the carrier.

If the system is in a steady state, the increases in tracer found in each compartment represent only "catch-up". If the system is changing (with components growing and changing its compartmental rates of turnover), the tracer uptake is partly catch-up and partly system growth or change. A simulation model including tracer and carrier is necessary to make sense out of the data. Figure 4.8 is an example of a model of carrier element, N and tracer, T_n. Each has its own state equations in the simulation. The model keeps the tracer and carrier compartments separate. The ratio of tracer to its carrier (specific activity) is readily calculated by the program, which may be easier than writing equations for specific activity based on various assumptions.

Simple Tracer Uptake in a Stream Microcosm

Cushing et al. (1975) added radioactive zinc to a stream microcosm containing a biomass of attached periphyton plants. They tested a model in which the zinc uptake in the periphyton compartment was a balance between inflow and outflow. Inflow was dependent on the zinc in medium and outflow of zinc to the water was in proportion to that in the biomass. An energy systems diagram for their equations is given in Figure 4.9a. The linear differential equation has an exponential integrated form, which is graphed in Figure 4.9b. The simulation has accounted for most of the uptake (Figure 4.9b). The small deviations from the model suggest a diurnal oscillation, possibly due to a small solar-driven uptake flow. This uptake pathway was added to the model diagram (K_3R).

For systems at steady-state, tracer catch-up is proportional to the distance from steady-state, and thus, linear simulation models correctly apply, though some pathways may be nonlinear. The use of linear models to represent tracer dynamics caused an early emphasis on linear models in systems ecology studies funded by the U.S. Atomic Energy Commission.

Tracer Uptake in a Terrestrial Microcosm Network

Patten and Witkamp (1967) studied radioactive cesium in terrestrial microcosms which were supplied with leaf litter already containing this isotope. Water, simulating rain, was poured through, leaching out some

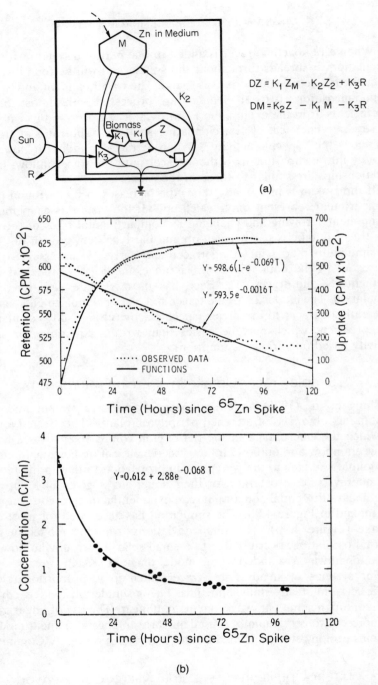

Figure 4.9. Zinc uptake by attached periphyton plants in a stream microcosm (Cushing et al. 1975). (a) Energy systems diagram to which a productivity pathway K_3R was added; (b) graphs of observed uptake and model simulation data on uptake and graph of equation for zinc uptake with time.

of the radioactive material. Other compartments of the microcosm, studied separately and together for tracer uptake characteristics, were soil, microflora, and millipedes. Concentrations of tracer in the litter declined as it increased in other compartments. The catch-up curves of tracer uptake in each compartment, separately and together, were measured and simulated by computer models. They were able to reproduce the tracer uptake curves with linear compartment equations summing the balance of inflow and outflow. The tracer outflows were proportional to tracer content in each compartment. Inflows were proportional to the amount of tracer in the compartment which was the source of the inflow. For the 30-day periods of study, tracer catch-up was the main process.

Effect of Carrier Flow on Tracer Flow

The amount of tracer which appears in a compartment depends on the amount of carrier in the system. For example, aquaria studied by Whittaker (1961) included examples with large and small phosphorus concentrations. With similar amounts of radioactive phosphorus tracer added to each aquarium, tracer was relatively more concentrated (with higher specific activity) where the carrier concentrations were small. Greater organism growth, incorporating more phosphorus in biomass, incorporated more tracer.

Errors in Radiocarbon Uptake Measurements of Photosynthesis

Many radiotracer studies of ecosystems were not accompanied by reference to the kinetics models necessary to interpret the results. Most of the use of radioactive carbon in ecological enclosures to measure photosynthetic uptake may have resulted in incorrect conclusions because the degree of catch-up of the tracer to its nonradioactive carrier was not calculated by means of the appropriate equations. Some published production rates are erroneous; the rates of uptake are partly the result of the catch-up process and partly due to net growth. Most radiocarbon photosynthesis experiments not making use of a model which incorporates tracer catch-up information do not provide insight on either gross assimilation or net assimilation, but something in between. The shorter and faster the turnover time, the larger the errors, so simply comparing radiocarbon uptake rates of large cells with that of small cells was erroneous. Wilson (1963), using microcosms in which he measured total carbon budgets, clarified the model and the relationships necessary for understanding radiocarbon uptake as a means to measure photosynthesis.

Because turnover times in different parts of an ecosystem vary, the errors in interpreting radiotracer distributions differ for each compartment. Whittaker (1961) determined radioactive phosphorus distribution within an aquarium microcosm at intervals of time after adding the tracer to the water, without the model necessary to infer rates correctly. Where a tracer is added to water and moves into other compartments, concentrations of the radioactive element peak and then decline as the tracer becomes more evenly distributed. However, one cannot infer the position of a compartment in the biogeochemical cycle by the timing of its radiotracer peak, because the turnover time depends on the storage and the pathway rates. Large storage takes longer for catch-up.

Steady-State Tracer and Carrier Distribution

Much later, when the tracer is distributed among all the compartments of the cycle, and the ratio of tracer to carrier (specific activity) is the same in all phases, one may use the tracer distribution to infer the carrier distribution. For example, radioactive potassium is a naturally occurring tracer in the biosphere (not due to fallout) and determining the potassium distribution with a gamma spectrometer is a method of analyzing the distribution of potassium within a system.

Small Differences in Isotope Processing

Although two isotopes of the same chemical element have the same chemistry, there are small differences in some of their flows due to their difference in atomic weight. Thus, the two isotopes become distributed slightly differently even when they are in steady-state relative to each other. Although very accurate measurements are required, ratios of isotopes of the same chemical element in compartments are used to infer the kinds of processes that contributed material to that compartment. The differences are so small that the custom is to move the decimal over three places. Thus the ratio 1.0235:1 is represented as a difference of 23.5. Gearing et al. (1984) studied the ratios of carbon-13 to carbon-14 in the MERL plankton microcosms (see Chapter 13). Sediments developing in MERL tanks had organic carbon ratios of -20.9 to 22.5 with more positive values (smaller negative values of -16 to -19) for carnivores. Land-origin carbon has a ratio of -26 and marine-origin carbon a ratio of -21.8.

Using Element Ratios to Study Two Coupled Chemical Cycles

In microcosms, as in studies of the biogeochemistry of large scale nature, ratios of two elements are sometimes used to understand the different behavior of chemical elements. For example, the movement of chemical elements from water to organisms to dead biomass to skeletons, etc. involve great changes in concentration, but for two somewhat similar elements traveling similar pathways, the element ratios indicate how the use of the elements differs. This is particularly useful where one is a common element with a familiar cycle and the other is a lesser understood trace element.

H.T. Odum's doctoral dissertation on Biogeochemistry of Strontium included experimental analysis of strontium in duckweed microcosms, with application of the Lotka principle, to explain the observed partition of strontium between water and snails in the ecosystem (Odum 1950, 1951). In Figure 4.10, a simulation model is given to represent the strontium

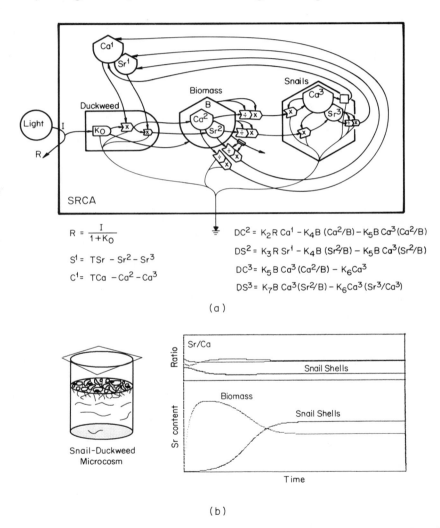

Figure 4.10. Simulation model of duckweed–snail microcosms and the cycles of calcium and strontium. (a) Systems diagram; (b) simulation results; (c) data from studies of duckweed–snail microcosms (Odum 1950, 1951, copright by the AAAS).

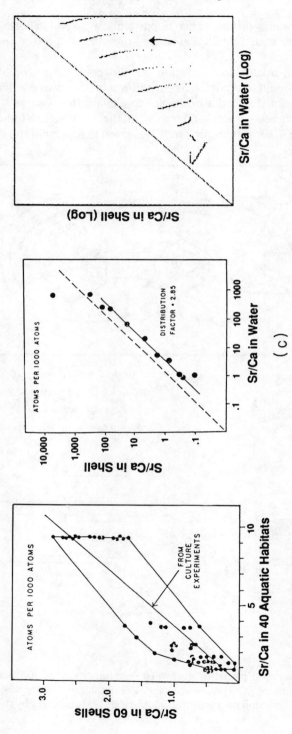

Figure 4.10. *Continued*

and calcium cycles within these microcosms. The model reaches a steady state (Figure 4.10b) for different strontium/calcium ratios in water. Different ratios are found in skeletons, with some exclusion of the minor element strontium when it is in low concentrations (Figure 4.10c).

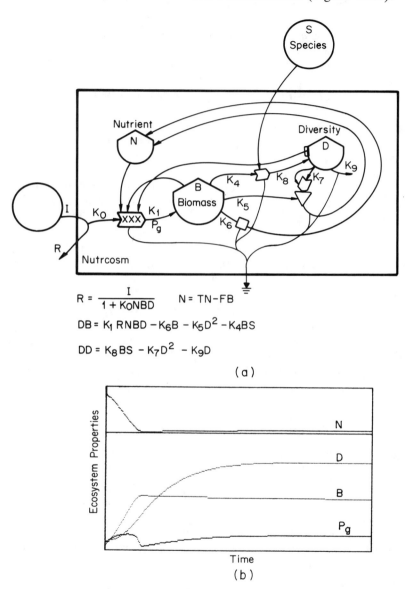

Figure 4.11. Model of succession minimodel from Figure 3.1 with nutrient cycle added (Program: NUTRCOSM) T_N and F as in Figure 4.7. P_g, gross production = $K_1 RNBD$. (a) Systems diagram; (b) simulation results.

Mineral Cycles Added to Succession Minimodel

In Figure 4.11, the minimodel NUTRCOSM is used to consider succession with a nutrient mineral cycle added (such as phosphorus). Here, primary production is a function of the concentration of the nutrient element, and the amount returned to the medium is proportional to the amounts incorporated into the biomass compartments. Including an element cycle in a holistic model is a better way of approaching a chemical element cycle than modeling it separately. One may still separate out a congruent diagram with the single element cycles alone (Figure 4.10a), but it should always be next to or overlaid on the network that is driving the flows.

The simulation of the model (Figure 4.11a) includes some properties often seen in nature. If TN, the total nutrient available, is small, gross production, biomass, and diversity are all small at steady-state. If TN is made large, so that it is not limiting, the resulting curves are similar to those of the model without nutrients (Figure 3.1), except that succession is quicker. The results in Figure 4.11b depend on an intermediate nutrient availability that causes the system to develop two sequential stages. The first involves nutrient binding and gross production going into biomass increase. Stage two occurs when excess nutrients, N, are incorporated, biomass increase stops, but gross production is restored by an increase in diversity and efficiency. This pattern of two stages has been observed in microcosms (Figure 3.15) as well as in natural succession.

Figure 4.12 (Fraleigh and Dibert 1980) shows rates of gross production and community respiration in 500-ml flask microcosms with different amounts of phosphorus (higher concentrations toward the bottom of the figure). Conditions in these microcosms involved limitations to available carbon dioxide as well as light. Gross production and total respiration were similar, increasing together during early succession. There were community changes, including some oscillation in chlorophyll quantities.

Adaptation for Maximum Power

The self-organization which occurs in ecosystems, reinforces species and varieties adapted to the energy regime. In other words, the coefficients for models such as that in Figure 4.11 may be changed toward higher input and transformation efficiencies during the successional period. For example, Smith (1980) adapted benthic estuarine microcosms to three intensities of light and then tested each microcosm at each light level to determine the amounts of primary production and community respiration. The microcosm with the longest period of adaptation had the highest performance at each light condition, compared to those more recently transferred from other light conditions. Figure 4.13 displays one of the results. Shortly after the change of conditions, there was 10 days of transition in which the P/R ratios were variable.

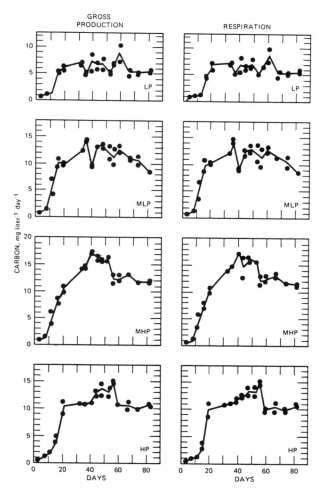

Figure 4.12. Gross production and total respiration during succession in which there were limitations in carbon availability (Fraleigh and Dibert 1980).

Exchange of Excess Material Resources in Lateral Organization

When an ecosystem receives an excess of a resource, it reorganizes to make more use of the material by stimulating development of those components which can process the excess. However, by exporting it to the next area, where that resource is not in excess, greater total performance may be obtained. Kelly and Levin (1986), reviewing high nutrient

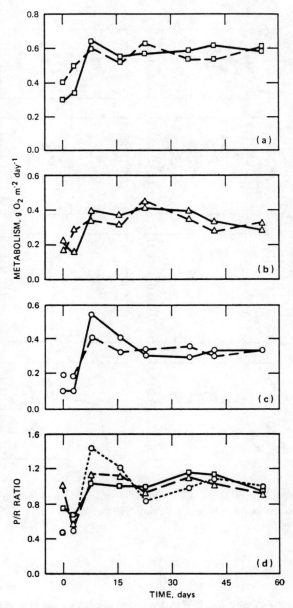

Figure 4.13. Role of previous adaptation of microcosms in maximizing gross production (solid line) and respiration (dashed line) when changed to a new light intensity (Smith 1980). Previous adaptation: squares, high light; trianges, medium light; circles, low light.

levels in some coastal marine waters, concluded these resources were in excess of the plants' ability to use. Further seaward, these nutrients do stimulate higher production levels. The observed pattern is consistent with good lateral self-organization for maximum regional performance. Good use of nutrients in the net production of food chains may benefit the larger system considered as a whole. In effect, it makes the inshore a waste treatment subsystem, which of course has other drawbacks.

Isolated microcosms cannot make lateral exchanges and may be incomplete for representing events in the world continuum. However, microcosms can be connected to simulate the process of exchange between ecosystem areas. See, for example, the salinity gradient microcosms of Copeland et al. (1972) and Figure 14.9.

State Hyperspace for Representing Normal and Disturbed Ecosystems

The normal range of state variables (properties like biomass, temperature, species numbers, diversity, phosphorus, etc.) for an ecosystem adapted to a set of environmental source influences can be represented on graphs of variables. Two state variables can be shown on paper at one time, where their normal range describes an area called the "state space." Such a graph has the same coordinates that one uses for phase plane representations of data or simulation results (Figure 2.4). A similar polygon may be described mathematically for many variables, even if it cannot be represented on a two-dimensional paper. Hutchinson (1978) called this polygon of characteristics an "n-dimensional hyperspace" and used it to define a niche for a species. Kersting (1988) uses these polygons to characterize normal microcosms and compare them with disturbed regimes where the polygon is changed. This method was suggested because, in this way, more of the interactive changes involved in ecosystem dynamics would be shown in one measure. Beyond three variables, sensitivity of the measure increases with the number of state variables included.

The perimeter of areas on the phase plane where an ecosystem oscillates is sometimes called an attractor. Microcosms usually change their space during succession and do so often after they adapt to a new condition. Microcosms may shift their spaces and attractors as part of chaotic oscillations (Figure 6.10f).

Product Index Versus Summation Indices of System State

Hyperspace is a linear concept where each variable adds equally. However, most state variables, being of different transformities, multiply their interactions as is necessary to maximize amplification of interactions between high quality and lower quality transformity factors. Thus, a multiplicative index might be more relevant for representing ecosystem performance.

Chapter 5
Diversity and Information

An ecosystem may be described as the biodiversity of its species plus the connecting flows of matter and energy by which they have become organized. Understanding the role of species is a major goal of ecology, and microcosms are one way of studying species' roles. Because containers limit access, lack of diversity often limits ecosystem development in microcosms. Even when continuously seeded, microcosms have lower diversity than the environments from which their species are derived or to which they are being compared.

In our systems diagrams, diversity and organization can be represented in two ways: (1) in detail, showing each species with a symbol and pathway lines representing functional connections to other species (Figure 2.20), or (2) aggregated, showing the quantity of species information as a single diversity tank (D, in Figure 3.1, S, in Figure 5.1). Adding diversity adds information. The word "information" in this book means description of the *components and connections of a systems organization*. Within each organism there is the inherited genetic information of its organismal system, the valuable storehouse of prior evolution introduced with each species. On a larger scale there is in the ecosystem information in the diversity of species, their proportions, and interrelationships. Indices of diversity are used to measure the genetic variety present and the complexity of connectance. For example, there are counts of species and counts of pathways of interactions.

In information theory, the index used is the logarithm of the number of components and connections. Information theory formulas by calculating the logarithm to the base 2 of the possible combinations (N) represent information (I) in bits ($I = K \log_2 N$). A bit is a yes–no decision. Diversity, expressed in bits, is the number of decisions required to des-

Diversity and Information

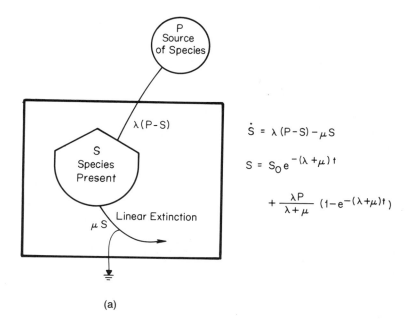

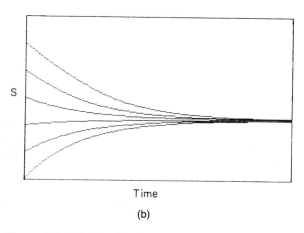

Figure 5.1. Model of balance of continuous introductions and linear extinction (MacArthur and Wilson 1967) SPECTANK. (a) Energy systems diagram and equations; (b) simulation results for microcosms with different starting conditions; (c) Example of colonization curves with pattern resembling the simulations (Maguire 1971. Reproduced with permission from the Annual Review of Ecology and Systems, Vol. 2, copyright 1971 by Annual Reviews, Inc.); (d) simulation results of different seeding rates. The lowermost curve has zero seeding.

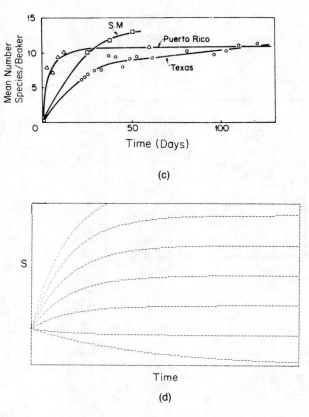

Figure 5.1. *Continued*

cribe the diversity. One popular expression, the Shannon-Wiener-Weaver equation (Wiener 1948; Shannon and Weaver 1949), is widely used to represent ecosystem biodiversity in bits per individual. Values range from 0, where there is only one species, to 7 bits per individual in some complex rain forests.

Emergy is the energy of one kind required directly and indirectly to generate a product (Chapter 3). The solar Emergy required for product formation may be used as a measure of information in ecosystems. The solar Emergy required to form a species is large, whereas the solar Emergy required to copy a species in reproduction, or transport a propagule is small. The Emergy per bit required for sustaining small species, such as microorganisms, is smaller than the Emergy per bit for larger organisms.

Some ecosystems have production, consumption, regular food chains, hierarchies, chemical cycles, etc. but only a few species. Others carry out the same functions, but with a great diversity of species having a large degree of complexity in their relationships. Lane and Collins (1985) applied loop analysis diagrams and derived matrix methods for representing the relationships of species in microcosms. This network analysis shows the distribution of mutualistic and competitive loops. Lane and Blouin (1984) using a similarity index found a 62% overlap between the networks inside lake enclosures and those in the free waters. More interspecies connections and stability were found in those with less acid. How many species functionally contribute to the major pathways, and how many more may be present without major functional roles? Having more species is a result of adaptation to change, but change also tends to eliminate species. Environments with less change can develop more complexity and diversity.

The maximum power theory of self-organization (Chapter 1) proposes that those species which are mutually reinforcing become functional dominants. It also suggests that the storing of many other species with only minor functions, may also contribute in the long term. By facilitating an efficient substitution of species, a gene pool helps maintain functions during times of change. After initial colonization has maximized what can be done with explosive exponential and exclusionary stages of growth, succession tends to develop specialization and diversity (Chapter 3). If productivity increases as microcosms self-organize, it is evidence of the maximum power principle. If consumers develop which are more useful to the microcosm system (by increasing interactions in order to amplify other processes, as in Figure 4.6b), this is more evidence. Consumers can reinforce a system and their own continuation by recycling materials more efficiently, helping with reproduction of other species, assisting with spatial organization, maintaining patches of diversity, eliminating destructive competition, etc. The MERL plankton-column microcosms (Chapter 15) were provided with continual seeding by pumped sea water. Smith et al. (1982), from studies in MERL, found similarity of species to be a good index for the results of mesocosm treatments. These studies showed the way a gene pool of little-used species provides for higher production and efficiency by providing quick adaptation to change. A pool of extra species allows continual fine-tuning of ecosystems in response to small changes. In ecological theory, as in economics, a greater diversity of processes and functions facilitated by a diversity of specialists, contributes to improved total system function. Although maintaining diversity information requires expending resources, diversity is self-reinforcing by increasing intake and efficient use of those resources. Simplified views relating diversity and function come with the simulation minimodels considered in this chapter. Kurihara, Shikano, and Toda (1990) introducing more microbial species increased growth rates by favoring competition and coexistence.

Diversity as a Balance of Information Inflow and Extinction

One factor in the maintenance of genetic information and diversity is the seeding from outside gene pools. The model of the balance of seeding and species extinctions provided by MacArthur and Wilson (1967) is given in Figure 5.1 (a variation on the single storage tank model in Figure 2.2). These authors included a "backforce" in their equation, with the inflow dependent not only on the external pool but on the difference between that pool and the diversity pool within the system. The model was important to experimental field testing. For example, Simberloff and Wilson (1970) measured colonization curves on tiny islands, and those curves resemble the first-order charge-up curves of the model. The phrase "Charge up" comes from the storing of electrical charge on a capacitor which has the same system design and mathematics as the model in Figure 5.1. Microcosms are tiny islands in which species seeding is controlled and often minimized. Maguire (1971) put open jar microcosms within the rain forest where they acquired species of microzoa from the environment. Charge-up curves (Fig. 5.1c) were like those in Figure 5.1b, leveling by the tenth day. Similar ones in Texas required more than 100 days. Those placed within the ionizing radiation field of the rainforest at El Verde Puerto Rico during a period of application of .66 MeV gamma radiation had as many species at 100,000 rads as did controls. Robinson and Dickerson (1984, 1987) and Dickerson and Robinson (1986) compared microcosms to islands for testing theories of island biogeography, especially the model in Figure 5.1. They seeded beaker microcosms from a larger, higher-diversity species pool with colonization curves similar to those in Figure 5.2. The regular exchange of species may counteract any tendencies for competitive exclusion, another way of maintaining useful diversity. Mitchell et al. (1971) maintained plankton microcosms with high diversities by introducing the population pressure of newly seeded individuals at regular intervals.

An old question in ecology is how much advantage is provided to a species that can get started first. When populations are in exponential growth, other things being equal, the one which starts ahead of the others rapidly expands its dominance over later arrivals. However ecosystems, including microcosms, do not have uncontrolled populations engaging in this kind of competition for long. How important is the time of arrival? Cairns and Yongue (1973) studied the effect of seeding on protozoan communities. Beyond an initial amount, increased seeding did not lead to higher numbers of species. Diversities temporarily raised by additions declined back to lower values (6 to 12 species per microcosm). They observed differences attributable to the timing of species arrivals. Half the invading species were successful. Robinson and Dickerson (1987) studied the effect of initial seeding on algal species developing in beaker

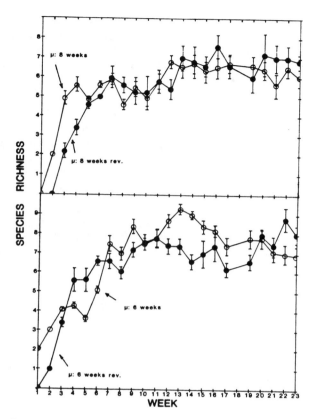

Figure 5.2. Colonization of species observed in microcosms by Robinson and Dickerson (1987) where sequences of species introduction were repeated every 6 weeks and every 8 weeks. Upper graph introduced species in reverse order of those in the lower graph.

microcosms. Differences corresponded to the order in which species were added. Figure 5.2 shows the effect of more frequent seeding and the effect of reversing the order of seeding. The graphs show that the general level of diversity achieved was about the same. It was slightly higher with more rapid seeding. However, different dominants resulted from different histories. In some cases, the organisms arriving first prevailed. If available species do not provide the best possible system for the resources available, the overall system may be displaced later, when other species are seeded. Microcosms synthesized with just a few species, such as those by Taub (1969a,b,c) may be in this category. Integrity is maintained when seeding is prevented. These microcosms are analogous to some ecosystems which evolved on isolated islands.

Microcosms, as outdoor ecosystems, all tend to be different. If a set of microcosms is started with each microcosm receiving similar seeding, the replications are often different in their prevailing species. However, if they are mixed together regularly, they will be more similar. Such mixing is a principal technique for preparing a set of replications for experimental treatments in which duplicates are desired. Robinson and Dickerson (1987) concluded that alternative ecosystems were operating in their replicate beakers. Mixing may tend to favor higher performances by providing better opportunities for combining the set of species with greater mutual reinforcement.

In another set of experiments by Dickerson and Robinson (1986), the more heavily seeded microcosms did not sustain their maximum diversities, but declined. When starting conditions are in excess of what can be maintained, a successional graph of the results goes through a maximum (Figure 5.3). When species are new, their maintenance requirement is less and greater numbers can be maintained than when the species are at an older average age. The relative importance of each species can be

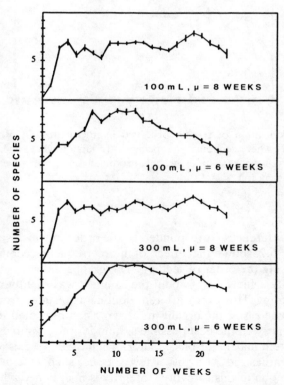

Figure 5.3. Colonization of species in beaker microcosms which showed maxima (Dickerson and Robinson 1986).

Diversity as a Balance of Information Inflow and Extinction

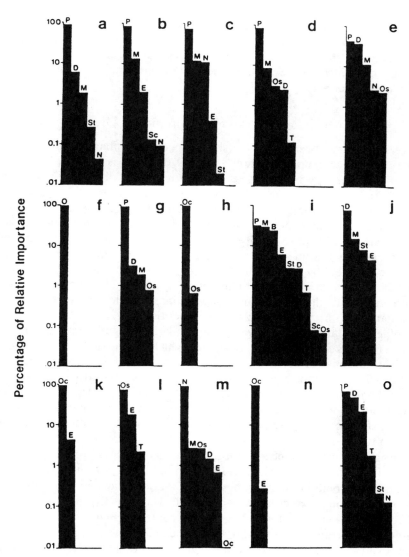

Figure 5.4. Rank-order diagram of species in beaker microcosms (Dickerson and Robinson 1984).

represented in rank-order graphs such as those in Figure 5.4 describing beaker microcosms. The rank-order graphs for microcosms are like those for the outer world except that they are truncated, with rarer categories represented less often. The shapes of these graphs can be generated by fractional division, as if they resulted from random influences, or by causal hypotheses, such as our hierarchical concepts of support. Hierarchical support is defined as a larger number of the first species being required for a lesser number of the second etc. (Odum 1983a, Chapter 18). There is an equivalence between species-area curves and rank-order curves, one calculable from the other, since one counts the more important species sooner.

A Model of Energy for Maintaining Diversity

A relationship between the diversity of species maintained and the available energy for support (Odum 1963, 1970, 1971) is modeled in Figure 5.5a. When energy is increased, there is a greater diversity maintained, but there is a diminishing return because of the greater requirements of the species interactions. Species either compete with negative effects, or expend energy on structures or behavior to separate their niche roles. Both paths may require energy in proportion to the square of the number of species. In this model, any reduction of energy available to species is accompanied by extinctions. The model shows why a considerable metabolic base is required to maintain species and their information. Pathway K_3 is the energy used up in inter-species interactions or in adaptations to prevent interspecies competition.

When available energy is varied, a simulation (Figure 5.5b) generates a curve (species verses energy available) which resembles the species-area curve traditional in ecology. For ecosystems operating on resources that are provided on an area basis (sun, rain, geological uplift to weathering), increasing the area examined increases the energy realm. If more than one kind of energy is contributing to the diversity, Emergy should be used instead of energy since Emergy expresses all kinds of energy in units of one type.

Insight on maintenance of information was obtained from experiments by Ferens and Beyers (1972). Ionizing radiation was used to destroy genetic information in microbes (both autotrophic and heterotrophic) prior to starting a microcosm with them. The growth–succession which followed was smaller than that of the controls (Figures 7.4 and 7.5). Holding the irradiated organisms for a time before initiating the microcosm improved their ability to develop growth and production (Figures 7.6 and 7.7). Apparently, there was some kind of repair and error removal process. The experiments show the way self-organizing systems can process the information for system function. Similar results were found with both autotrophic and heterotrophic organisms.

In microcosms, after the abundant dominant species are established, there appear to be few of the rarer species. Curves, of species present as a function of area sampled, flatten after reaching a shoulder. The absence of the appearance of additional species after early samples are taken may indicate the role of the container in eliminating those which were initially present but, requiring a larger minimum area, failed to survive. See for

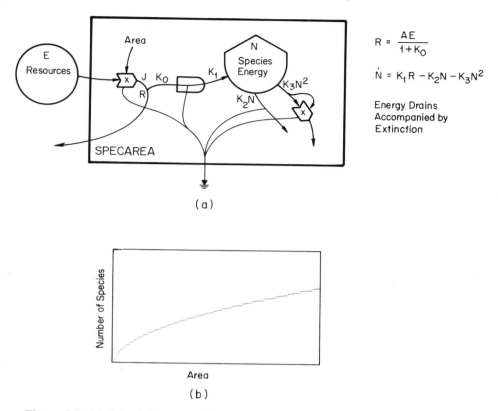

Figure 5.5. Model of the energy basis for diversity as a balance between input and quadratic requirements for information maintenance (Program: SPECAREA). (a) Energy systems diagram; (b) results of simulation of a species-area curve by increasing the area A in the model.

example the species diversity counts in our flowing stream microcosms (Odum and Hoskin 1957). In microcosms, small size may reduce the number of species to those that can remain in small densities, available for future contingencies. Even with frequent seeding, only four to nine species were maintained in the beaker microcosms of Robinson and Dikerson (1984, 1987). These researchers, using an index from Berger and Parker (1970) found 76 to 85% of the individuals to be of the first dominant species.

Although not typical of less constrained ecosystems, the flat curves make counting species in microcosms easier. For example, Dickerson and Robinson (1985) sampled their phytoplankton microcosms successively, and obtained the cumulative species curves in Figure 5.6. This kind of evidence suggests that enclosed sanctuaries are no means for protecting the world's gene pool. The situation is much different in the sea, with its

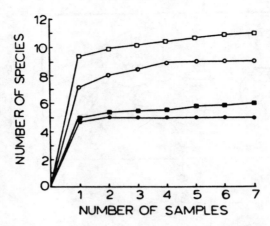

Figure 5.6. Species sampling curves from aquatic microcosms with restricted species access (Dickerson and Robinson 1985).

restless tidal exchanges of species, compared to land, which is being converted into isolated islands by private property practices.

Combining Mechanisms to Make a Successional Minimodel

In Figures 3.1 and 5.7 the two important aspects of short-run diversity are combined (Odum 1983a). The energy required for support, from Figure 5.5, is combined with the model of introductions and extinctions from Figure 5.1. Figure 5.7b–e has simulations of the self-organization that results from the model's interplay of information and energy. Of course, there are important factors beside genetic seeding and resource inputs in real world systems, some of which are included in models discussed below.

The changes in biomass, functional relationships, and diversity which occur in ecosystems with time, are called succession (Chapter 3). Patterns during succession depend partly on the initial starting conditions and partly on the inflow of resources and diversity. Although including only several primary factors, simulations of the minimodels (Figures 3.1 and 5.7) predict some of the main properties of succession. These include growth in biomass, temporary high net production, and rising diversity (Figure 5.7b–d). There are two gross production pathways, k_1 and k_8. The first is maximized by high diversity, reinforcing photosynthesis. The second is amplified by maximizing autocatalytic feedback from the biomass. Figure 5.7b begins with the system low in biomass and species but with available resources and species seeding. Rapid organization with more species develops a kind of climax (Figure 5.2). Both in the outdoor world and in microcosms, major substitutions of dominants are observed with succession. For example, microcosms developed by Kelly (1970), operating with one set of species, after a year, suddenly replaced their species with a different set. The conditions generated by one set make possible another.

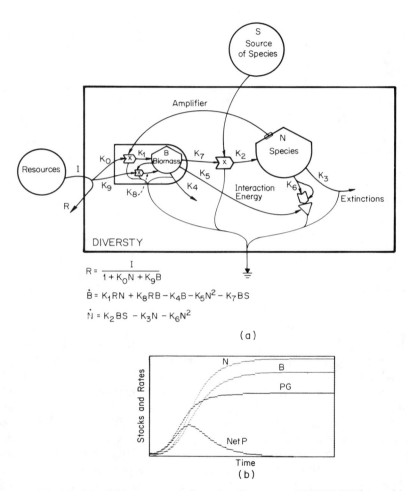

Figure 5.7. Model of biomass and diversity (Program: DIVERSTY) combining the introductions–extinction mechanisms from Figure 5.1 and the energy-support mechanisms from Figure 5.2. P_g = gross production = $K_1RN + K_8RB$; Net P = dB/dt. (a) Energy systems diagram and equations; (b) results of simulation of system organization starting with low diversity, $N = 1$, low biomass, $B = 100$, and available species seeding, $S = 1$; (c) results of simulation with initial large species introduction and restricted seeding thereafter; (d) and (e) results of simulations comparing low levels of input resources with high levels.

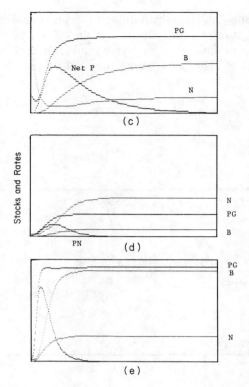

Figure 5.7. *Continued*

The simulations in Figure 5.7 show what is possible if the species present are able to utilize the opportunities provided. Species adapted for generating high net production in a microcosm's early stages are likely to be replaced with those which can generate higher gross production, more lasting products, and better recycling. The model helps determine when the typical successional pattern may be different. Figure 5.7c shows the results of a simulation which starts with a heavy seeding of species and organic matter, but undergoes no later seeding. Many microcosms have been processed in this way. The result is an initial precipitous drop in the number of species, excess respiration over production, followed by rapid growth in biomass and low levels of diversity. There is a high net production, partly because lower diversity requires less respiratory consumption. Initial efforts to start microcosms with a high diversity have usually failed. Most species disappear, but after substantial biomass and metabolism develop, the system gradually recovers some diversity and performance. The model predicts that if more species are added initially than can be supported, diversity declines to a sustainable level, as seen in the simulation results in Figure 5.7c. See also the example where Silver Springs periphyton was added to a flowing water microcosm (Figure 10.7).

When a model such as that described in Figures 5.1 and 5.7, has species introduction cut off, extinction causes diversity to go to zero. However, extinction is not what is observed in small ecosystems if resources are made available. There are many testimonials about sealed microcosms that have gone on for years with a minimal diversity but without extinction. See examples from Folsome in Figure 20.3. Apparently, there are extinction resistant species specialized to prevail with high reliability in conditions of little species competition. A threshold was added to the model DIVERSTY (Figure 5.7), stopping extinctions at a low threshold (Figure 5.5d).

Nutrients, Diversity, and Longitudinal Succession

Everything else being equal, the more resources provided an ecosystem, the more life it can support. When models such as that in Figure 5.7 are given more input resources and allowed to reach a steady state, higher levels of diversity result (Figure 5.7e). In Chapter 3, questions were raised about high nutrient levels (eutrophy) and diversity. Minimodels can generate high productivity without nutrient inflow or organic outflow (Figures 2.3, 2.11, and 2.12) if nutrients are in abundance and recycled. Closed-to-nutrient systems with high productivity based on recycling, are eutrophic in the sense of high photosynthetic production and respiration (older definition), but not eutrophic in the sense of having excess nutrients in their waters. These systems have high gross production, but little net product. Diverse species facilitate good recycling. Thus, on theoretical grounds and from observations in microcosms and nature, high diversity is associated with tight-cycle productivity. The diversity facilitates high gross production, thus reinforcing that system over any other designs with less gross production and recycling. This is apparently the best system design for maximum power (productivity) when nutrient inputs are small.

In an ecosystem or microcosm started with an excess of nutrients, a first stage is the low-diversity, weedy, net production phase, which binds excess nutrients and prepares the system for a subsequent high-diversity, tight-cycle type of climax. There is low-diversity net production first and later high diversity without much net production. Margalef (1969), studying external environments and microcosms, found higher diversities in oligotrophic waters. He found fertilized algal cultures with low diversity. Many of his microcosm cultures went from a low-diversity successional state, where nutrients were still in excess, to a state with tightly bound nutrients and higher diversity (low net production and probable lower gross production).

Where nutrient materials are maintained in great excess by inflows, a climax eutrophy may develop. It has great net production (as well as gross production), with a rapid deposit of organic matter. Yount (1956) assembled evidence that species diversity was reduced when resources were in excess, because a few species would overgrow others during competition. In other words, high nutrients favored low diversity. His study of diatoms on glass slides in Silver Springs, Florida showed relatively low diversity in an environment where excess nutrients were flushed through and production was limited only by available light. But when a dominant species has excluded other competitors for its principal role, it has interfered with the pool of other species the system has in reserve, and therefore has reduced the information in the gene pool. Sometimes, when the counts of dominant species are subtracted, what remains has fairly normal diversity indices.

Mitchell et al. (1971) added inorganic fertilizer (Figure 5.8) to simulate high levels of eutrophy. Net growth of phytoplankton cells was much stimulated at the intermediate level of 40 mg/l. The Shannon-Weaver index of diversity of fertilized microcosms was much lower than controls, consistent with a net production phase. They found their eutrophic microcosms, seeded regularly with lake waters, maintained a stable, moderate diversity of species, but less diverse than the unfertilized and equally seeded controls. Having regular seeding from outside helped maintain the high species level, and the restricted dimensions assisted in protecting the systems from external fluctuations. This example is evidence that stable conditions, high energy availability, and frequent species seeding favor the maintenance of diversity. The diversities were fairly similar at low and medium concentrations. At ten times higher levels (400 mg/l), the fertilizer was toxic and the Shannon-Weaver diversity index approached zero. This behavior may fit the model in Figure 7.2.

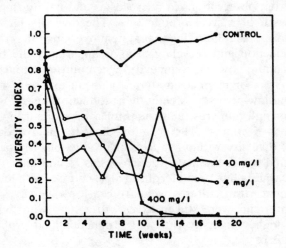

Figure 5.8. Effect of eutrophic fertilization on microcosm diversity (Mitchell et al. 1971).

Both in nature stressed by pollution and in microcosms, ecosystems with excess unused resources tend to have low diversity. Under these conditions, high rates of gross production do not require much cycling or division of labor. Maximum net production is favored by the most adapted net producer overgrowing the rest. Intense competition occurs with simplicity and low diversity, and without many control organisms. The most rapid converters produce organic products of low quality, high turnover, and little permanence, because all their energies go into maximizing net conversion rather than organizing structure. Thus, power is maximized with a simple design, if there are large inflows of nutrients. (Strictly speaking, in discussing maximum performance of an ecosystem, "Empower" may be a better measure than power. Empower is the flux of Emergy production and use. Empower is the available-energy use expressed in units of one kind of energy.)

The sequence of changes which comes about in zones arranged in series, due to linking of ecosystem outputs and inputs, is sometimes called longitudinal succession. Each zone can be in a constant state. With a high level of nutrients entering from upstream, a zone of low-diversity weeds is followed by a zone of higher diversity recyclers downstream. Microcosms can be adjusted to represent the series by adjusting the turnover time of the inflows and outflows.

The ratio of carotenoids to chlorophyll has been used as an index for succession in microcosms. Early stages with net production are very green, whereas mature stages are often yellow. Margalef (1969) used the ratio of the optical densities of community extracts, tested at 450nm to 665nm, as a convenient measure. The green-to-yellow shift accompanies a change from net production, to more nearly balanced consumption and production. It represents a change in the type of photosynthesis predominating, a change in the chlorophyll required, and sometimes at least adaptive photorespiration to improve cycling. Thus, the green to yellow color is an index of P/R ratio, indicating net production, not necessarily succession. It is an index of succession only for the common general case which starts with high nutrients and later develops a tighter cycle with fewer free nutrients. In our 1955 flowing bluegreen algal mat microcosm (Figure 8.5) (Odum and Hoskin 1957), the upper end first receiving the nutrient rich waters was green, whereas two inches further down the tube the mat was bright yellow. We reversed the tube. Within a few days the green zone had reappeared in the upper position, which was receiving the regenerated nutrients from the dark part of the circulation cycle. The succession which others observed in color change with time was, in this system, laid out in space.

Low-diversity eutrophic ecosystems are useful to other ecosystems which may receive the products of the net production. They are successionally arrested and part of longitudinal succession. A few dominant species can perform a necessary low-species role that is a stage in the movement toward prompt development of high diversity. Low-diversity systems convert nutrient excesses into organic pools from which a more complex web of consumers and producers can develop. As in early succession, low diversity is an adaptation to the stage that leads to rapid development of high diversity.

Because our human-dominated landscape has a great excess of consumption over photosynthetic production (a fossil fuel based economy), outflows from the society (wastes, storm runoffs, exhausts, agricultural fertilization runoffs, etc.) have changed more and more of our environmental systems from tight-cycle type to input-net producer types. These changes, with decrease in diversity, are often seen as pollution and loss of quality. Yet, the low-diversity net-production systems are nature's way of correcting excess nutrient conditions, using weed dominance to rebind the excesses, making conditions downstream suitable for the higher diversity systems. Often, diversity is used as a general synonym for a good ecosystem, yet eutrophic systems with lower diversity have their place and are needed in order to bring others into high-diversity stability.

Larger Scale Influences, Pulses and Patches

The influences on ecosystems from higher levels of natural hierarchy are those of larger-sized units, having sharper pulses, greater effects, and longer intervals between pulses. The factors shown in the models (Figure 7.2) coming from the right are examples with high transformity such as the introduction of information and species.

Large-scale and longer period influences may add to diversity through seeding, regulation, developing patches, or eliminating dominants. The action of toxic substances (Chapter 7) usually causes a decline in diversity. Other influences may reduce diversity through generalized stress, e.g., by draining energies in excess of energy available, by restarting succession, or subtracting information (Fig. 7.2). See Chapter 6 for the effects of larger scale influences. Both positive and negative effects of higher levels of hierarchy are included.

Large-scale actions may be important to most ecosystems, but large-scale influences on microcosms are often nearly absent because of container walls. Special arrangements are required to include them. Considered over a longer scale of time and space, the maximum gross production may require pulses of consumption that generate patches of succession and climax in a sort of mosaic. The pulsing alternation of production and consumption is simulated in Figure 2.15. Pulsing consumer actions may cause patchy microstructures to develop within microcosms (Figure 6.2). Increasingly larger animals are found managing their realms of smaller components, controlling patterns, diversities, and introducing larger scale events. A study of tropical tide pools by Weaver (1970) provides some insight. Unlike laboratory beakers, tide pools are provided access to larger scale species. Weaver found great heterogeneity of tide pools associated with the great heterogeneity of fishes found in the pools, at least temporarily. The Shannon-Weaver diversity index ranged from 0.63 to 2.8 bits per individual, which is fairly high for microcosms. This is an indication of the contribution open access, however temporary, brings to small realms.

To include more of the influences of the larger scale, other, additional pathways can be added to the model in Figure 5.7. For example, carnivores may be added which contribute reinforcing service in proportion to consumption (Figures 6.5 and 6.6). The action of larger consumers not in symbiotic relationship was discussed in Chapter 3. Controls from the larger scale outside the ecosystem can be provided in the form of temporary additions, human actions, or technological controls. The container barrier is simulated by turning off the cross boundary pathways. In this way diversity is restricted. The unfertilized phytoplankton microcosms of Mitchell et al. (1971) were more stable, and had higher diversities on the average than the lake from which they were drawn (Figure 5.9). At any one time, outside diversity may reflect the large-scale, longer-period cycles of the larger systems with oscillations of higher animals and terrestrial influences. The microcosm in a container was separated from these influences and kept on a smaller hierarchical scale. See discussion in Chapter 2.

Many studies have suggested that the size of organisms tends to be diminished when populations are held in small containers. The ability to adapt to smaller realms by adjusting size can allow a population to survive when larger members would use too many resources, reducing

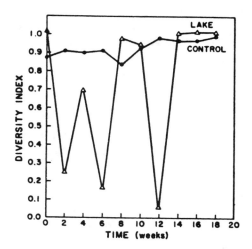

Figure 5.9. Effect of microcosm isolation on stabilizing long-period species oscillations generated in larger scale realms (Mitchell et al. 1971).

the number of members below the minimum necessary for species continuation. Becoming smaller is tantamount to being at a lower position in the hierarchy, with smaller territories, resource needs, and turnover times.

Very high-diversity ecosystems of the world, such as some coral reefs and rain forests seem, to have a pool of surplus species. These are species beyond those necessary to operate the ecosystem's main functions. Elsewhere, in areas with fewer species, similar ecosystems occur without the surplus pool, yet show parallels in principle structures and functions. For example, insular rainforests have lower diversity but functions similar to those found in South American rainforests. Atlantic coral reefs have lower diversity but functions similar to those found in the reefs of the tropical Pacific. When present, a species pool is a contingency for maintaining fine tuning and adaptation. The similar species do not seem to drive each other out by competition. Apparently, the dominant species of the ecosystem are prevented from competitive exclusion by the existence of organized relationships. The many minor species of the surplus pool in similar niches, do not have enough resources to allow any one to overgrow the others. Most microcosms may be without much of a surplus species pool. Some microcosms have been operated so as to maintain high levels of species availability. This is tantamount to providing some of the equivalences of larger size. McCormick and Cairns (1991) found "limited and unlimited membership" of microbiological species and several years to develop full complement.

Transformity is defined as the Emergy per unit energy in emjoules/Joule (Odum 1986). Since the energy content of the carrier of information is small (DNA, memories, seeds, etc.) and the Emergy to make the information is large, transformities of information tend to be large. By definition, items of high transformity are high in the energy hierarchy, associated with structures and functions of large scale and long period. Transformities may be useful in comparing large-scale influences. Species may be arranged on a scale of transformity that involves their size, complexity of internal structures, and other properties.

Diversity and Instability

Whereas the models of this chapter are based on a mutualism between diversity and productivity favoring sustainability, there are opposing views. Many ecologists including Hutchinson (1978) have associated diversity with instability with the idea that changing conditions inhibit competitive exclusion. Wilson and Botkin (1990) applied the species modelling approach to plankton microcosms that was previously used for forest trees. A production function with physiological requirement was arranged for each species while sharing the pools of light and nutrients. The model generated emergent properties since species were not maximum according to their individual optima. Runs using 5 species were more stable than those with 10 species that went extinct. They suggested that the emergence of specialization accompanying complexity made the 10 species system more vulnerable.

Chapter 6
Hierarchy, Control, and Oscillation

Ecosystems developing in microcosms, like the larger unconfined systems, soon develop hierarchical networks with many small organisms at one level converging support for fewer, larger organisms at another level (left to right in energy systems diagrams). The higher level units provide controls over the units at lower levels in the form of animal services and chemical substances (right to left in energy systems diagrams). With hierarchical population structure, "prey–predator" type oscillations appear. With higher energy levels, more complex "chaotic" oscillations may appear. In this chapter we consider the hierarchy of populations and processes, spatial aspects of hierarchical organization, population oscillations within microecosystems, and the effect of containers in isolating a system from large-scale influences. As elsewhere in this book, we keep in mind the compelling intuitive hypothesis that the observed hierarchies and oscillations occur because they reinforce maximum power.

Hierarchy in Space and Time

In Figure 6.1 is a diagram of spatial territory and replacement time, which is a common way of representing hierarchy in most fields of science. Aquatic ecosystems tend to represent a faster, smaller realm than many of the land ecosystems (Steele 1985). Microcosms, by their size limits, are forced to be smaller windows than the systems about which they are often intended to give insight.

Figure 6.1 helps interpret events at one level in terms of another on a different time–space scale. Things are in comparable proportion if one multiplies both territorial size and renewal time by the same factor. For example, a computer simulation model can serve at more than one

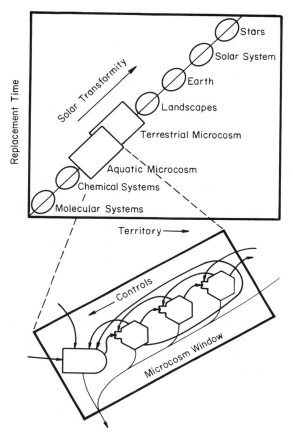

Figure 6.1. Size of territory, support, and influence as a function of replacement time for various systems, showing the window of concern of ecological microcosms in the hierarchy of sciences.

size–time level by declaring that a second of computer time represents a larger unit of time. Of course, the size amplitudes must be changed by the same factor. Cross-level comparisons are appropriate where the systems are similarly organized. Perhaps self-organization tends to generate similar designs, models, and system mathematics at different levels, because reinforcement of these designs maximizes power at all size levels. Scaling factors are needed where microcosm results are to be used for management of ecosystems on a larger scale. Uhlmann (1985) reviewed ways of scaling for several purposes, concluding that no one scaling factor is appropriate for all purposes. He recommended empirical comparisons of microcosm toxicity with large-scale tests to obtain scaling factors. Some of the parameters given for making scaling comparisons are:

Surface to volume ratio of bottom and sides
Turnover times (respiration/weight; water flushing; cells/retention)
Turbulence indices (Reynolds number; Froude number; dispersion number)
Total metabolism; inverse production depth relation

After self-organization, many microcosms show spatial hierarchy with centers where inward flows and transformations converge, and outward processes diverge. For example, a reef is a center of convergence and recycling (Chapter 14). Such spatial organization apparently facilitates the mutual reinforcement of hierarchical levels. In brine microcosms self-organized at 1,000 ft-candles and 10% salt in growth chambers, Nixon (1970) found bluegreen algal clusters had developed on the bottoms of the containers. These clusters had been sculptured by brine shrimp feeding around the edges (Figure 6.2). Kurihara (1978) provides a photograph of the clumps of clustered algal–bacterial colonies consisting of *Mougeotia* and *Schizothrix* that include concentrations of *Chlorella* and bacteria, being grazed around the edges by ciliates. He described the colonies as equidistant from each other. Kawabata and Kurihara (1978a) found the presence of an oligochaete (*Aeolosoma*) was necessary to the formation of colonies and increased algal biomass. They inferred that its presence increased productivity. Spatial organization was being controlled by animal behavior, from which a successful reinforcement of colonies and animals resulted in a hierarchical pattern, apparently an example of organization for maximum power. Energy transfer networks are accompanied by patterns of spatial hierarchy. Oscillations in time contribute to spatial variation when pulses of consumption generate "gaps." A mosaic of patches in different stages of micro-succession results.

An energy measure of hierarchy, the transformity, was defined earlier (Odum 1976, 1985, 1986, 1987) as the energy of one type required, directly and indirectly, to generate another kind of energy (see Chapter 3). Now applied to all forms of energy, the transformity concept is a generalization of the trophic pyramid concept of ecology, except all kinds of energy contributing are evaluated, each traced to the solar energy required. Table 3.1 uses solar transformities to evaluate the direct solar energy, physical stirring energy, and chemical potential energy of nutrients contributing to a microcosm. Transformities of items in the geobiosphere include the solar equivalents of solar, tidal, and geologic heat energies, all of which contribute to the system.

Because the solar transformity is the energy which is required at one level of hierarchy to make a unit of a higher level, it is a useful scaling factor for comparing small things like microcosms with larger outdoor systems. Populations of small things can be compared to populations of larger scale by multiplying by the ratio of their transformities. Note the increasing solar transformity on the size-time graph (Figure 6.1) within

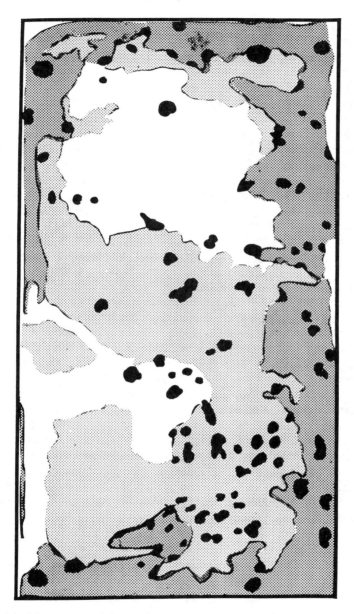

Figure 6.2. Map prepared from photograph of bottom of brine microcosm 125 days after seeding (Nixon 1970). Black areas are bluegreen algal mounds, shading indicates density of sediment cover.

and between windows of concern. Larger animals with larger areas of influence have higher transformity and more influence. Findlay and White (1983) added sand dollars to marine bottom microcosms, and found the animal role in this system was to increase the zone of oxidized sediment and change the lipid signature used as a chemical signature of these ecosystems. Animal control of chemistry is apparently a major mechanism of reinforcement and hierarchical organization. Terami and Watanabe (1989) using mirrors found Guppy's distributed according to perceived space in their receipt of visual information. Role of higher animals may often be organized to feedback to a larger area. Gorman 1988 using 75 liter aquaria found species distributions different according to other species present, a tendency that allows fishes to cover space in ecosystems.

Many heavy metals which have high transformities (large Emergy required for their concentration in geologic and biospheric process) will tend to assemble at high trophic levels and act as controls. These are beneficial when adapted, and detrimental when not adapted. Knight (1980) considered the heavy metal cadmium, estimated its transformity, and studied its role in microcosms, including the experimental streams at the Department of Energy's Savannah River Plant at Aiken, South Carolina, and in tube microcosms submerged in Silver Springs, Florida (Figure 10.16). Evidence was found in self-organized microcosms for low concentrations of cadmium concentrating in upper food chain levels. There it had beneficial effects. Copper acts on algae as well as small consumer animals, and less on the larger ones. Microcosm studies show copper inhibits algal growth early in succesion. In later stages, when algal growth in normally held down by consumer populations, copper, which is toxic to the consumers, stimulates net gain of algal growth (Taub et al. 1989; Taub 1984). Large and bulky, leaves of aquatic plants are higher in hierarchical position than fast replacement microorganisms. Murray and Hodson (1986) showed that chemical substances from decomposition of *Nymphea odorata* can alter the growth rate of bacterioplankton in Okeefenokee Swamp microcosms. Similarly, Alongi (1987) used microcosms to show that "tannins" from mangrove leaves controlled meiobenthos.

There may be a general principle here, that self-organizing system, through interunit reinforcement, develop a network which provides greater benefit. One of our referees called this "ecosystem immunology."

Hierarchy and Minimodels

Controversies concerning modeling revolve about aggregation and the window of interest. Science traditionally has sought rigor by concentrating effort at one hierarchical level of size and time, which is appropriate when the emphasis is on looking at one part or mechanism of a system. However, since items at upper levels control lower levels, modeling of ecosystems and microcosms requires inclusion of two or more levels of hierarchy, such as producers and consumers. One view seeks to understand the species as much as the system, and this bottom-up modeling starts with the parts. It is hoped that if the parts are perfectly known, the whole can be understood through combination. However, when the approach to simulating a system is to add equations for parts, there is the pitfall of carrying forward population details not relevant to a larger scale of time and space, and leaving out the main system mechanisms which are on the larger system scale. The other viewpoint is whole-system oriented and sees the parts being controlled and organized by mechanisms at a larger scale. These include nutrient cycles, information controlling behavior, reinforcements of choices, and control by selection. Top-down modeling emphasizes these larger scale processes. Most species in a functional guild may be aggregated into a single unit that represents them collectively. So-called noise may be due to dynamic oscillations at a lower level, too small and rapid to concern the window of interest. The pitfall in top-down modeling is that some unrecognized small-scale feature may have a major role in long-term process. At the extreme, holistic modeling uses a generalized network which is believed to apply to any ecosystem because it contains the mathematical and energy-maximizing constraints of any self-organizing system. Energy systems diagrams which have been found to be common to most ecosystems are used to initiate model development.

In the succession of many ecosystems, small, fast components appear first with larger and longer-acting components developing later. Kurihara et al. (1990) found changes of growth characteristics within microbial species increasing their hierarchical level within microcosms. Kemp et al. (1980) represented the role of small and larger microcosms as experimental means for understanding the worlds of several size–time scales (Figure 6.3). They suggested that intermediate-size experiments were compromise tradeoffs between the controllability and minimal expense of the small units and the realism of the large scale. Papers reviewed in this book illustrate the conceptual problems faced by those with different windows of interest and the differences between the models they build.

Steele (1977) recognizing the smaller time-space scale of smaller populations in the sea showed that there was an appropriate time-space scale of sampling. In considering patchy spatial distributions, and recognized the role of larger-longer animals in impressing larger scale dimensions on the smaller-faster populations. Populations were also thought to be coupled to turbulent oscillations, on several scales.

Populations and Microcosm Oscillations

Microcosms like outdoor systems, develop food chains and prey–predator type oscillations, especially when energy levels are high. Does oscillation provide more long-range stability than level steady-states? Do self-

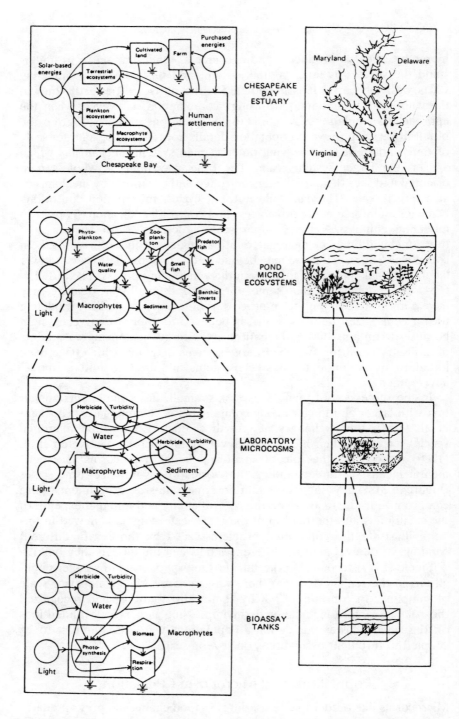

Figure 6.3. Role of cultures, microcosms, and models in studying larger ecosystems (Kemp et al. 1980).

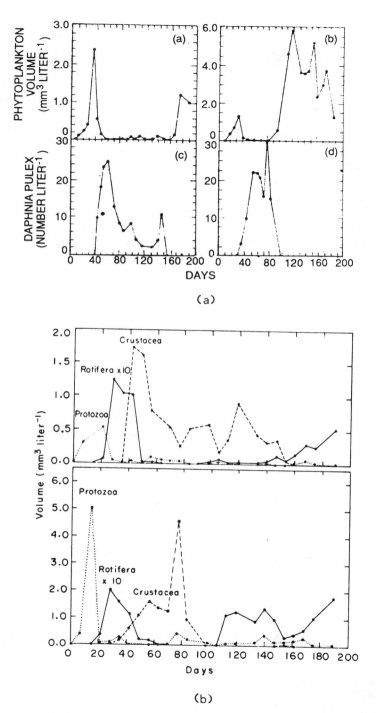

Figure 6.4. Oscillations observed during succession in microcosms (Jassby et al. 1977 given by Rees 1979). (a) Phytoplankton; (b) zooplankton.

116   6: Hierarchy, Control, and Oscillation

organized oscillations maximize resource intake and use (i.e., provide maximum power)? Kok (1956) and Grossman (1990) found alternate light and dark more efficient than continuous light. Although laboratory studies of one or two isolated species are outside the scope of this book, population phenomena, such as competition and prey–predator oscillations, are an integral part of hierarchy in ecological microcosms where they interact with production, consumption, recycling, diversity, and other system properties. For example, prey–predator type oscillations occur in microcosms (Figure 6.4). Oscillations were observed among the smaller animal populations in the estuarine sewage mesocosm ponds at Morehead City, N.C. especially during initial succession (Odum et al. 1985; Odum 1989). See also Figure 13.11.

Whereas a typical study of populations might try to eliminate all extraneous factors, even to the point of testing pure cultures, researchers studying populations as part of ecosystems often consider the population in conjunction with as much of normal ecosystem complexity as possible,

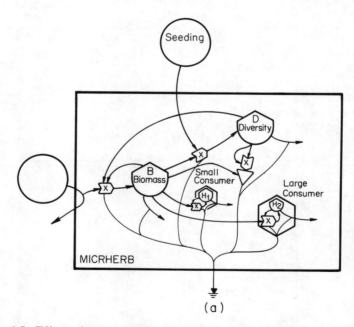

Figure 6.5. Effect of autocatalytic consumers on the general successional minimodel (Figure 3.1). (a) Energy systems diagram; (b) two consumers in parallel: the larger one with long turnover time replaces the smaller, faster one which predominates at first (Program: MICRHERB); (c) after steady-state is reached (without the consumers), a large biomass of long-turnover consumers is added, which produces overgrazing and consumer extinction, followed by restoration of a productive regime, reminiscent of snails added to macrophyte aquaria (Figure 9.7) (Program: CONSIZE).

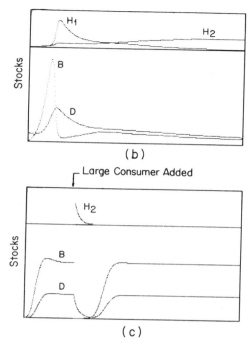

Figure 6.5. *Continued*

changing some variable of interest. An intermediate kind of experiment starts with a complex microcosm, but introduces factors which eliminate some part of the system in order to reveal that part's effect on a population in the system. Exclusion experiments can be done in a microcosm as well as in he field. We may gain some insight on population–system interactions by adding population units to ecosystems minimodels. Two populations with different replacement time were added to the biomass–diversity model in Figure 6.5 and a prey–predator pair was added to the model described in Figure 6.6. Figure 6.5 shows the successional, biomass, and diversity minimodel from Figure 3.1 to which consumer populations were added, but without the positive feedbacks often observed. The model places the two consumers in parallel, competing for a common food source. The simulation results in Figure 6.5b illustrate the effect of turnover time on the timing and prevalence of consumers during succession. The smaller, faster consumer grows first, but is displaced later by the larger one with longer renewal time. The simulation results in Figure 6.5c show the effects of suddenly introducing a large numbers of a large consumer (long renewal time). Overgrazing decimated the biomass and diversity, causing the large species to go extinct, after which regular functions returned. The addition of snails to plant-filled aquarium micro-

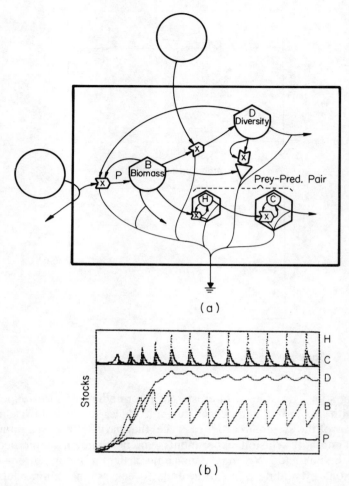

Figure 6.6. Prey–predator pair inserted into the general successional model in Figure 3.1 (Program: VIBRATOR). (a) Systems diagram; (b) simulation results.

cosms (Figure 9.7) illustrates such overgrazing in a microcosm. In microcosms, the usual larger, top carnivores are not present because of the restricting size of the microcosm, and so smaller species become the highest carnivores the system can support. Food chains may converge on the top dominant consumer so that the population may reflect the whole trophic base of the ecosystem. Goodyear et al. (1972) studied *Gambusia* as the top carnivore in circular plastic pools (2.44 m wide by 0.45 m deep). By application of different concentrations of fertilizer (8:8:8), a five-fold range of gross primary production was developed. Gambusia production was proportional (Figure 13.2).

Oscillating subsystems within an ecosystem may cause the whole system to oscillate, which is a form of control action, although many prey–predator oscillations in microcosms die out or approach a level steady-state. The oscillations of small realms occupy short periods of time and small areas, and variations are readily absorbed in the convergence of their products to the next higher levels on the larger scale of time and space. To detect oscillation, the measurements must be more frequent than the time and space scale of the phenomena. Unsynchronized microbial oscillations may be damped out or homogenized in a sample on a larger scale. The higher levels operating on a longer time scale filter out the more frequent inputs, which are not important at that scale. Moving in the opposite direction, the oscillation of a larger unit may pass its long-period pattern to many smaller units through the feedback of service, materials, or population control.

The usual experience of many who have studied microcosms is that larger consumers overconsume and become extinct after one or two population surges. See Figure 6.5c and papers by Taub and McKenzie 1973 and Taub et al. (1981). Adding a container boundary eliminates those species which require a larger area for sustainability. Since the period of oscillation is determined by the replacement times of the larger members, eliminating larger members of the population eliminates the longer period oscillations. After elimination of larger components, microcosms viewed from the larger scale window tend to be steady. Kawabati and Kurihara (1978a,b) found steadier patterns in microcosms that in wild prototype ecosystems. Steady patterns may improve replicability and make microcosms more useful for some kinds of experiments. In Figure 6.6 a consumer pair has been included in the general minimodel for a microcosm containing biomass, diversity, and material cycling. Hierarchy is represented if the model if calibrated to give the consumer to the right a much larger turnover time. Simulation results show that such a pair can cause the whole system to oscillate or not, depending on other conditions.

Figure 6.7 relates an oscillator configuration to the input energy source. Figure 6.7a presents a model with energy supplied from a steady inflow. The prey–predator subsystem cannot increase much without becoming damped at the source. Because the consumer pair is attached to a renewable inflow which does not have appreciable storage, the system cannot sustain an oscillation (Figure 6.7b). If, however, as in Figure 6.7c there is a large accumulation of food before the introduction of the first of the populations, the system can maintain a sustained oscillation. In microcosms, as in nature, ecosystems generally develop a pool of detritus organic matter during their self-organization. This pool may have many useful mechanisms besides supporting oscillation (if indeed oscillation can provide greater long-range performance). If oscillation generates more power than a steady system of production and consumption, then systems that develop a pool of food for consumers can prevail. Alongi and

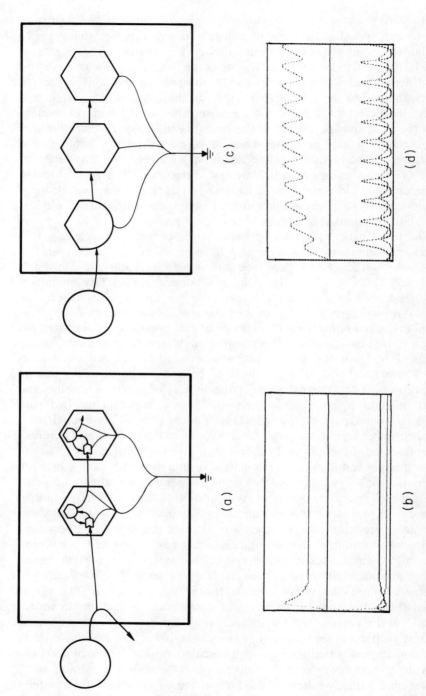

Figure 6.7. Effect of source energy storage on predator–prey oscillation. (a) Energy based on fast turnover resources without much storage (Program: LIMTPAIR); (b) simulation results of model in (a); (c) energy drawn from a storage with steady balance of inputs and linear outflows (Program: PREYPRED); (d) simulation results of model in (c).

Hanson (1985) observed oscillations of capitellids (micro-annelid worms) in microcosms with added detritus, and Alongi and Tenore (1985) found competitive interactions with smaller meiofauna. Alongi (1985b) found wide ranging oscillations in bacteria, flagellates, and ciliates (three orders of magnitude), but did not attribute these to meiofauna or physical disturbance of the bottom. Salt (1967) observed prey–predator oscillation of the carnivorous protozoan *Woodruffia* consuming *Paramecium* which, in turn, fed on the organic nutrition (bacteria) from boiled wheat grains. The *Woodruffia* population pulse reduced *Paramecium* population to low levels; the *Woodruffia* encysts hatched when the *Paramecium* population numbers once again returned to higher levels. Cultures were frequently restarted, thus maintaining a steady pool of detritus for the oscillation. Taub et al. (1981) started microcosms with different initial densities of algae. With larger initial food storage, the consumers increased in number sooner and kept algal populations steadier than when initial algal food storage was lower.

Where larger consumers are absent, the organisms which in their natural habitat would have been a food source may, in the microcosm, maintain higher levels, being regulated by their own food resources or other crowding effects. These may be irregular. Bell and Coull (1978) adapted tidal salt marsh microcosms (Figure 6.8) to function without shrimp, measuring substantial but irregular populations of meiofauna. After shrimp were added, there was a reorganization to smaller, steadier meiofaunal populations. At this scale of time and space, the shrimp served as a steady control of the smaller, faster regime. Vanni and Findlay (1990) found fish affecting phytoplankton, not only through there feeding control of zooplankton, but also by recycling nutrients and increasing zooplankton excretion.

Chaos

At high energy levels, complex kinds of oscillations develop in many kinds of systems, and these are said to be "chaotic." The oscillatory patterns are described with deterministic equations, but the graphs they produce are sometimes so complex that they seem without principle, hence the misleading name "chaos." As often in science, a word with a general meaning is taken and provided with a special meaning that is different from the general one. In this case, the general meaning "to be without order," is almost the reverse of the scientific meaning of the word, particularly as concerns causality.

In contrast to a regular oscillation like Figure 6.7d, "mathematical chaos" is an oscillation that shifts its position over time. In Figure 6.9 are simple models which generate mathematical chaos when their energy sources are increased and pulsed. Figure 6.9a is a model that generates logistic growth with time. Figure 6.9b has two prey-predator oscillations

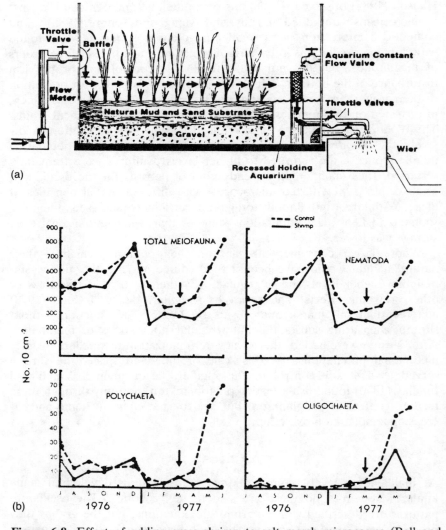

Figure 6.8. Effect of adding grass shrimp to salt marsh microcosms (Bell and Coull 1978). (a) microcosm view; (b) comparison of tanks with shrimp and controls.

on different time scale one driving the other (predator stock of the rapid oscillator above supplying the energy for the slower oscillating pair below).

In Figure 6.10 are the results of simulations of different kinds of steady states, represented with time graphs and phase plane plots. The first case (Figure 6.9a) is the typical steady state which develops according to the logistic equation model following a growth phase. In our use of the

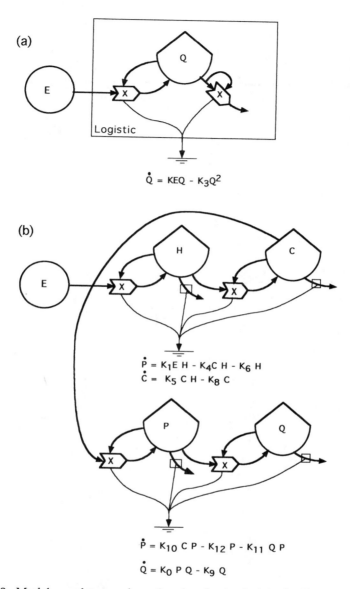

Figure 6.9. Models used to produce the chaotic simulations in Figure 6.9. (a) Logistic model used to relate chaos to increasing energy from source E; (b) Prey–predator model at one level of hierarchy driving another prey–predator cycle at another level of hierarchy (calibrated with different turnover time).

(a)

(b)

(c)

(d)

Chaos

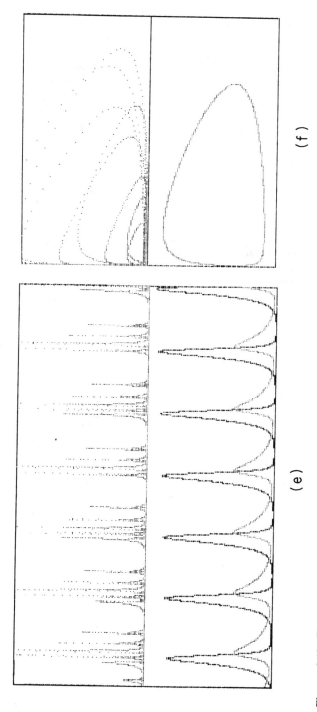

Figure 6.10. Types of steady states represented on time plot and phase plane. (a) Logistic growth (Program: LOGCHAOS) with simple level steady state; (b, c, and d) logistic bifurcations with successively higher energy source; (e) time plot of fast prey–predator oscillator (above) driven by slow prey–predator oscillator (below); (f) phase plane of the simulation in (e) showing attractor expanding as amplification from the large-scale cycle increases.

logistic model, the energy source is explicitly identified. The phase plane is a dot. Following May (1974), the logistic model is often used to show how jumping occurs, but the results are chaotic only if the operation is performed in interrupted steps, as on a digital computer. In nature, logistic growth does occur with separate steps. When the input is rhythmic. For example, each day there is an increment of light which is incorporated into plant growth, followed by night interruption in which there are losses, this cycle being continued again the next day. The model in Figure 6.9a was simulated and the results are shown in Figure 6.10 with part (a) depicting a steady state. When given more energy relative to the time step, the plant storage fills on one iteration and discharges on the next, repeatedly. The result is a steady pattern of two states jumping back and forth (Figure 6.10b). An abrupt change emerging from a continuous function is called a bifurcation. As the energy level is increased, the simulation develops increasingly complex bifurcations (Figure 6.10c and d). With any simulation model, bifurcation is often observed when the increment of flow into a storage compartment is larger than the storage. Almost any model can be made to display bifurcation by increasing the time step in Euler integration, as used in the minimodels of this book. We call the jumping "artificial chaos," since it is an artifact of the simulation procedure that uses time steps which are not in the real system being studied.

Simple regular oscillation is a kind of steady state. For example, the simulation results in the lower part of Figure 6.10e have textbook prey–predator patterns which are in an oscillating steady state. Such an oscillation is a simple closed loop on the phase plane graph (Figure 6.10f, lower) which it approaches under some conditions. Another type of steady state the upper part of Figure 6.10e has an oscillation that shifts its phase plane loop back and forth (upper right). The locus within which the oscillations shift is called an "attractor," and such behavior is a mild form of chaos. The system in Figure 6.10b shows the slow prey–predator cycle (of higher levels of hierarchy) draining the smaller, fast prey–predator pair which represents lower levels of hierarchy. The result is an irregular but repeating pattern of oscillation.

Oscillation on one hierarchical level, linked to another oscillation on a different time scale is represented on a phase plane graph as a shifting loop. Models that jump back and forth between steady states or switch their oscillations from one locus to another, are said to be chaotic. At high energy levels, they switch among so many states so quickly that the result seems without order. Chaotic oscillations have not been much noted in microcosms. Many microcosms are relatively low energy, operating in growth chambers, often with much less light than is available outdoors. However, with the insights from chaos studies, we can find signs of chaotic behavior in earlier studies, such as those in which microcosm oscillations were observed by Tsuchiya et al. (1972) (Figure 6.11).

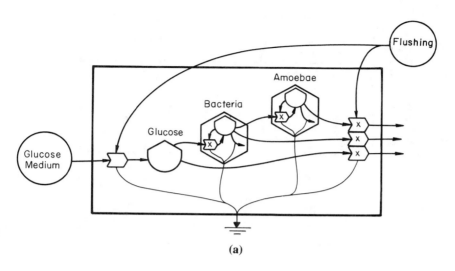

(a)

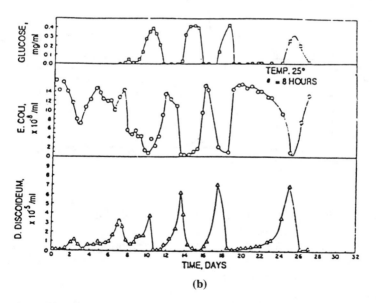

(b)

Figure 6.11. Simulation model of glucose–bacteria–slime mold in chemostat microcosm (Drake et al. 1968; Tsuchiya et al. 1972). (a) Energy systems diagram of the model equations; (b) and (c) simulation results and data.

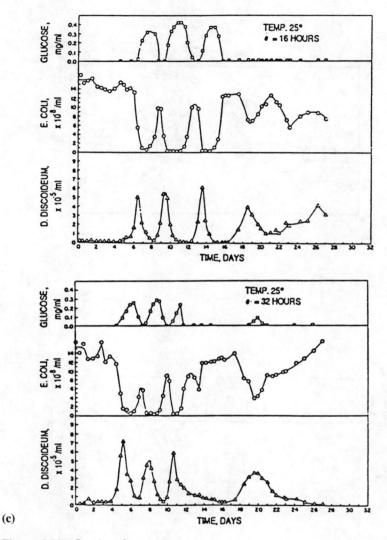

Figure 6.11. *Continued*

Especially when the flux of the energy source was large, there were shifts in the attractor of the oscillation.

Accompanying bifurcations and attractor shifts, are large gradients driving the filling and discharging of storage tanks. As with pulsing light and photosynthesis, more performance may be obtained from the same energy source with a pattern that maintains higher gradients, than one that maintains a smooth, steady state. In other words, it may be that chaotic ways of maintaining high-efficiency are effective because this mechanism reinforces productivity and use.

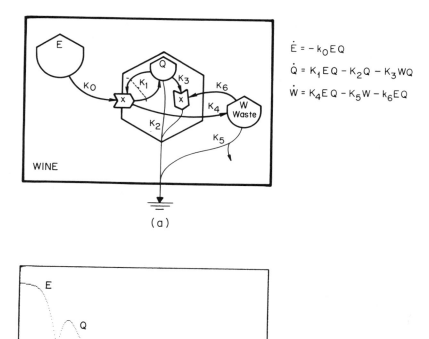

Figure 6.12. Minimodel of declining inhibition due to accumulating waste product. (a) Energy systems diagram; (b) simulation results (Program: WINE).

Inhibition in Microcosms

Of ancient origin is the domesticated microbial ecosystem used for making alcohol through yeast fermentation. This batch microcosm accumulates a byproduct which eventually terminates the microbial action. The system has a network configuration that is found in many ecosystems where metabolism is concentrated and wastes can accumulate. Simulation model WINE (Fig. 6.12a) generates the kinds of curves for the rise and decline of microbes observed in such cultures. In this model, the rise and decline are due to depletion of substrate and the destruction caused by accumulating waste. Kawabata and Kurihara (1978a, 1978b) include several of these waste-control configurations in simulation models to facilitate their understanding of the results of their experiments with microbial microcosms. Measurements included assays of organic matter, filtered after accumulations of wastes, of *Chlorella* (Mc), bacteria (Mb), and protozoa (Mp). One microcosm contained four species of bacteria (B) growing on polypeptone (N) and metabolic products, *Chlorella* (C) using recycled byproducts, and the protozoan, *Cyclidium*, (P) growing on bacteria. Another microcosm included an oligochaete. These microcosms presented successional transients followed by a steady state without much oscillation (Figure 6.13). Experimentally, and thereafter in modeling, they examined system components first in isolated pairs, before combining them. They measured effects and wrote equations for predation relationships and density dependent inhibitions by metabolic wastes. They presented the

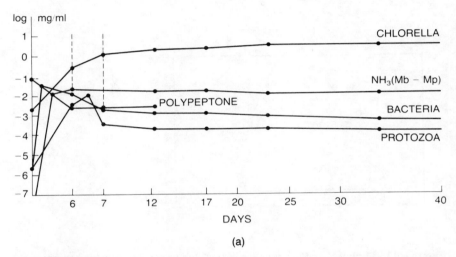

Figure 6.13. Succession and simulation of microbial microcosms (by Kawabata and Kurihara). (a) Observed succession for system of *Chlorella*, bacteria and protozoa (Kawabata and Kurihara 1978b); (b) growth with higher animals (Kurihara 1978a); (c) successional pattern and simulation results of transients, using more complex equations (Kawabata and Kurihara 1978c); (d) diagram of steady-state model representing the following equations:

$dN/dt = -k1NB$

$dB/dt = -[k3N/(k3 + N) + k4Mc - k6Mb/(k5 + Mb)]B - k8BP$

$dMb/dt = k9B - k10MbC$

$dP/dt = [k12B/(k11 + Mb) + k21Mp/(k20 + Mp) - k22Mc]P$

$dMp/DT = K16P - K17MpC$

$dC/dt = [k19Mb/(k18 + Mb) + k21Mp/(k20 + Mp) - k22Mc]C$

$dMc/dt = k23C - k24McB$

Inhibition in Microcosms 131

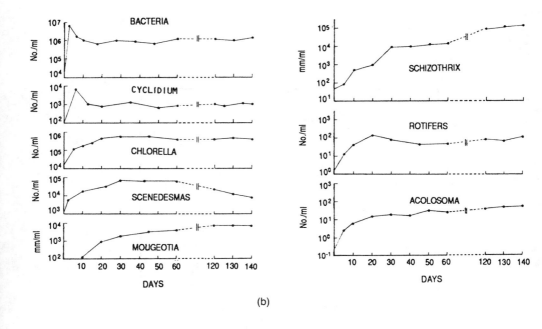

(b)

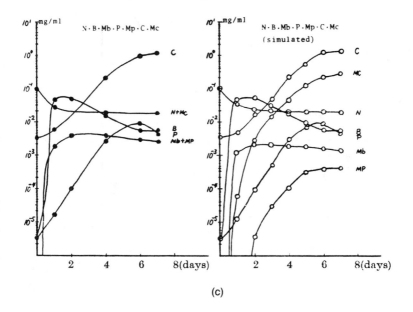

(c)

Figure 6.13. *Continued*

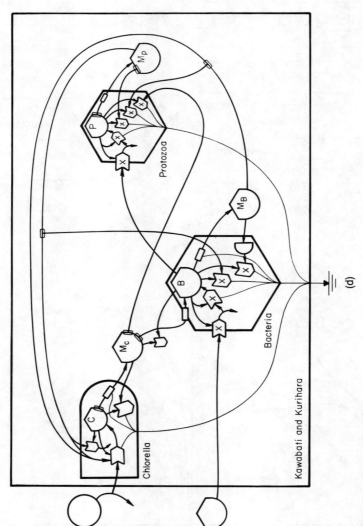

Figure 6.13. *Continued*

following matrix of relationships between the "elements" of their microcosmic system:

```
     N    B    Mb   P    Mp   C    Mc
N    -
B    +    -    -         -    -         +
Mb        +                        -
P         +    -         -    -         -
Mp                  +              -
C              +              +    -    -
Mc        -                   +
```

In their approach to simulation modeling, Kawabata and Kurihara (1978a, 1978b) considered the relationships of the above matrix, developing an equation for each: predation, growth, waste generation, waste inhibition, etc. Since models with very simple mechanisms (linear and prey–predator pathways) did not reproduce observed growth curves, other models were tried. Mathematically innovative expressions were applied, including quadratic denominators for density dependent action where Y is a population:

Density dependent inverse quadratic drain $D = k_1/(k_2 Y^2)$

and inhibition effects with limiting factor kinetics which decay with time:

Inhibition by decaying chemical $J = YMy \exp - [KMy/(k_1 + My)]t$

where K, k, and k_1 are constants.

Michaelis-Menten algebra was included to allow the representation of limits within the populations during succession, but in the final runs these terms were dropped without affecting the steady state. The equations fit the observed curves for isolated interacting pairs, although explanations were not given for why these mathematical relationships were appropriate. Then, when these mechanisms were included in a model for the whole microcosm (N, B, Mb, P, Mp, C, Mc), the curves resulting from the simulation were a reasonable fit for the data collected over seven days of succession. Because this system started with an unrenewed batch of polypeptone, an overshoot could be expected. When the model projected results for a longer time, all the variables went to zero with mutual inhibition, whereas the real system went into a steady state (Figure 6.13). Apparently, the real system organized itself enough to metabolize the byproducts, counteracting the inhibition. A new set of simpler equations was developed, and these did support a steady state, but did not track the successional stages very well. Figure 6.13d has a diagram inferred from the equations. Their idea was that interactions "decreased" in a steady state, so that the equations that adequately represented succession were no longer suitable. This theory opposes the more usual idea that in early succession, separate struggle among species acting individually is replaced with more complex interactions and more stability.

To create a simulation generating results which fit the steady state observations, Kawabata and Kurihara (1978a,b) changed coefficients and the structure of some equations. According to some schools of thought, the changing of equations during simulation is a useful procedure, but others regard the need for this as a failure of the model. If a coefficient is changed, it is really a variable and should be so represented. As self-organization changes a system, the changes should be explained by dynamic mechanisms, not by empirical substitution of coefficients which fit observed curves without being understood. If an item is in a model without a reason, there is no way to predict when the model will apply. When equations for parts are simply added to a computer program, coupling may not be consistent. For example, diagraming of the equations (Figure 6.13d) showed flows leaving one component with a product function, but going into a receiving component with a linear function. Waste compartments were arranged to act as inhibitors, but without any of the contents being used up. A better approach is to draw the system diagram first and then write the equations for the variables, referring to the diagram to be sure that all pathways and couplings are consistent.

Chemical and Population Effects in Synthetic, Standardized Microcosms

Frieda Taub and her associates have used standardized, synthetic microcosms. These were established with defined media and pure cultures of bacteria, algae, and small animals to test effects of toxic substances, not only for pollution studies, but in order to understand ecosystem dynamics, using the differential toxicities to eliminate various ecosystem units (Figure 6.14). Called SAM, the "standardized aquatic microcosm" contained 3 liters of chemical nutrient medium, ten algae (*Selenastrum*, *Chlorella*, *Scendedesmus*, *Ankistodesmus*, *Chlamydomonas*, *Lyngbya*, *Nitschia*, *Selenastrum*, *Stigeoclonium*, and *Ulothrix*), five animals (*Daphnia magna*, the amphipod *Hyalella*, the ostracod *Cyprinotus*, the protozoan *Hypotricha*, and the rotifer *Philodina*. *Enterobacter*, which was included as food for *Philodina*. Other bacteria were introduced from conventional cultures of zooplankton. Reinoculation was performed at intervals to prevent extinctions and allow recovery from temporarily toxic conditions. A protocol for standard use of the microcosm for toxicity testing in many laboratories has been provided (Taub 1985). Simultaneous replicate experiments were performed at four laboratories, with general similarity but some differences in timing (Taub et al. 1989).

Swartzman et al. (1984, 1987, 1989) in a series of papers and a book on simulations (Swartzman and Kaluzny 1987) developed a moderately large ecosystem model for the SAM system (Figure 6.15a and b). The models

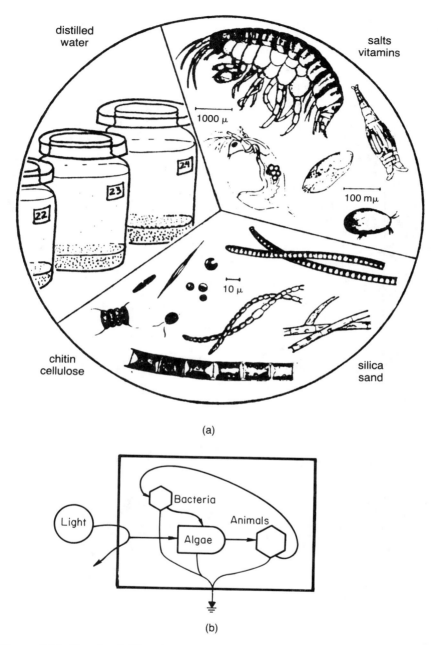

Figure 6.14. Sketch of "Standardized Aquatic Microcosm" (SAM) for toxicity testing (Taub 1989. Reprinted with permission from Aquatic Toxicology, Vol. II, copyright 1989, CRC Press, Inc.). (a) Sketch of chamber and contents; (b) summary energy diagram with microbes in appropriately low hierarchical position.

6: Hierarchy, Control, and Oscillation

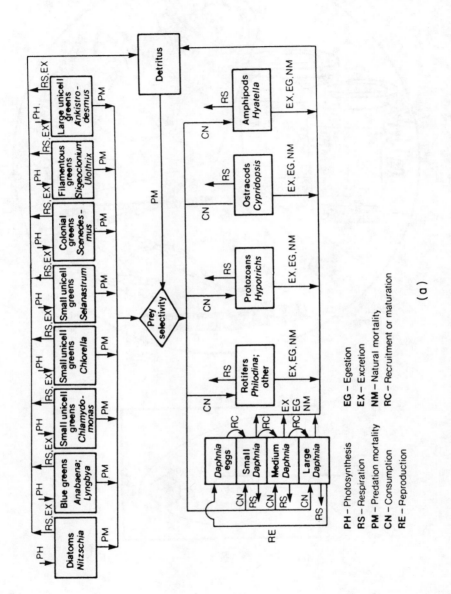

(a)

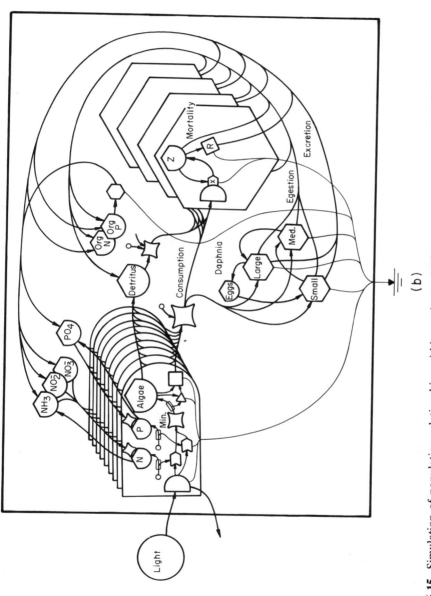

Figure 6.15. Simulation of population relationships within a microcosm (Swartzman and Rose 1984; Swartzman and Kaluzny 1987). (a) Summary diagram; (b) energy systems diagram derived from equations given by Rose et al. (1988). Included are eight similar phytoplankton species equations, four animal consumer population equations plus *Daphnia* represented by 4 equations each for life history stages (eggs, small, medium, and larger individuals). Plants have internal unassimilated nutrient pools as well as nutrients in biomass structure exchanging with media pools.

for each of the plant species were similar but with different coefficients, each determined separately from laboratory measurement in isolation and pairs. The plants had internal reservoirs where nutrients such as nitrogen and phosphorus could be sequestered in advance of exchange with nutrients from the storage pools in the aquatic medium. In the animal population models, animals drew their food from the plants and the detritus, according to coefficients of food selection determined experimentally. Both plant and animal models had limiting factor kinetics. The modeling procedure involved calibrating the populations as parts and putting them together in simulation just as the populations had been combined in the microcosm (Figure 6.16). The main self-organizational mechanisms provided in the simulation were the phosphorus and nitrogen cycles, and the reinforcement that occurs automatically when there are several coupled producers and consumers, so that those with complementary coefficients can be reinforced. The graphs plotting data over time revealed surges during periods of succession stabilizing later. Coefficients were not changed to make a better fit between the simulation and the data, although the real system may have changed its coefficients with "learning," as part of self-organization adaptation.

Taub et al. (1981) added streptomycin to defined synthetic microcosms, causing temporary reduction in algal productivity. After production was undetectable (28 days), these systems continued to have changed algal species (more *Scenedesmus*, *Gomphonema*, and *Nitzschia*, and less *Ankistrodesmus* and *Lyngbya*), lower diversity, and shifts in abundance of ostracods and *Daphnia*. Swartzman et al. (1989) simulated the streptomycin effect as linear, decreasing plant production. The simulation did represent the observed surges, although it was slightly off in the timing of some blooms (Figure 6.15b). When these authors added malathion, the *Daphnia* and ostracods were removed from the top of the food chain. A few dominant algae proliferated, generating a large, stored biomass. This was later followed by recovery and a bloom of the animal populations. These effects, possibly to be expected from consideration of ecosystem hierarchical structure, would not have been predicted from single species bioassay data. Figure 6.17 shows a prey–predator oscillation in the SAM system in which control was shifted as a result of the introduction of malathion, causing animal extinction and subsequent population blooms (via reinoculation) on accumulated algae.

Chemostats

Continuous culture methods have been adapted for ecosystem-level study of multiple species, some together in the same chamber, and others separated in compound microcosms of connected chambers. Chemostats (actually a misnomer for "biostats") have a flushing in of media, balanced by a flushing out of culture waters at the same rate. Margalef (1958,

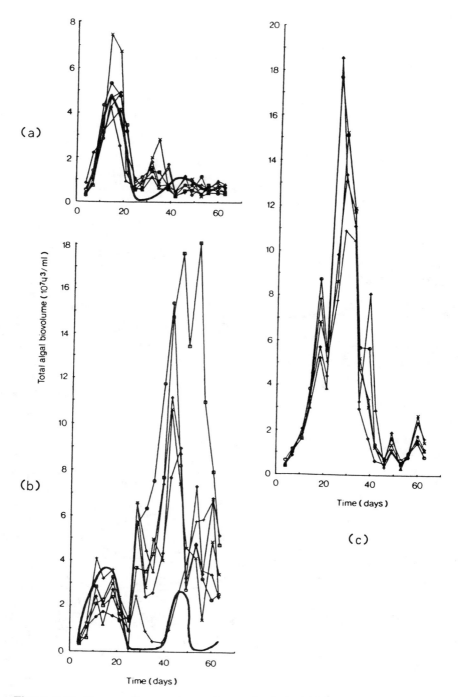

Figure 6.16. Curves of succession and growth of algal biomass in aquatic standardized microcosms (SAM) including curve (heavy line) from simulating the model in Figure 6.14 (Swartzman and Kaluzny 1987). (a) Control microcosms; (b) streptomycin; (c) copper.

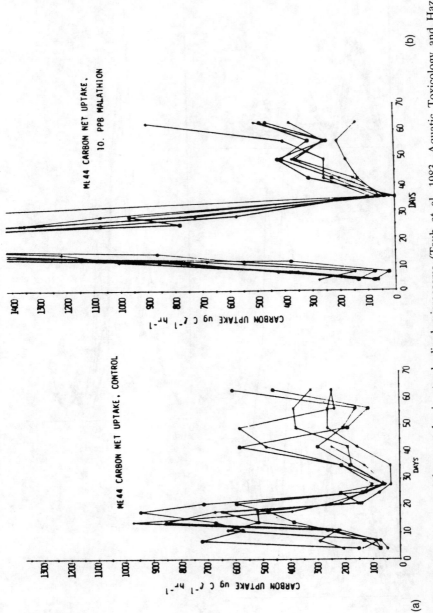

Figure 6.17. Experiments on net carbon uptake in standardized microcosms (Taub et al. 1983, Aquatic Toxicology and Hazard Assessment, 6th Symposium, ASTM STP 802). (a) Control; (b) malathion effects.

1967), in his early experiments in using continuous cultures to simulate ecosystems, described using vessels like that in Figure 7.3. Increasing the flow rate flushed out species, reducing diversity, but in a chain of chambers, diversity increased downstream. This was especially true when the organisms present were able to attach to chamber walls and thus had less mortality than unattached species. Chambers at the head of the chain where nutrients were in higher concentration produced more chlorophyll and algal cells, but the chambers at the tail of the chain had equally high rates of photosynthetic production. If operated with appropriate flushing rates (in and out), a coupled predator–prey type microecosystem can be sustained, although the original chemostat of Novick and Szilard (1950) was designed to flush out all but one species, as a form of competitive exclusion. Prey–predator coupling constitutes a form of cooperative unity. Although chemostat chains were operated to simplify study of plankton dynamics, Margalef wrote: "Brutal competition for dominance based on the rates of increase has given way to more subtle and interminable processes, and the chemostat is prevented from attaining a stationary state. The situation is interesting as an example of development of more organization that the experimenter desires" (Margalef 1967).

In a chemostat, Drake et al. (1968) and Tsuchiya et al. (1972) studied oscillation in a system of slime mold amoebae (*Dityostelium discoideum*) feeding on glucose-consuming bacteria (*Aerobacter aerogenes*) (Figure 6.11), and related the observations to a simulation model. An energy diagram of their equations is given in Figure 6.11a and some of their simulation results are compared with data in Figure 6.11b.

The living oscillation produced populations out of phase within concentrations of glucose, bacteria and amoebae. When part of the system oscillated, the whole chain oscillated. The simulation model (Figure 6.11a) produced oscillatory curves which fit the data well for some time, eventually drifting out of fit, apparently due to lags and other factors. At times, the system stopped oscillating and shifted to a steady-state regime. Oscillations were absent at low rates of food intake (glucose), damped at intermediate rates of feeding, and periodic at high energy availabilities. In other words, oscillations were dependent on environmental resource conditions. Among the bacteria, a hierarchical size distribution was found, with many small individuals, fewer intermediate, and a few large. These researchers also studied a system comprised of glucose, two bacteria, *Azotobacteria vinelander* and *Eschericia coli*, and a predator, *Tetrahymena pyriformis* (Figure 6.18). Although these organisms were in parallel in competing for glucose, some mutualism was found. Oscillations were observed (Figure 6.18b). The limit cycle on the phase plane was independent of the starting conditions. Growth constants of the species were not sufficient for modeling without considering the differing energy supplies.

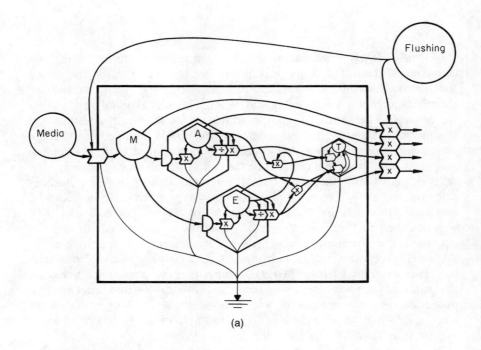

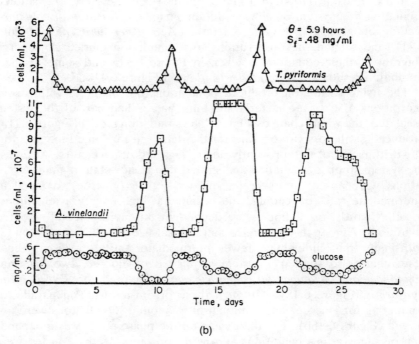

Figure 6.18. Oscillations of system of *Azotobacater*, *Escherichia*, and *Tetrahymena* in chemostat microcosm (Tsuchiya 1972). (a) Energy systems diagram derived from equations; (b) data and simulation results.

Compound Microcosms and Chemostats

Many engineered environmental systems consist of unit processes which are separate but connected by valve-controlled flows. This kind of construction echoes the "take apart and control" aspect of our education. In this methodology, understanding separate units is accomplished first and then the units are connected and tuned to form a system. The "split-level" microcosm uses these ideals of separate unit processes to isolate and understand the separate processes of an ecosystem and their connections to other compartments. We operated a class demonstration of trophic levels in our NSF Microcosm project at Duke in 1956 with three cultures: an algal culture, a *Daphnia* culture, and a tank with *Gambusia*. Each day we dipped a cup of algal culture and added it to *Daphnia* tank, and a cup of *Daphnia* went into the fish tank, thus the trophic level transfers were controlled and known. But ecosystems generally have their hierarchical components occupying the same space without the rigid pipeline, roads, and wires which characterize some technological hierarchies. Considering how many pathways of interaction there are even in an ecological microcosm, there would be no room to move if all components were connected by permanent conduits. The issue of developing a human

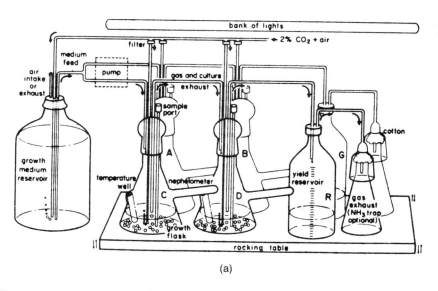

(a)

Figure 6.19. Continuous algal and protozoan cultures in series for study of trophic dynamics (Taub and McKenzie, copyright 1973, Munksgaard International Publishers Ltd., Copenhagen, Denmark). (a) Apparatus with two series: series A-B-G without protozoa; and the other with protozoa in chambers D and R, A and C in the light; (b) algal densities in flasks A-D; (c) algal densities predicted from simulation model; (d) experimental and predicted protozoan densities.

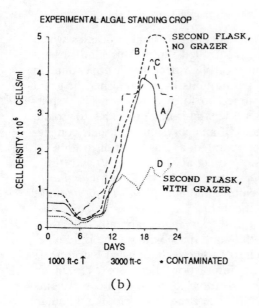

(b)

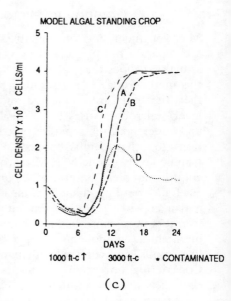

(c)

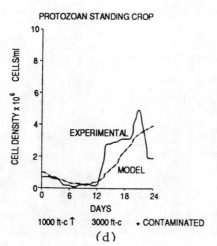

(d)

Figure 6.19. *Continued*

microcosm for space involves the controversy between strings of unit processes and the all-in-one self-organizing ecosystem. See Chapter 19.

A large area of research concerns compound continuous culture units (compound chemostats), which have more than one chemostat culture connected to form a system (Herbert 1964). See the example in Figure 6.19 where Dollar et al. (1964) and Taub and McKenzie (1973) studied continuous cultures as a separated trophic level microcosm. In other words the chemostat cultures, when connected to form compound chemo-

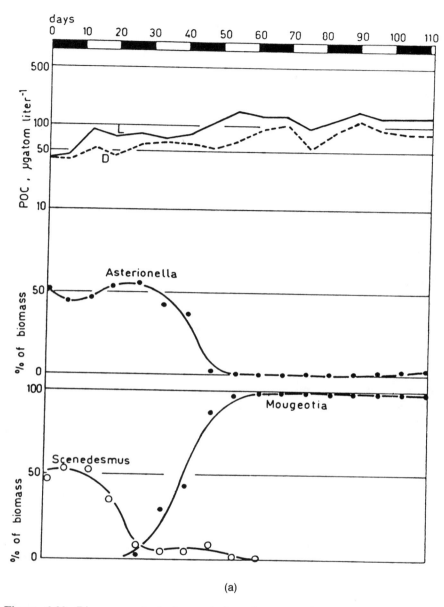

Figure 6.20. Biomass and algal populations in a two-chambered recirculating microcosm (L, lighted chamber for algal photosynthesis and D, dark chamber with zooplankton) (Sommer 1988). Particulate organic biomass, POC.

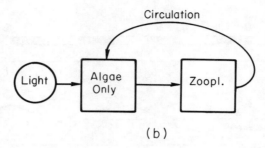

(b)

Figure 6.20. *Continued*

stats, have characteristics of a dissected microcosm. In the experiments using the apparatus of Figure 6.18, there were two chains of flasks, one with algae only and the other with algal consuming protozoa, *Tetrahymena*, in the downstream flask. These cultures were all in glass-contained, bacteria-free systems, thus achieving some of the ideals of organismal biology for defined, known conditions. One of many growth series is shown in Figure 6.19b–d. Up to a point, increasing light intensity or inflow of nutrient resources increased protozoan populations; increasing the flow rate or decreasing the light intensity could cause extinction of protozoa if their death rate exceeded the birth rate. Sommer (1988) circulated medium between a light chamber where photosynthesis took place and a dark chamber with zooplankton. As shown in Figure 6.20 organic matter was maintained at higher levels in the photosynthetic chamber and at lower levels in the consumer chamber. In the course of the experiments, there were substitutions of species with a succession to

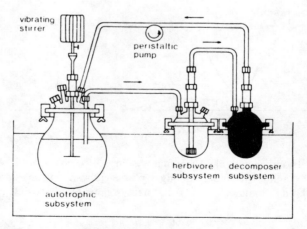

Figure 6.21. Three-level microcosm for study of ecosystem state space. (Ringlberg and Kersting 1978).

dominant algae, less edible by the *Daphnia* consumers, with phosphorus bound in tissues with high carbon–phosphorus ratios. The patterns had fluctuations, and most experiments did not demonstrate steady states within the 100 or more days of study. The presence of the grazers resulted in cultures that were more sensitive to the control of ratios of silicon to phosphorus, causing substitution of the diatoms. Ringelberg and Kersting (1978) used the three-chambered split-level microcosm with recycling (Figure 6.21) to study the disruption of system properties by toxic substances and other influences. During the 142-day stabilization phase, microcosm replicates were mixed weekly. Zieris, Feind, Huber (1988) used "the split pond" means of getting comparable model ecosystems responses to chemicals as the basis for environmental regulations.

Summary

Some microcosms with hierarchical populations oscillate. Some do not. The microbial microcosm of Kawabati and Kurihara (Figure 6.13) did not oscillate, nor did their models, but these systems had strong dampening effects from quadratic density dependency and waste product inhibition. The multispecies microcosm of Taub et al. (Figure 6.16) did oscillate, as did their model. This microcosm provided choices for coupling, production, and consumption, so that self-organization in the microcosm and in the mathematical model, resulted in a number of possible pathway coefficients. With many species providing many coefficients, it was possible for oscillation to develop if the oscillation pattern provided greater reinforcement power. By seeding to prevent extinction of larger animals, the upper levels of hierarchy were sustained. Generally, when one major component oscillates, all components oscillate (Figures 6.11, 6.16, 6.18). Many simulation models behave in this way. See the microcosm simulation model by Swartzman and Rose (1984) in Figure 6.15, for an example. In some models, a slight change in a coefficient may switch the results from all components oscillating to all components at steady state. Others shift their patterns chaotically.

The multispecies models and microcosms, even more than the simple ones isolated for study and simulation, show overall coexistence and mutualism. Wilson (1984) found infaunal populations with similar niches coexisting in marine sediment microcosms. Observed hierarchies, division of labor, and pulsing are consistent with the general hypothesis that self-organizational reinforcement prevents competitive exclusion and generates higher systems performance.

Chapter 7
Stress, Toxicity, and Adaptation

In the biosphere, there are large and long-interval phenomena that shock ecosystems with infrequent destructive pulses. Examples are storms, floods, volcanic actions, fires, or surges of long-period consumers, such as snowy owls. When conditions change, ecosystems reorganize. When outside influences cause severe disruption, new succession follows. To better understand these processes of stress, environmental impact, and recovery, agencies of environmental protection have supported extensive research on the effects of chemical, physical, and biological factors on microcosms.

When ecosystems come under stress due to detrimental factors and toxicity, components and functions are lost, the ecosystem changes, and new patterns emerge. Microcosms have some of the same responses to stress as the larger ecosystems. Consequently, there has been a large volume of research employing the microcosm as a means of testing chemicals and other environmental influences.

Let us define stress as some factor for which species are not adapted to profit. Stress, therefore, is only temporary, for new species and self-organization can usually provide a new system that derives some utility from the former stress. The ecosystem adapts to new conditions, using species with the special characteristics adapted to the altered condition and capable of being relatively stimulated.

External Stress

First, let us consider the stresses that operate from outside the system, the actions being dispersed afterward as heat. Some of these are natural impacts such as passing carnivores, winds, earthquakes, landslides, etc. If a stress is steady, it results in the development of ecosystems for which the stress is no longer a stress, such as when beach crabs become adapted to the surf. If the stresses are from small-scale events, short in time and magnitude, they are easily absorbed by the longer period components of the ecosystem. These small-scale events are seen as noise. The stresses that are important are the pulsing energies from the large-scale events. Consequently, relative to the smaller stresses, they have high transformity, high amplifier action, and are infrequent.

Pulsed Stress on Microcosms

Microcosms are sometimes perturbed by outside destructive pulses either by accident or as part of a study. For example, Richey (1970) applied flame to terrestrial microcosms and ultrasonics to aquatic microcosms. The initial effect of a pulse is to reduce metabolism because of the disabling of functioning components, but there is a quick return and overshoot as the system returns to its unperturbed pattern.

Model of External Stress

The basic production-consumption model of ecosystem metabolism was arranged for application of external stress (Figure 7.1). The destruction acts to increase consumption and recycling. If the stress is a pulse, the effect is a surge of consumption, recycling, and subsequent production a short time later. In Figure 7.1b is the result of a simulation of the basic production-consumption model including an external pulse which drains consumers, C. The model simulation, like the real microcosm, responds with rapid regrowth, using the store of disordered fragments as raw materials to stimulate the restoration. If the stress is steady and without compensating adaptation, the effect is the maintenance of a lower level of structure. The results in Figure 7.1c were obtained by successive simulations run with a steady stress, S, increased with each run. The steady-state value of consumer stock was plotted as a function of the increasing stress. Note the inverse relation of stress and stock at steady-state.

External Stresses and the Biomass Diversity Model

Having previously considered the way diversity varies with resources and succession (Figures 3.1 and 5.7), we next add stress removal. Most studies of external stress in microcosms show decreased diversity. In order to represent outside stresses and the larger scale, in other words, influence from outside a microcosm, the model in Figure 7.2 was created. Two stress sources were added to remove information and diversity from the system diagramed in Figure 3.1. Since gross production in the model is proportional to both the biomass and diversity, any drain on either tends to reduce production at first.

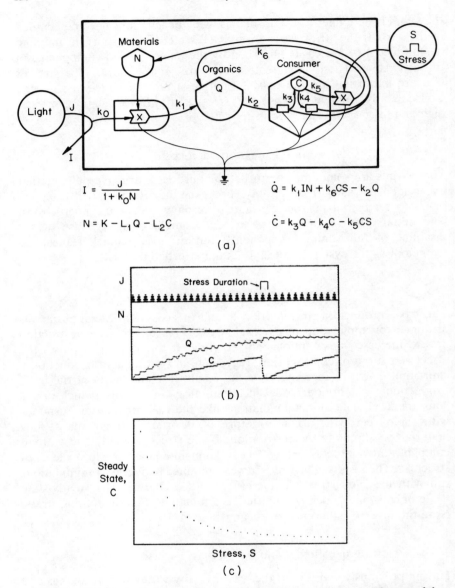

Figure 7.1. Response of a basic production and linear consumption model to an external destructive pulse (Program: STRESSPR). (a) Energy diagram; (b) simulation results with large initial concentration of nutrient; (c) steady-state consumers, C as a function of stress, S.

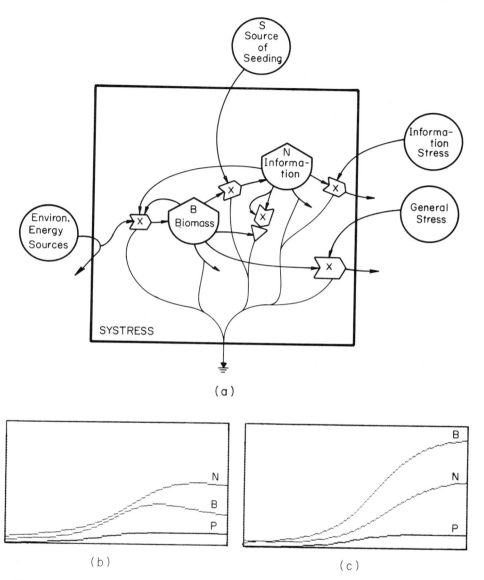

Figure 7.2. Model of succession and diversity from Figure 3.1 with two levels of external stress (Program: SYSTRESS), one draining biomass, the other, like irradiation, draining information. (a) Systems diagram; (b) simulation results with removal of stress on biomass; (c) simulation results with removal of stress on species information.

As in Figures 3.1 and 5.5, the number of species is represented as a state variable. Collectively, species interactions are in proportion to the square of the species number. Symbiosis-favoring functions may be in proportion to the square, and negative effects, such as competitive exclusion of species, may also be in proportion to the square. In Figure 7.2a, the stress on the lower right drains biomass, affecting information and diversity indirectly. The stress on the upper right is targeted selectively to eliminate information, affecting biomass indirectly. Ionizing radiation is an example of this kind of stress.

Substitution of Biomass and Diversity

Already well known in economics is the substitution of one factor for another and the effect on productivity when both are necessary. In the simulations (Figure 7.2b and c), both biomass and diversity are autocatalytic. A decrease of either increases the other's role so that the gross production which develops at steady-state is about the same as usual, but production is ultimately restricted by the flow-limited external light source. The drain of biomass (Figure 7.2b) reduces both, but diversity can increase relative to biomass and sustain gross production. The drain of species favors the biomass, partly by decreasing the amount of storage, since less storage is needed to maintain diversity, and partly because of the autocatalytic substitution. Microcosms which receive toxins sometimes generate a large amount of low diversity biomass, sometimes consisting of only one species.

Chemostat Flushing Rate and Diversity

The chemostat is a culture with media flushing through and was originally used to study single-species populations in high-turnover net-production states. To species not properly adapted, which includes the preponderance of species that might otherwise grow in the rich medium, the flushing is a stress which fits the model in Figure 5.8. By flushing media through with high turnover, all but the adapted species were flushed out. Others investigators, such as Margalef (1967), have operated chemostats in ways that maintained more diversity (Figure 7.3). Continual removal by flushing is a type of stress requiring continual replacement and this replacement occurs under conditions that provide an excess of the resources necessary for regrowth. These are the conditions for low diversity. The faster the flushing, the smaller the organisms that prevail, because turnover times must be faster to prevent population wash-out. It is the larger organisms that are large-scale organizers in time and space. Thus, general flushing tends to remove the control system and keep the system arrested in uncontrolled population competition. In an abstract way, the chemostat is a model for the decrease of diversity under stress.

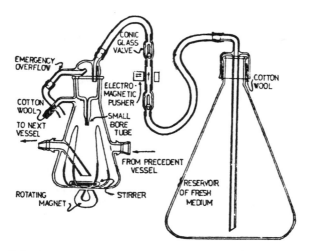

Figure 7.3. Chemostat microcosm (Margalef 1967, Copyright AAAS).

Generalized stresses which remove structure are included in the model shown in Figure 5.6. Species must be replaced if diversity is to be sustained. Another kind of stress is the diversion of previously available resources. In the simulations, the effect of adaptation is provided by reducing the biomass coefficient of stress removal. The reduction of species causes resources to be channeled to those remaining, sometimes with an increased net production. The mechanisms of action included in Figure 7.2 generally reduce diversity. About this there is controversy.

Stress as a Diversity Generator

The Hutchinsonian "paradox of the plankton" concept (Hutchinson, 1961) suggests that a general stress such as physical turbulence and dispersal could increase diversity by preventing any one species from developing dominance. This theory suggests that there are no competitive exclusions because the species are all maintained in small populations. If true, the energy of the stress would be an example of energy being used to maintain diversity, a special case of the generalization already discussed, that more resources can sustain more diversity. Energy for this work can be supplied from outside the system, or from within, perhaps by action of managing carnivores. On the other hand, Margalef (1958) found that turbulent waters contained low diversity. That a generalized turbulence did not result in higher diversity, does not support the disturbance theory of diversity in the environment for which Hutchinson first discussed it.

A stress is an energy influence for which the system is not adapted. After exposure to stresses, especially chronic ones, the emergent system is one which has reorganized to use the previously negative influence in a positive way. It is always an advantage to an adapted species if the stress is deleterious to others, thus giving the adapted species the benefit of more of the available resources. The result is lowered diversity which may require larger areas for species' long-range support.

Species Diversity and Uniformly Distributed Stress

When stress is applied uniformly, as for example, when a whole microcosm experiences a heat stress or toxicity, diversity of species is usually reduced. The aforementioned theory suggested that diversity during early successional stages is increased by a general stress which prevents any species from over-growing any other. Yet we find no case in microcosms where this has been demonstrated. Generalized stresses, such as cutting the vegetation or irradiating a community, invariably reduced the diversity. However, when stress is applied in patches, as for example when areas of an ecosystem are removed, the larger realm which included those areas may have its diversity increased, provided that there has been appropriate seeding. Pioneer species develop in the bare spots while the undisturbed areas continue to maintain the mature species.

Disruption and the Larger Scale

If disruption is examined in a larger context, a patch of displacement may constitute additional variety in the larger mosaic. Much of the research on microcosms in recent decades has dealt with environmental impact measurement, often using diversity and information measures as indicators.

Microcosm Boundaries Affecting Outside Stress Actions

Either in microcosms or in the biosphere, sharp disruption by nature or by human hands reduces diversity of the directly affected system. Some of this loss is in the removal of species, causing temporary local extinction. Active seeding is required to reestablish the removed species. In microcosms, unless specifically supplied by some regular input process, the reseeding may be prevented or delayed. The isolation effect may result in stressed microcosms having lower levels of diversity than the more easily seeded outdoor environments.

Under conditions of stress, larger species are more sensitive than smaller ones. One reason is that they take longer to be replaced. Continual or intermittent stress tends to favor those that can come back quickly. These are the species that often thrive early in succession, since their energy goes into replacement rather than developing long-lasting structures. Some kinds of stress are cumulative, such as ionizing radiation which causes genetic damage. Organisms that accumulate toxic substances also accumulate damage. Those species such as small microorganisms with short lifetimes and turnover times, replace their structures more rapidly than larger and longer-lived units. Thus, they can not accumulate as much damage from cumulative type stresses. Microcosms with a preponderance of smaller components may be more resistant. Some loss of species may be caused by the removal of controls and integrating mechanisms which normally prevent competitive exclusion. Larger species may be the controls that pollution or other stress may remove. Without some kind of controlling mechanism, two species drawing on the same resources each tend to grow exponentially. The better adapted one outgrows the other. It is in the mathematical nature of exponential growth that the competetive edge one species has over the other becomes larger and larger.

Species which have means for symbiotic interactions are set to use energy for these control actions, with less reserve for simple replacement. Such species tend to be displaced during stress by the successional species, sometimes called the weed species. Mature systems may be set back to less organized stages, and the recovery may involve replacing multiple species with a few types which are appropriately specialized to adapt to the impact. Especially when there is repeated (chronic) interruption, advantage goes to the particular species that have special means for thriving under that kind of stress. For example, *Scenedesmus* has been observed to predominate in growth on copper metal test plates in a microcosm of Silver Springs ground water.

Energy, Emergy and Stress

Energy can be used to put stresses on a common basis, including the energy resources diverted, the processes interrupted, or the storages or structures diverted. Energy is required to maintain the processes, structures, and controls for high diversity. By diverting resources or causing resources to support more replacement, energy is diverted from the support of species diversity. Energy is required to maintain a diversity of species, whether it is only a question of supporting more units or whether it is also a question of supporting more controls. Because energy flows of more than one type are involved in ecosystems and other natural hierarchies, the effect of different types of energy and different levels of hierarchy can be removed by expressing stresses in Emergy units. This is done by multiplying the energy storage or flow by its solar transformity, thus expressing the energy flows and stresses in comparable terms. It expresses what is required by the system in the terms of one kind of energy.

Stress and Transformity

Stresses may also be measured by their transformities, with larger values associated with larger effects. High transformity items are scarce because more Emergy is required for their development. High transformity items surviving prior self-organization may be those which have effects commensurate with the resources they require.

Ionizing Radiation

Ionizing radiation is a generalized external stress with the ability to destroy information to a greater degree than functional structures. Irradiation effects are greater where the informational storages provide larger targets, e.g., genetic materials in cells with larger nuclear volumes. Irradiation has a delayed effect since the information damage is not significant until normal repair and replacement fails because of faulty genetic templates. Genetic information is shared information with a high transformity and long period, consistent with the observed facts about time of action. The role of ionizing radiation in ecosystems has been studied in microcosms as a way to investigate genetic disruption and to make inferences about ecosystems in space or nuclear holocaust. Included are the effects of ultraviolet light, gamma radiation penetrating from outside the system, alpha particles, beta particles, and gamma radiation issuing from within the biotic components of microcosms.

Radiation studies provide insights concerning what disrupts the information content of an ecosystem. Most disruptions of ecosystems leave much of the genetic information intact, which facilitates reconstitution, but radiation has the reverse effect, attacking the genes and leaving the rest of the systems less directly affected. See the model for two kinds of stress in Figure 7.2

Irradiation of Species Prior to Starting Microcosms

The previously unpublished experiments by Fraleigh and Beyers used ionizing radiation to destroy genetic information in microbes prior to staring a microcosm with them. The growth–succession that followed produced less biomass than the controls (Figures 7.4 and 7.6). Holding the irradiated organisms for a time before initiating the microcosm improved their ability to develop growth and production (Figures 7.5 and 7.7). Apparently there was some kind of repair and error removal process. The experiments revealed the way self-organizing systems can process the information necessary for system function. The microcosms in Figures 7.4 and 7.5 were heterotrophic, running on a supply of organic matter. Those in Figures 7.6 and 7.7 were autotrophic.

Example of an Externally Radiated Aquatic Microcosm

Microcosm techniques permit investigation into the effects of gamma radiation upon aquatic ecosystems which would be almost impossible under field conditions. The high dosages required to elicit a response

External Stress 157

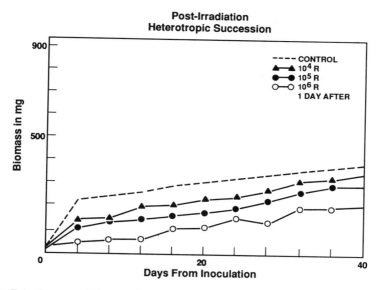

Figure 7.4. Course of biomass increase with time in heterotrophic succession after irradiation of inoculum with 10^4, 10^5, and 10^6 rads. Inoculation was performed on the 2nd day after irradiation.

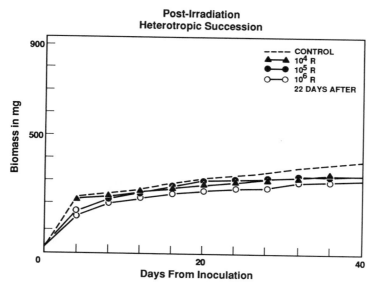

Figure 7.5. Course of biomass increase with time in heterotrophic succession after irradiation of inoculum with 10^4, 10^5, and 10^6 rads. Inoculation was performed 22 days after irradiation.

158   7: Stress, Toxicity, and Adaptation

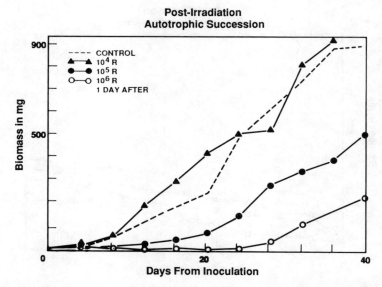

Figure 7.6. Course of biomass increase with time in autotrophic succession after irradiation of inoculum with 10^4, 10^5, and 10^6 rads. Inoculation was performed 1 day after irradiation.

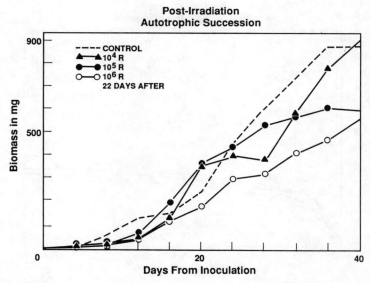

Figure 7.7. Course of biomass increase with time in autotrophic succession after irradiation of inoculum with 10^4, 10^5, and 10^6 rads. Inoculation was performed 22 days after irradiation.

from small aquatic organisms are impractical and dangerous to apply to natural bodies of water. The microcosms derived from a sewage oxidation pond, described in Chapter 3, were the subject of several experiments with external radiation.

Experimental Procedures

Microcosms were used which had undergone succession and reached the climax stage. They were irradiated using an 11,500-curie ^{60}Co source. The dose levels were 10^4, 10^5, 10^6, 2×10^6, 3.5×10^6, and 5×10^6 rads. These doses were delivered a within 0.6 min, 6 min, 1 h, 2 h, 3.5 h, and 5 h, respectively. Survival of the climax system and its ability to initiate a new succession were observed. With the exception of the death of the ostracods at 10^5 rads, there was no visible radiation effect on the climax system at or below 10^6 rads. At 2×10^6 rads the particulate fraction of the system lost its color 40 days after irradiation. At 3.5×10^6 rads it bleached after 4 days, and at 5×10^6 rads it bleached after 2 days. The loss of color, of course, indicated the death of the algae. Climax systems subjected to the three higher doses lost their ability to initiate a succession. However, an estimate using the serial dilution technique indicated that a small number of the bacteria (0.00017% of the number present in the controls) were still alive and capable of cellular fission after a dose 5×10^6 rads. This latter finding illustrates the extreme tolerance of these prokaryotic heterotrophs to gamma radiation. Partly because their target DNA content is small and partly because of their rapid metabolic and reproductive processes, these damaged units can be rapidly replaced.

Succession of Irradiated Systems

With the range of sensitivity having been established by the experiments described above, climax systems were used to investigate the effects of gamma irradiation on the autotrophic and heterotrophic successions. Three dose levels were used. Six unirradiated climax systems were each divided into two parts. In each case, one part served as the experimental system and the other as the control. Two of the replicate systems were irradiated at 10^4 rads, two at 10^5 rads, and two at 10^6 rads. The control systems and the irradiated systems were used to inoculate control, heterotrophic, and autotrophic successions. These inoculations took place at 1, 8, 15, and 22 days after irradiation. Data for both the autotrophic and heterotrophic successions inoculated on day 1 and day 22 at all three irradiation levels are shown in Figures 7.4–7.7. The irradiation was found to depress both types of succession in proportion to the amount of radiation received. However, this effect lessened with time, suggesting a recovery of the component organisms of the system.

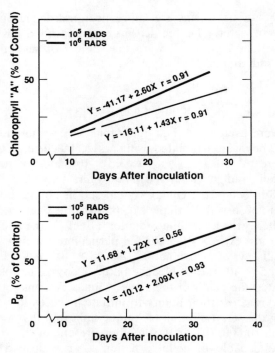

Figure 7.8. Linear regressions against time as a percentage of its control for the microcosms irradiated with 10^5 and 10^5 rads. (a) Chlorophyll; (b) gross productivity (after Ferens and Beyers 1972).

Metabolic Experiments

In a set of similar experiments, Ferens and Beyers (1972) irradiated the same system at 10^5 and 10^6 rads. They demonstrated that chlorophyll a content in the succeeding heterotrophic systems was depressed in relation to the dose received (Figure 7.8). However, the amount of gross productivity was greater at the higher doses, a fact which may reflect the greater radiosensitivity of the *Chlorella* as opposed to the *Schizothrix*. The death of the *Chlorella* at 10^6 rads removed it from competition with the *Schizothrix*.

Toxic Stress

Toxic stresses are detrimental effects from substances incorporated within an ecosystem. Heavy metals, radioactive substances, and toxic organic compounds are examples. These substances have large effects per unit mass and high transformity. Their effects are often at the control level of ecosystems where turnover times are slow. These substances are processed in chemical metabolism and cycle along with the normal metabolic materials, so that effects depend in part upon the ecosystem's stage and organization. When a toxic substance is added to an ecosystem, in may reduce some storages and interfere with some pathways. It may also be circulated as part of the material cycles. It may decay or be denatured. Microcosms have become very important in the study of toxic substances on the environment, since hazardous materials can be contained and test replications are relatively inexpensive.

Toxicity Tests with Organisms Versus Those with Microcosms

There has long been a controversy as to whether environmental measurements of toxicity should make use of microcosms or organisms. Investigators using standardized animals, such as white rats, *Drosophila*, or *Daphnia*, point to their known biochemistry and the benefits they contribute toward statistical calculations, and the comparing of results worldwide. Those advocating microcosms say that processing of a substance by an organism is entirely different from the processing accomplished with a full production–consumption system. A whole ecosystem shows what happens as a result of the normal self-organizational processes of selection and adaptation. Some progress has been made in duplicating standard microcosms either by the gnotobiotic culture approach, where everything else is kept out, or with the multiple seeding method, where everything is super seeded for a standard set of conditions so that self-organization produces a standard mix. The many sewage plants of the world operating with the same living components are a world-wide standard microcosm. See the discussion of the "white mouse" eutrophic microbial microcosm in Chapter 3. The organisms chosen to become standard test organisms are generally the tough generalists which are easily cultured and tend to be more resistant than the specialist members of complex ecosystems. As we have seen in Chapter 5, microcosms also tend to contain the more resistant species capable of operating with a less complex web. The two approaches are not necessarily incompatible. Lassiter (1983) suggests that data on the toxicology of individual species may be added to operating ecosystem models as a way of ascertaining the ecosystem-level consequences of a particular substance whose toxicity has been determined for organisms.

Sensitivity of Organisms Versus Microcosms

Whether self-organized systems are more or less sensitive than their organisms is also controversial. Highly organized ecosystems, such as tropical forests, achieve some resistance to stress through their gene pools and diversity. Their organisms, individually, are less resistant to stress in the same way that specialists have poor survival on a frontier. In some cases, experiments have shown the test animals to be more resistant and in other cases, less resistant.

Simulated Role of a Toxic Substance in a Production–Consumption Microcosm

The model in Figure 7.9a is a combination of the basic production–consumption model (Figure 2.3) and the pattern which results when a tracer is added (Figure 4.8). The toxic substance flows into the medium,

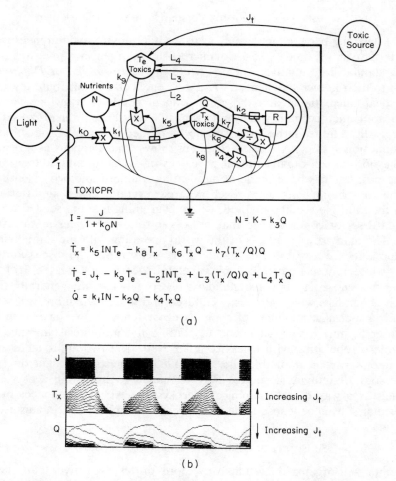

$$I = \frac{J}{1 + k_0 N} \qquad N = K - k_3 Q$$

$$\dot{T}_x = k_5 INT_e - k_8 T_x - k_6 T_x Q - k_7 (T_x/Q)Q$$

$$\dot{T}_e = J_t - k_9 T_e - L_2 INT_e + L_3 (T_x/Q)Q + L_4 T_x Q$$

$$\dot{Q} = k_1 IN - k_2 Q - k_4 T_x Q$$

(a)

(b)

Figure 7.9. Simulation of a toxic substance within a basic production–consumption microcosm (Program: TOXICPR). (a) Energy diagram and equations; (b) simulation results of toxic uptake, T_x and organic stock, Q for successive simulation runs, each with a higher inflow of toxic material, J_t. k_3 is nutrient fraction of Q.

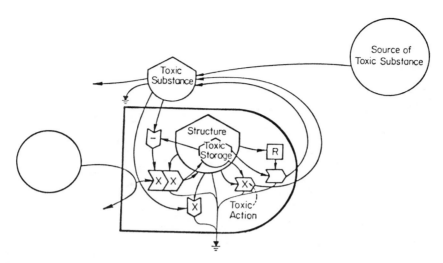

Figure 7.10. Concept of cadmium within a flowing-stream microcosm (Knight 1980, 1981).

T_e, where it is incorporated into biomass in proportion to primary production. While there, it causes extra consumption of biomass, Q, and is recycled to the medium with biomass consumption. Successive simulation runs in Figure 7.9b show increasing quantities taken up in response to increasing inflowing quantities, J_t. The effect is a decreasing quantity of biomass, Q, the toxic action removing it as fast as it is formed. The toxic material has a linear decay rate and so a balance of inflow and decomposition develops. In the simulation results, increasing quantities of toxicity produced higher levels of the toxin within the biomass, greater toxic action, and reduced production and consumption.

Knight (1980), studying cadmium microcosms, provided a unit model of the interactions of cadmium and an ecosystem (Figure 7.10). The toxic substance was circulated and had its effect on the components in which it was contained as well as on others. Knight, in reviewing the effects of increasing toxicity on ecosystems, found the model to produce the often-observed shapes of dose-response graphs.

Simulation of a Toxic Substance in a Mesomodel

The principles illustrated with the model in Figure 7.10 were applied by Knight (1980, 1981) to a more complex model of the contents of flowing microcosms (Figure 7.11a). The simulation is compared with observation in Figure 7.11b. As in the simpler version, the cadmium was processed through the compartments of the food web and recycled, while exerting toxicities in each compartment. In these flowing-tube microcosms, there

164   7: Stress, Toxicity, and Adaptation

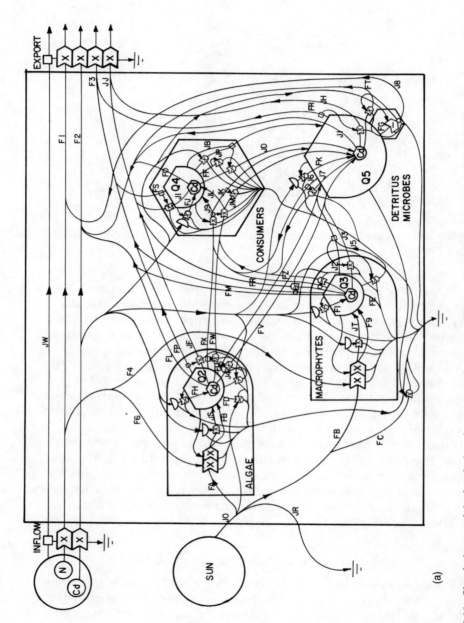

Figure 7.11. Simulation model of cadmium in stream microcosms (Figure 10.16 Knight 1980, 1981). (a) Energy systems diagram; (b) simulation results and observed data.

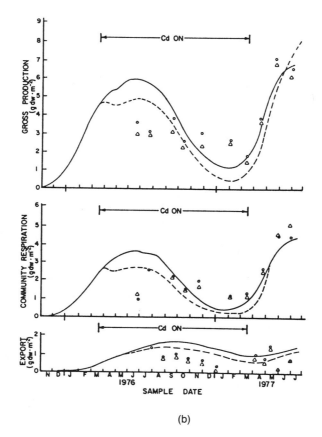

(b)

Figure 7.11. *Continued*

were steady substance inflow and outflow. The simulation results were more accurate in simulating metabolism than quantity of detritus.

Optimum Concentrations of Toxic Substances for Maximum Benefit

Knight (1980) in a literature review and in his own flowing-tube microcosms within Silver Springs, found that toxic substances caused some positive stimulation, when at very low concentrations, becoming highly toxic at higher concentrations. In this behavior, microcosms resemble organisms in their pharmacological responses to various chemicals.

Transformity and Position in Ecological Hierarchies

Many toxic substances cause their effects by interacting at the higher levels of the ecological web. This may be an example of the principle of

Toxic Stress and Reduction of Diversity

Generally, the stress of toxic substances, by reducing biological functions also reduces diversity. The species remaining have inhibited production, divert energies into repair and replacement, and, even if adapted, divert more energy into adaptive work and therefore have less to spend on the interspecies relationships necessary for maintaining diversity. When radioactive substances are within the ecosystem, the biogeochemical cycles control their distribution. A circulating tracer catches up with its carrier (nonradioactive equivalent), so that at steady-state the tracer becomes distributed among phases of the ecosystem in the same proportions as the carrier. Thus, the ecosystem determines the distribution of the radiation impact. For example, some heavy metals tend to concentrate in animals

matching energies. Items of high transformity develop interactions with items having transformities one or two orders of magnitude smaller, since in this way items which have large requirements for their manufacture, through amplification have effects commensurate with their formation requirements.

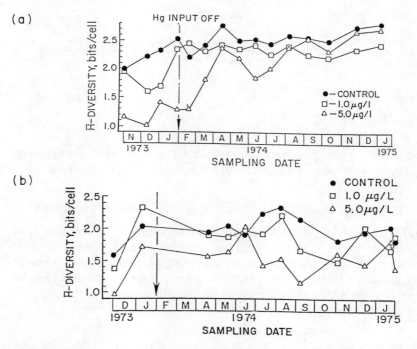

Figure 7.12. Increase in species diversity after termination of mercury input in large-stream mesocosms (after Kania 1981). (a) Plants; (b) insects.

high in the food chain and radioactive versions of these elements would concentrate their internal physiological damages there.

Figure 7.12 displays the way diversity returns when inflows of mercury are stopped (Kania 1981). At least for this class of stresses, microcosms do not show diversity to be augmented by stress. Even when applied in patches, the net effect is not an increase in diversity. Pratt and Bowers 1990 studied selenium effects on microbial communities in laboratory microcosms. Diversity and carbohydrate were reduced, but chlorophyll and the ratio of production to biomass were stimulated. Microcosm results were similar to those in outdoor streams.

Organic Toxicity in Microcosms

Because of the high toxicity of organic chemical substances, many experiments have been done with organic substances added to microcosms. The way these are processed, transformed, and affect the ecological systems can suggest a basis for policies concerning the larger outdoor system. For example, Mitchell (1971) found algicide strongly reduced the diversity of algal plankton in microcosms. Although a number of researchers have attempted to investigate the effects of toxic organic compounds on the ecosystem level using microcosms, criticism has been leveled at these efforts. Metcalf (1977) has written, regarding this approach, that microcosms "... have well-defined limitations, including limited applicability to long-term effects (i.e., multigeneration studies, time for ecosystem stability), uncertain predictability from one model system to another, limited predictability as to vital biological components and vulnerable biological processes, and the fact that replicability and reproducibility need careful evaluation." Moriarty (1977) concurs: "The model ecosystem is too simple to give a realistic representation of a field situation, but too complicated to yield useful quantitative data." However, these opinions have not deterred others from continuing this line of investigation.

Dieldrin Uptake in Flowing Microcosms

Rose and McIntire (1970) studied dieldren and benthic algae in laboratory stream microcosms. Wooden troughs with paddle wheels similar to that in Figure 10.3, were used with a current of 28 cm/sec. Dieldren uptake was measured from flows with dieldrin concentrations ranging from 0.05 to 7 ppb. Accumulations in algae ranged from 0.1 to 200 ppm, which is approximately 30,000 times the dieldren concentration in the water. Faster current caused more filamentous forms and higher dieldrin uptake. Filaments forms may be able to take better advantage of turbulent energy because they may extend further into the water.

Toxic Heavy Metal Elements in Microcosms

Cosmically and globally, light chemical elements are generally more abundant and more involved as normal components of living systems. Many of the heavier elements, especially the heavy metals, are scarce but biologically active, either as biochemical trace requirements or as toxiins. As one of the main classes of wastes which can have a large impact on environments, the heavy metals, their movements, transformations, and effects have been important points of research in microcosm work.

Zinc and Plankton Fed Flowing Microcosms

Williams and Mount (1965) prepared four stream microcosms by lining outdoor channels in the ground (4 m by 0.3 m) with plastic sheeting. Flow of uniform hard spring water was passed into ponds, which developed plankton blooms, passing next into the microcosm channels. Each channel contained a different concentration of added zinc (0, 1, 3, and 9 mg/l). Periphyton on glass slides in the microcosms was studied. The long-term effect was a buildup of a zooeal slime mat made up of fungi, bacteria, and bluegreen algae feeding on dead, settling plankton from the incurrent. Species variety was greatly decreased with increased zinc concentration. The industrial use of heavy metals has led to increasing pollution of the environment with these elements. Public awareness of the problem has, in turn, led to a demand for research on the fates and effects of these substances on ecological systems. Since widespread additions of these frequently toxic metals to natural systems for research purposes are both expensive and environmentally and politically unsound, the microcosm technique has been an excellent method for studying the subject.

The following heavy metals have been used as single stressors in microcosm experiments of one sort or another: Arsenic (As), Sanders and Osman (1985); Cadmium (Cd), Niederlehner et al. (1986); Copper (Cu), Harrass and Taub (1985); Manganese (Mn), Hunt (1983); and Zinc (Zn), Stewart et al. (1986). Many investigators have included suites of heavy metals in various studies in attempts to trace several cycles simultaneously: Cesium (Cs), Cobalt (Co), Ragsdale et al. (1968); Zn, Cerium (Ce), Americium (Am), Giesy and Geiger (1980); Thorium (Th), Iron (Fe), Ce, Co, Scandium (Sc), Mercury (Hg), Santschi et al. (1983); Cu, Pb, Mn, Cd, Hunt and Smith (1983); and Lead (Pb), Cu, Zn, Silver (Ag), Nickel (Ni), Mn, Hardy et al. (1985).

Effects of Dose

Heavy metals often have a Jekyll-and-Hyde aspect to their role in ecological systems. They may be essential nutrients, required as coenzymes by many organisms. In this aspect they are often limiting in the Liebig sense. On

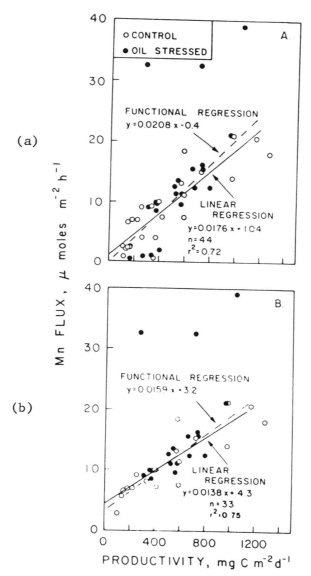

Figure 7.13. Relationship between the benthic manganese flux and average daily primary productivity in the water column in MERL mesocosms. (a) Results for annual cycle; (b) results for fluxes obtained when the water column was less than 14°C (after Hunt 1983).

the other hand, in higher concentrations, they may accumulate in various tissues and act as poisons. Thus, special care must be taken in the design of experiments and the interpretation of results. As a practical matter, heavy metal concentrations in experimental systems should approach those concentrations expected to be found in nature, either naturally, or as the result of human activity (Pratt et al. 1987). The essential nature of trace heavy metals is illustrated by the data of Hunt (1983) in Figure 7.13. In the MERL microcosms, the flux of manganese is correlated with primary production. Although there is some doubt as to which is cause and which is effect, the close coupling of a community metabolic parameter to the availability of a single metal demonstrates that, in concentrations approaching nature, these divalent elements may be critical indeed.

Response and Recovery

On the other hand, a more typical ecosystem response to heavy metals is the one shown by the *Elodea* dominated microcosms of Ausmus et al. (1980 and Giddings & Eddlmon 1978) (Figure 7.14). As in response to almost any kind of stress, the primary production and community respiration decrease and then increase with time, demonstrating system recovery and adaptation, perhaps through species replacement or through physiological accommodation. This is the same pattern shown in microcosms subjected to thermal stress and overgrazing (Beyers 1962b, 1963b). Response is often proportional to dose, and recovery does not take place prior to cut-off of the contaminant (Hardy and Apts 1984, Hedtke 1984) (Figure 7.15).

Ecosystems exhibit another phenomenon associated with heavy metal (and other forms of) contamination. This is recovery as evidenced by elimination of the pollutant both from the system components and the system as a whole (if the system has outputs and inputs). Giesy et al. (1981) found in the large (91.5 m) Savannah River channels described in Chapter 8, that cadmium disappeared rapidly from the rooted macrophytes after the input of the metal ion was halted (Figure 7.16). This is another variation of the famous oxygen sag curve of Streeter and Phelps (1925). In classic sewage science, the pollutant is a slug of organic matter which is eventually oxidized by the microbes present in the system. Recovery can be measured by the restoration of dissolved oxygen levels within the system after the oxygen sag.

Yet another example of this recovery was demonstrated in the same systems with another heavy metal, mercury, and another ecosystem parameter, species diversity (Ferens 1974; Kania 1981). For over a year, mercury was added to the Savannah River site channels at the 1 and 5 µg/l levels. After termination of the mercury input, sampling was continued for an additional year. The Shannon-Weaver diversity index (H bar) increased for insects and plants (Figure 7.12). Gladyshev, Gribovskaya,

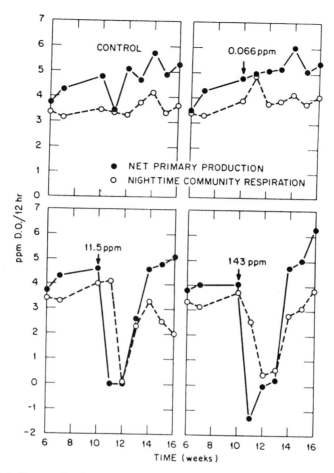

Figure 7.14. Net production and nighttime community respiration in controls and at three arsenic levels before, during, and after arsenic addition (Giddings and Eddleman 1978, Ausmus et al. 1980).

and Shchur (1990) found autocatalytic model required for Phenol decontamination in a microecosystem.

Heavy Metal Response of Organisms in Microcosms

The microcosm technique has proved useful in studies where investigators desired to evaluate the effects of specific metal ions on individual species or types of organisms. The obvious advantages of reproducibility and control of the experimental situation, and lack of addition of pollutants in field situations has led several investigators to pursue microcosm studies

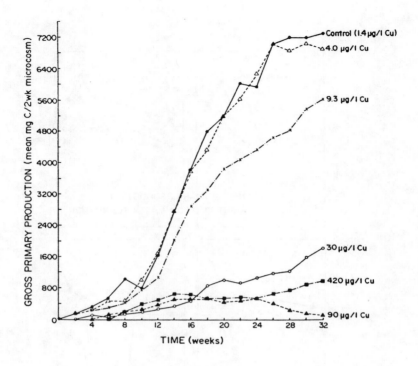
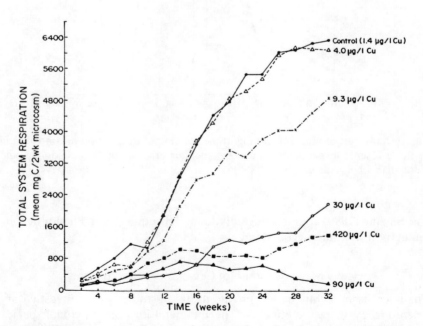

Figure 7.15. Biweekly gross primary production and total system respiration of microcosms exposed to various concentrations of copper (after Hedtke 1984, Aquatic Toxicology 5:227–244).

in which the focus was on one type of animal or plant, or a specific trophic level.

Giesy (1978a) utilized the Savannah River channels to conduct studies of the effects of 5 and 10 µg/l concentrations of cadmium on the decomposition of leaf material by fungi and bacteria. Mixed packets of pine, sassafras, oak, cherry, and maple leaves were used. Microbial activity was determined by scanning electron microscopy and by loss of weight from the packets of leaf material placed in the channels. Both levels of cadmium inhibited decomposition, but the higher level of cadmium did not show a significantly greater inhibition than the lower. Since the detritus food web is the energy basis for most lotic environments, any reduction in decomposition of allochthanous material could have deleterious effects on higher trophic levels. Also studying cadmium effects, Sanders and Cibik (1985) used 500-liter fiberglass tanks with flowing Patuxent River, Maryland (estuarine) water. The object of the experiment was to determine the effects of 5 and 15 µ/l cadmium concentrations on the growth of a natural assemblage of diatoms, and on spore formation of the dominant diatom, *Chaetoceros debile*. The phytoplankton exposed to chronic low cadmium doses showed little change in species composition, growth rate, cell densities, or particulate carbon to nitrogen ratios. However, the production of spores by the dominant was greatly reduced, the concentration of cadmium compatible with spore reduction being approximately an order of magnitude lower than the amount required to suppressed cell growth. Since spore production and germination is a mechanism for maintaining dominance in diatom species, such reduction could lead to long-term alteration of species composition and dominance. Diatoms are important primary producers at the base of estuarine food webs. Thus, alteration in species composition of the base could lead to alteration of the species composition higher in the food webs. The work of the group at Virginia Polytechnic Institute and State University (Niederlehner et al. 1986; Pratt et al. 1987) is also concerned with the effects of heavy metals on microorganism species composition. Using a polyurethane foam technique to trap protozoa and other microorganisms, they investigated the effects of varying concentrations of zinc and cadmium on species reduction. In these cases, species reduction was in inverse proportion to heavy metal concentration (Figure 7.17). Phelps et al. (1985) investigated the effects of increased copper concentration in pore water on the burrowing behavior of the clam, *Protothaca staminea*. Since burrowing is the primary method of predator avoidance for this species, any reduction in the speed of burrowing can have unfortunate consequences for the population. They found a direct correlation between distance of burrowing and the concentration of copper in the pore water of the sediments in which the clams were digging (Figure 7.18).

All of these examples illustrate the subtle effects a heavy metal pollutant can have on an ecological system. Since an effect on one organism can be

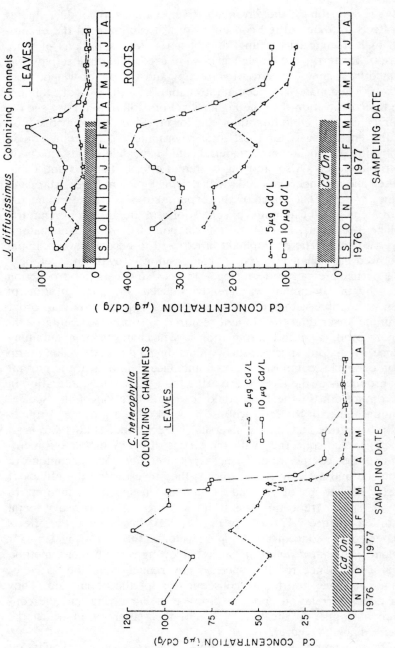

Figure 7.16. Cadmium content in *Callitriche heterophylla* and *Juncus diffusissimus* in Savannah River channels during and after cadmium addition. Concentrations expressed on a dry weight basis (after Giesy et al. 1981, Copyright 1981, Pergamon Press).

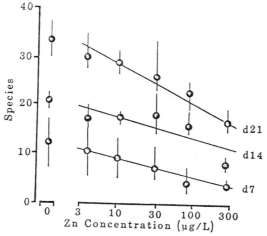

Figure 7.17. Log-linear regression of species versus zinc concentration on days 7, 14, and 21. Increasing zinc diminished the log of the number of species present (after Pratt et al. 1987).

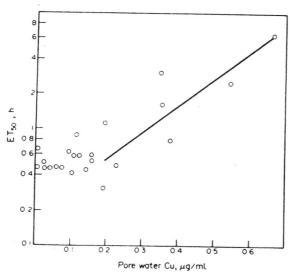

Figure 7.18. Clam burrowing time (ET_{50}) versus pore water copper concentration in sediments (after Phelps, Pearson, and Hardy, 1985, Copyright 1985, Pergamon Press).

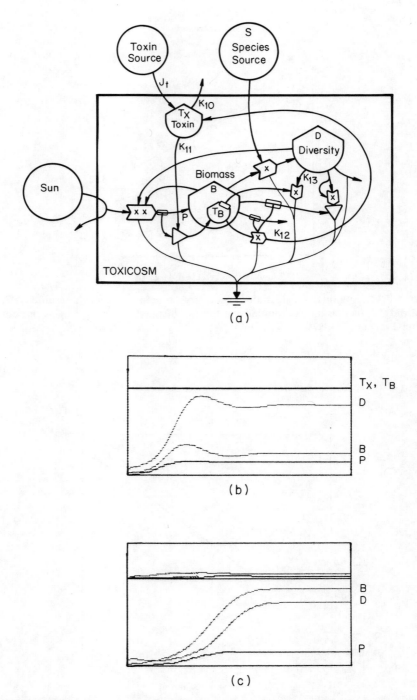

Figure 7.19. Biomass and diversity minimodel from Figure 3.1 with a toxin added (Program: TOXICOSM). (a) Systems diagram; (b) simulation results without toxin; (c) simulation results with inflow of toxin.

amplified throughout the food web, a whole ecosystem can be deleteriously affected through this mechanism. These facts also demonstrate the advantages of testing pollutants on the ecosystem level rather than on the single organism level, as such testing in microcosms or enclosures in the field is more likely to pick up ecosystem effects than single organism studies including such techniques as LD_{50} or indicator organism experiments.

Toxic Substances and the Minimodel of Diversity

Figure 7.19 depicts the diversity minimodel (Figure 3.1) including a toxic substance. With the toxic substance, diversity is reduced and more energy is directed into biomass.

Lichens with their symbiotic organization of photosynthetic and heterotrophic consumer tissues have self contained metabolism and are like small microcosms (Ahmadjian 1967; Hale 1983). They may be useful for microcosm studies of toxicity, since field observations suggest sensitivity to air pollution. Models like that in Figure 7.9 may be pertinent. Steele (1979) studying enclosure results from Loch Ewe and Saanich inlet suggests that simplified food chains in enclosures as the best compromise between natural conditions and experimental control for toxicity testing.

Chapter 8
History

Modern biological science tends to regard ecological microcosms and other similar objects as relative newcomers on the research scene. Indeed, most of the serious scientific literature dates no earlier than 1960 and the technique was not named microcosmology until 1980 (Giesy and Odum 1980). However, the philosophy behind the microcosm concept can be traced back to the ancient Greeks and some very good science was done on containerized living systems prior to 1960. In this chapter, we trace the history of the concept and consider some of the early work.

Robert Warington Introduces the Balanced Aquarium

On the evening of Friday March 27, 1857, Mr. Robert Warington gave an address to the Royal Institution titled: *On the Aquarium*. One-hundred-thirty years ago, the home aquarium was just becoming popular (Atz 1949) and Warington was generally considered to be the originator of this "new pleasure." A similar concept in the terrestrial realm was the dish garden (Beard 1930) which was also becoming popular in England at that time, although it had been known in Japan since antiquity.

As every contemporary owner of a fish tank knows, there is a great deal of misinformation extant about the subject. This was also true in Warington's day, and his purpose was to put some of the then current misconceptions to rest. However, in the process of giving, and subsequently publishing his talk (Warington 1857), he produced one of the first scientific papers on microcosms. He began his work with a review of the literature. He quoted such famous names as Hooke, Boyle, and Priestly, referring mainly to their work on respiration and gas exchange. He went on to discuss the containers used to hold aquatic organisms, a subject much discussed by present writers in the field (Gearing 1989). His equivalent of a methods section consisted of his description of how he had placed two goldfish and some *Vallisneria* in a twelve gallon aquarium. The purpose of his experiment was to demonstrate that the "growing vegetation would counterbalance the vital functions of the fish." However, as so often happens, his experiment went awry. The surface of the water became covered with "green confervoid mucus," the plant leaves decayed, and the fish became sickly. He then introduced five or six snails of the genus, *Limnea*. After this addition, Warington writes:

> Thus was established that wondrous and admirable balance between the animal and vegetable kingdoms, and by a link so mean and insignificant as almost to have escaped observation in its most important functions. The principles which are here called into action are, that the water, holding atmospheric air in solution, converts the oxygen constituent into carbonic acid. The plant, by its vital functions, absorbs the carbonic acid, and appropriating and solidifying the carbon of the gaseous compound for the construction of its proper tissues, eliminates the oxygen ready again to sustain the health of the fish. While the slimy snail, finding its proper nutriment in the decomposing vegetation and confervoid mucus, by its voracity, prevents their accumulation, and by its vital powers, converts that which would otherwise act as a poisonous agent into a rich and fruitful pabulum for the vegetable growth (Warington 1857).

Warington thus pointed out the interaction of the three trophic levels: producers, consumers, and decomposers, and also laid the basis for the concept of the aquarium, with its contained organisms functioning as a unit. This holistic concept underlies all microcosm work. Warington writes further: "Reasoning from analogy, it was evident that the same balance should be capable of being permanently maintained in sea water and thus a vast and unexplored field for investigation opened to the research of the naturalist." Again, he expressed a concept fundamental to microcosm research, namely that the results of microcosm experiments can be carried over to other situations and other levels of organization. Such analogical reasoning is known in philosophy as "microcosmic theory."

Microcosm Concepts in Philosophy

Microcosmic Theories in the Ancient World

Conger (1922), in his monograph *Theories of Macrocosms and Microcosms in the History of Philosophy*, states that according to microcosm theory, "portions of the world which vary in size exhibit similarities in structures and processes, indicating that one portion imitates another or others on a different scale." He refers to such generalized schools of thought involving the concept of analogy as "microcosmic theories." Most often such theories dealt with an analogy between man and God, man and the universe, or other similar religious themes. Although there are indications that such ideas may have even earlier origins, Conger traces their earliest expression to the works of Plato, and attributes the first authentic occurrence of the term "microcosm" to Aristotle. He then goes on to trace the persistence of the microcosmic idea through the writings and teachings of the Stoics, the early Jewish philosophers, the Neo-Pythagoreans, the Eclectics, and the Neo-Platonists. Therefore, the microcosm concept was firmly established in pre-Christian thought.

Microcosmic Theories in the Medieval Period

Medieval philosophers could be generally grouped into one of three categories Christian, Jewish, and Islamic theological scholars. The early Christians took over much of the classical Grecian thought and with it some vestiges of microcosmic theory. Although never greatly in evidence, some microcosm ideas can be found in the writings of such famous early Christians as Clement of Alexandria, Pope Gregory the Great, Thomas Aquinas, and Augustine. Microcosmic theory was much more prevalent among Jewish and Islamic medieval writers. Of course, as with the Christians and classical Greeks, much of the microcosmic thought revolved around man as a microcosm of God or the universe. As Conger puts it:

> The medieval Christian writers transmitted, while the Jewish and Mohammedan writers elaborated in some fantastic details, the ancient views of man as a microcosm.... The three traditions apparently developed in considerable independence of one another, although some cases of interaction, especially between Jewish and Arabian writer, are fairly evident. In general, during the middle ages, the microcosmic theory served as a convenient and uncritical method of reconciling religion with the natural sciences, which even then were beginning to raise questions and difficulties for the faithful (Conger 1992).

Renaissance to Modern Times

As so many other medieval ideas, the microcosmic theories suffered an eclipse during the Renaissance and Reformation. However, as the importance of science increased, the idea of trying to understand natural phenomena by arguing from analogy, the basic microcosmic theory, regained its prominence.

Early Modern Times to the Beginnings of this Century

Much impetus was given to the idea of analogic reasoning by the work of Haeckel (1874) whose famous "ontogeny recapitulates phylogeny" was the ultimate in analogical thought. Haeckel's "recapitulation theory," also called the "morphogenetic theory," the "biogenetic law," the "doctrine of parallelism," or the "repetition theory," became widely popular among philosophers and scientists in the nineteenth century, and was carried over with various modifications into such diverse fields as psychology, sociology, geology, physics, chemistry, and astronomy. For a discussion of the roots of Haeckel's ideas and a critical discussion of them, see pages 272–274 in Hyman (1940).

As an illustration of the popularity of this idea, Conger finishes his monograph with a discussion of what he calls "monads," what ecologists would call trophic levels, and system scientists call hierarchies. These hierarchical series, drawn from physics and astronomy, biology, and psychology are postulated, by analogy, to possess similar structure and function to the other members of the series. Conger writes:

> The analogies are ... scattered, and uncertain, and uncoordinated; there are, however, enough of them so that from this point of view the suggestion presents itself that if the world could be interpreted in terms of a concrete and realistic monadism whose individual units, as studied in laboratories and observatories, are seen in their interactions with one another to exhibit essentially similar structures and processes, the way might be open for a modern microcosmic view ... (Conger 1922).

Microcosms and Science from the Turn of the Century to 1960

With the exception of a few scientists such as Warington, microcosmic theory in the 1800s was the realm of the philosophers. However, in the first part of the twentieth century, the history of microcosmic thought shifted from philosophy to science, mainly biological science. In the first decade, Woodruff (1912) did his classic experiments on succession in hay infusions. Figure 8.1 describes the well known replacement of one species by another with time, as well as the model of the organisms involved and their interactions. In the 1940s and 1950s limnologists, marine biologists, fisheries scientists, wildlife managers, foresters, and agronomists became increasingly interested in measurements of primary production in natural and man-managed systems. The obvious advantages of the microcosm method in studying community metabolism and related phenomena were recognized by some pioneering scientists. Edmondson and Edmondson (1947) studied the effects of fertilizing salt water in concrete tanks at Wood's Hole, Mass. They demonstrated the correlation of gross primary production with nutrient concentration (Figure 8.2). In similar experiments, with outdoor tanks and with battery jars in the laboratory, Raymont and Adams (1958), while attempting to mass culture *Phaeodactylum*, demonstrated the inverse relationship between that phytoplankter and a flagellate herbivore, *Monas sp.* (Figure 8.3). Odum (1950, 1951) studied strontium cycles in duckweed microcosms (Figure 4.10). In a series of

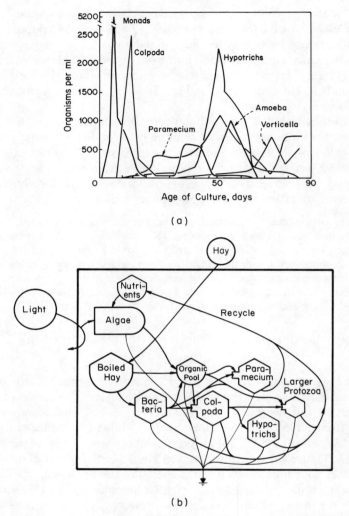

Figure 8.1. Woodruff succession in hay infusion microcosms. (a) Rise and fall in population numbers with time; (b) functional model (after Odum 1983).

papers, Kuhl and Mann (1951, 1955, 1956a, 1956b) investigated the diurnal chemistry and nutrient biochemical cycling of a marine aquarium and published the first diurnal oxygen curve (Figure 8.4) of the typical shape subsequently shown by many investigators to be present in most microcosms and many natural bodies of water (Beyers 1963b). Over approximately a ten year period, Kurihara undertook a study of the ecological systems within man made containers. Some of these were stone basins and bamboo vases in Japanese cemeteries, while others were

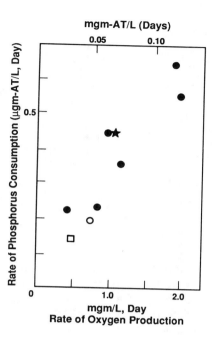

Figure 8.2. Correlation between total oxygen production and phosphate–phosphorus consumption in fertilized salt water in a concrete tank. Square symbol represents rates in raw sea water, open circle, rates after nitrate fertilization, and starred dot, after nitrate and phosphate fertilization. Other points refer to determinations on subsequent days (after Edmondson and Edmondson 1947).

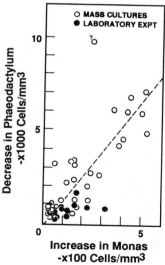

Figure 8.3. The decreases in density of *Phaeodactylum* and corresponding increases in *Monas sp.* in tank, battery jar, and laboratory experiments. The dashed line represents the relationship: 1000 phytoplankter decrease per 100 herbivore increase (after Raymont and Adams 1958).

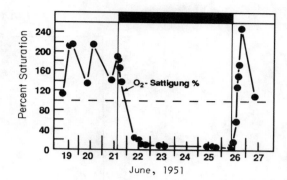

Figure 8.4. Diurnal changes in percent oxygen saturation in a marine aquarium containing *Ulva, Polydora ciliata, Balanus improvisus,* and *Nereis diversicolor.* Dark bar represents period of no light (after Kuhl and Mann 1951).

bamboo microcosms fashioned by the investigator. These efforts were undertaken to understand the production of mosquitos in these ubiquitous containers. The result was eleven papers (Kurihara 1954a, 1954b, 1957a, 1957b, 1958, 1959a, 1959b, 1960a, 1960b, 1960c; Kurihara and Kato 1952) dealing with succession and community structure within the mosquito breeding microcosms.

With the increasing interest in space generated during the 1950s, some thought began to be given to the biological support of man in space. The first speculative papers in this area were published (Myers 1954; Golueke et al. 1959; Rich et al. 1959). In all of these investigations, the researchers recognized the basic necessity of some form of photosynthetic gas exchanger to provide oxygen for humans. The other support functions of the biosphere were also acknowledged. These papers mark the beginning of the interesting area of microcosm research revolving around humans in closed systems (Chapter 19).

During the first half of this century, the interest in aquaria as homes for fishes and other aquatic organisms had not waned. Richardson (1930), much in the style of Warington seventy three years previously, published instructions on methods of keeping fish alive in aquaria, based mainly on practical experience, while Breder (1931) discussed the balance of respiratory gases and the activity of bateriophages in conditioning aquarium water. Interest in the aquarium as an ornament in the home also continued with the periodic publications of various handbooks on the subject (Simpson 1946). During this period, most aquarists sought to achieve "balance" in the aquarium. By that was meant the achievement of a condition similar to that described by Warington (1857) in which the autotrophic and heterotrophic activities of the organisms counterbalanced each other. However, since the principal interest of most aquarists was in the animal component of the system, the population of these organisms was usually increased, with their energy and matter requirements being met by supplemental feeding. If plants were present, and the fish survived under this kind of regime, the aquarium owner would proudly proclaim his tank to be "balanced." In a reaction to this kind of thinking, Atz (1949) published his attack on the possibility of closed ecosystems. While he correctly decried the overfed, heterotrophic dominated system as not being balanced in Warington's sense of the word, he also implied that it was not theoretically possible to have a balanced system which was closed energetically, materially, or both. To quote Atz: " 'A vessel of water containing plants and animals must be looked upon as a little world,' wrote Edwin Lankester in 1856. We can now just as categorically state that it must not be so considered". Although Atz's contentions were not widely accepted by aquarists or school teachers, who continued to construct thousands of sealed carboy microcosms in schoolrooms throughout the country, based on earlier studies of duckweed microcosms (Figure 4.10), Odum and Johnson (1955) published a rebuttal. They characterized the Atz type of aquarium as follows:

1. The light intensity is too small to support the community. The energy for the community comes primarily from added food rather than from the processes of photosynthesis.
2. The oxygen is maintained above a lethal level by diffusion from the air. The accumulation of carbon dioxide is likely to be serious, especially if the water is soft and does not have buffer capacity.
3. With plant photosynthetic metabolism a relatively minor factor, dangers from the accumulation of toxic wastes are great.
4. The ratio of plants to animals is not pertinent since the animals do not get most of their food from the plants, but would have to be fed.

On the other hand, a materially closed, but energetically open balanced aquarium would possess the following characteristics:

1. There must be strong enough light so that the resulting photosynthesis will support the animals and plants day and night.
2. The community must possess a large mass of plants to provide food for a smaller mass of animals.
3. The aquarium must regenerate its own plant nutrients from its wastes, such as the fertilizer elements phosphorus and nitrogen.
4. It must not develop toxic conditions of any kind due to wastes.
5. It must maintain enough oxygen for the animals and enough carbon dioxide for the plants through day and night.

In order to demonstrate the feasibility of these theoretical considerations, Odum and Hoskin (1957) constructed the first microcosm designed to study the growth, succession, and community metabolism of the aufwuchs community of a flowing stream in the laboratory (Figure 8.5). This construction was the forerunner of many more elaborate and highly instru-

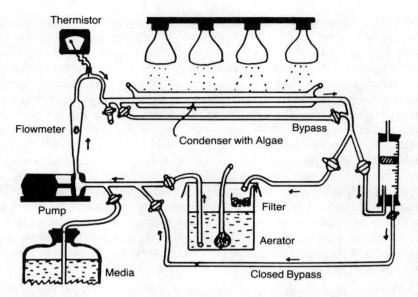

Figure 8.5. Diagram of the flowing stream microcosm. The system was normally open to gas exchange but closed during metabolic measurements. Temperature and stream velocity were controlled (after Odum and Hoskin 1957). Also see Figure 10.7.

mented types of apparatus such as chemostats, laboratory streams, etc. Odum and Hoskin showed that the communities developing in the microcosm during succession resembled communities and nature with respect to production to respiration ratio, chlorophyll content, assimilation number, species diversity, and nitrate and phosphate uptake. The microcosm and the macrocosm were indeed similar.

Summary

In our western culture, the concept of the microcosm was the contribution of classical Greek thought. The Greeks developed microcosmic theories which embodied the ideas of hierarchy and analogy: that one level's (monad) structure and function was similar to another's, whether the levels involved man and God(s), animals and man, or any of the monads of nature. Such ideas were seldom compatible with medieval Christianity but were developed by Jewish and Islamic philosophers of that period. With the Renaissance, European scientists revived the concept and applied it to various fields of biology, particularly embryology. The popularity of the home aquarium in nineteenth century England led the scholars of the time to study it. Earlier work with gases and whole animal physiology was applied by analogy to the aquarium. Concepts such as autotrophy and heterotrophy, although not under those names, came to be understood. An equalization of these two functions was desired and means to achieve it were discussed under the name "balance." Scientists began to conceive of aquaria as little worlds or microcosms.

The advantages of the microcosm method for some forms of ecological research were recognized by a few workers in the first few decades of the twentieth century. However, the concept of balance and the entire microcosm philosophy were called into question by Atz (1949). The defense was led by Odum and Johnson (1955) who answered on grounds both theoretical and, with Hoskin (Odum and Hoskin 1957), practical. The Odum and Hoskin (1957) set of experiments did indeed demonstrate that a microcosm could be a little world unto itself, and that little world would emulate, in many respects, the greater world around it. Thus the long history of microcosmic thought was proven to have a basis in fact, and the stage was set for rapid expansion in the quantity, types, and objectives of microcosm research.

Part 2
Kinds of Microcosms and Mesocosms

In Part 2 are chapters on different kinds of microcosms, including those large enough to be called mesocosms. The purpose of Part 2 is to give examples that further represent the principles given in Part 1 and at the same time introduce as many kinds of microcosms from the fascinating literature as possible. Some of the creative varieties of microcosm structures are shown in order to help those planning new studies. Ideas and interpretations are included here as suggested by these examples and their authors.

Chapter 9 includes aquaria and small tanks; Chapter 10, flowing water microcosms; Chapter 11, terrestrial microcosms, including soil ecosystems; Chapter 12, experimental pools and ponds; Chapter 13, aquatic microcosms dominated by encrusting life and reefs; Chapter 14, tall aquatic microcosms with conditions favoring plankton; and Chapter 15, microcosms adapted to high temperature, high salinity, and other extremes.

Chapter 9
Aquaria

Small, still aquatic microcosms such as aquaria, small tanks, and flasks, have a high ratio of wall and bottom surface to volume and small exchanges with the outside (Fraleigh and Dibert 1980). The most relevant outdoor ecosystems are small, shallow aquatic ecosystems with little plankton and the littoral zones of larger bodies. Aquaria were the microcosm type first set up by Robert Warrington (Chapter 8). If one counts the aquaria in schoolrooms around the world and in the homes of hobbyists, probably millions of these microcosms have been set up, but relatively few have been seriously studied.

Strategies for Maintaining Aquaria

The vast majority of enclosed volumes of water, from the aquaria of the tropical fish hobbyist, to the commercial seaquaria scattered around the world's tourist coasts, are used to hold fishes and other large aquatic animals for the pleasure of the visitor. Of necessity, the large animal biomass must be supported by food and oxygen from the outside, and wastes removed by mechanisms other than the activities of in situ decomposers and autotrophs. Thus, we find that the simple under-the-gravel filter and the external fiberglass-and-charcoal filter of the hobbyist are scaled up to the elaborate filtering and recycling systems found in the large fresh- and salt-water aquaria of the world. Oxygen is supplied by aeration with air or bubbling with pure oxygen. Carbon dioxide is swept out with the excess gas, or absorbed by plants in the aquarium. Solid and dissolved wastes are decomposed in the filters. Dissolved nitrogen and phosphorus compounds produced from the wastes are taken up by the plants and result in rapid growth. Also, some of the plant nutrients are removed from the system as aerosol droplets. The end result is an input–

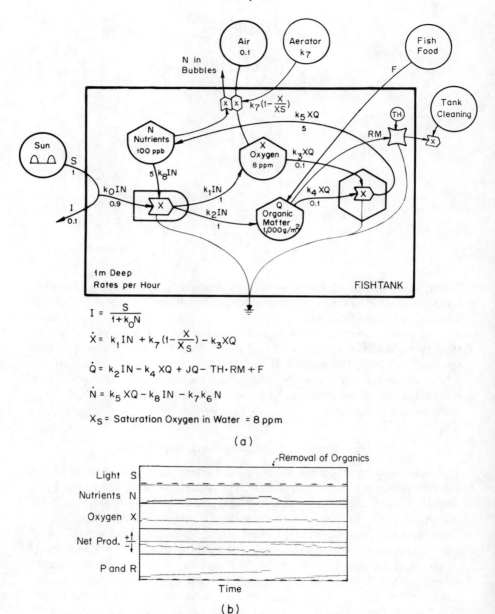

Figure 9.1. Model in Figure 2.14 adapted for typical fishtank with low light intensity, regular additions of fish food, aerator bubbling air, and occasional removal of excess organic matter. (a) Systems diagram; (b) simulation results.

output system designed to meet the needs of the animal populations. The simulation model of this kind of aquarium is presented in Figure 9.1. Although this type of system is designed to accommodate animal specimens, it often just barely does so. Most neophyte fish tank owners lose their pets through overcrowding, a synonym for lack of sufficient input or output. These facts underlay the positions of Atz (1949) and Breder (1931) when they pointed out the fallacy of calling most hobbyists' aquaria balanced (Chapter 6).

The opposite extreme to the input–output system is the sealed, truly balanced aquarium, such as represented by the model presented in Figures 2.3 through 2.5. The grammar school teacher who takes pond sediment, water, plants, and animals, puts them in a carboy, seals it, and places it in a sunny window is striving for such a system. However, once again, it is an input–output system, with the sole input being thermodynamically high grade photons and the output being low grade heat. There are no material inputs or outputs. All matter is recycled within the system. Usually the experiment is considered a failure because most of the large organisms die. The volume of water and the energy input is insufficient to support them. What is left is both autotrophic and heterotrophic microbes carrying on their life processes. Were the teacher able to observe the microscopic events occurring within the carboy, he or she would indeed be able to demonstrate the ecological phenomena of biogeochemical cycling, production, and respiration.

Most aquaria used for scientific investigations fall somewhere between these two extremes. They are usually buffered by some kind of aeration and are thus open to atmospheric gas exchange. The main advantage of this procedure is that if the experimental treatment should result in an excess of either production or respiration, the biota is not stressed by either a lack of carbon dioxide or oxygen. The pH, eH, and other chemical parameters of the water are also stabilized. Depending on the experimental design, various material inputs may be added, either on a one time basis or continuously. Outputs, too, can be continuous or periodic. A few examples of various kinds of aquaria established as components of experimental designs are presented below.

Phosphorus Movement in Aquaria

Whittaker (1961) exploited the greater safety and control of the microcosm technique in order to follow the movement of radiophosphorus in an aquatic community. He used 60-gallon tanks, under artificial illumination, which were filled with Columbia River (Hanford, Wash) water and seeded with pond organisms. Only four compartments of the system were considered, the water, the sediments, the plankton and seston, and the algae on the walls. Phosphorus tracer was added to the water. Initially, it

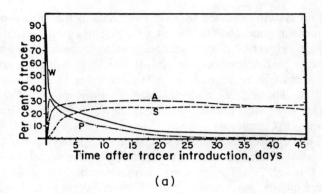

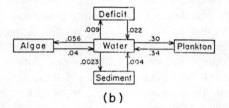

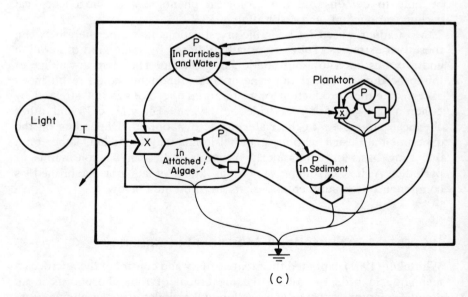

Figure 9.2. Radiophosphorous distribution in Columbia River water microcosm in relation to time after tracer introduction (Whittaker 1961). Ecosystem compartments: W, water; A, algae attached to aquarium surfaces; P, plankton or seston; and S, sediment. (a) Observed tracer uptake with time; (b) Whittaker's model of tracer flows; (c) energy systems diagram of the same pathways.

disappeared rapidly and it continued to disappear, more slowly, following the pattern of a simple drain model (Figure 2.2). The tracer entered the plankton and then was gradually lost over the period of the experiment. Phosphorus was taken up more slowly by the side wall algae and the sediments, but tended to remain there (Figure 9.2).

Marine Carboy Microcosms

In an effort to simulate a marine estuary, Abbott (1966) filled 19-liter glass carboys with estuarine water and sediments from a tributary of Biloxi Bay, Miss. The bottles were stirred, but not aerated. Like most other aquarium microcosms, the systems were maintained under artificial illumination which rose suddenly to a constant level at dawn and then fell to zero twelve hours later. By recording diurnal oxygen rate-of-change curves, he demonstrated the afternoon depression of photosynthesis and post nightfall acceleration in respiration (Figure 9.3). With similar microcosms and similar regimes, McConnell (1962, 1965) obtained similar results (Figure 9.4). Abbott also concluded that aquatic microcosms show replicability comparable to that found among replicated biological units in various types of statistical trials. In later experiments on the same systems, Abbott (1967), studied the effects of single doses of nitrates and phosphates. Like Confer (1972), who did similar experiments with continuous flow systems, he concluded that: (1) high nutrient levels lead to development of luxuriant algal floras with species changes in the dominant primary producers (Chapter 6); (2) primary productivity achieves high levels; and (3) phosphate and nitrate consumption occur independently.

Nutrient Enrichment Microcosms

After observing large differences in productivity in two antarctic lakes located only 500 m apart, Samsel and Parker (1972) conducted a series of microcosm experiments to determine the cause of these differences. Lake water analysis revealed there were differences in the phosphate, ammonium, nitrate, and calcium ion concentrations between oligotrophic lake number one and eutrophic lake number two. They used 600-ml Erlenmeyer flasks, 3,000-ml beakers, and 5-gallon aquaria in a constant temperature room under alternating high and low light intensities simulating antarctic summer. It is interesting to note that they recorded essentially the same results with systems varying in size to such a large degree. In the microcosms, specially designed synthetic defined media simulated the ion concentration of the two lakes in question. The microcosms were inoculated with net plankton from the lakes. The productivity of the second lake's control microecosystems, as measured by ^{14}C techniques, greatly exceeded that of the first lake's controls. Lake one replicates were enriched with ammonia, phosphate, nitrate, silicate, and calcium ions in various com-

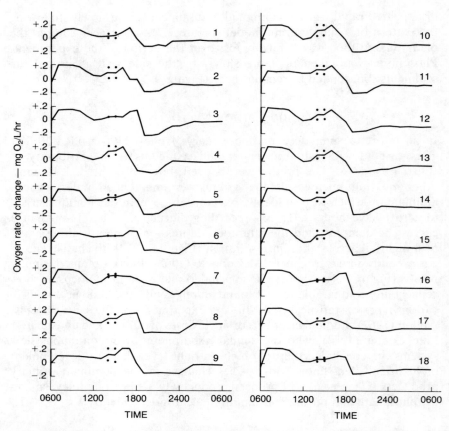

Figure 9.3. Oxygen rate-of-change curves for marine carboy microcosms exhibiting afternoon depression of photosynthesis and post nightfall acceleration in respiration (Abbot 1966).

binations to the same level as the lake two controls. Ammonia was demonstrated to be the principal nutrient causing the differences in primary productivity between the two lake microecosystems (Figure 9.5). Phosphate also induced a small but significant stimulation of primary productivity, especially in combination with ammonia. Other enrichments were ineffective.

Estuarine Microcosms Treated with Sewage

Copeland (1967) transplanted bottom sections of a shallow Texas estuary to microcosm containers in constant temperature chambers under 3,000 ft-

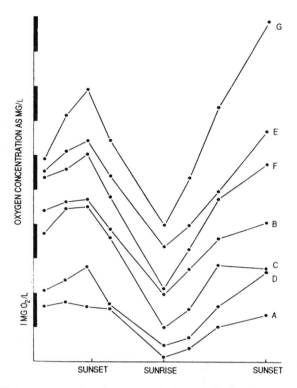

Figure 9.4. Diel oxygen pulse (not rate of change) in freshwater carboy microcosms showing typical pattern (McConnell 1962).

candles of artificial light with a 12-h photoperiod. After a 45-day acclimation period, he dosed them with various levels of artificial sewage containing high organic loading, and measured the community photosynthesis and respiration. Both the photosynthetic and respiratory rate increased with increasing concentrations of sewage (Figure 9.6). He postulated that the community adjusted to the disturbance by merely increasing the rate at which it cycled materials.

Herbivory Experiment in Benthic Systems

In one of a series of experiments with 3-gallon freshwater aquarium microcosms dominated by *Vallisneria*, Beyers (1963b) introduced an exotic herbivore into well established systems which had been in existence for

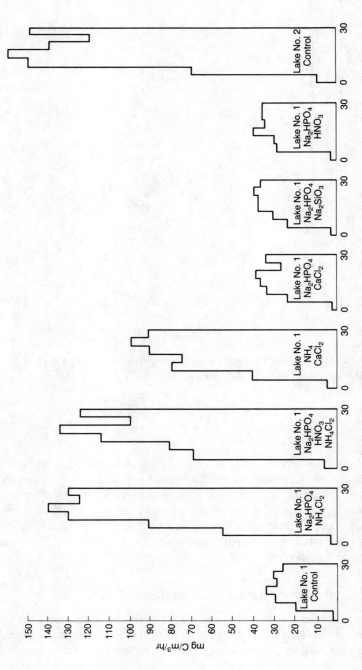

Figure 9.5. Oligotrophic (Lake 1) microcosms continuously enriched with ammonia, phosphate, nitrate, silicate, and calcium ions, and eutrophic (Lake 2) control. Ammonia appeared to be the limiting factor (Samsel and Parker 1972).

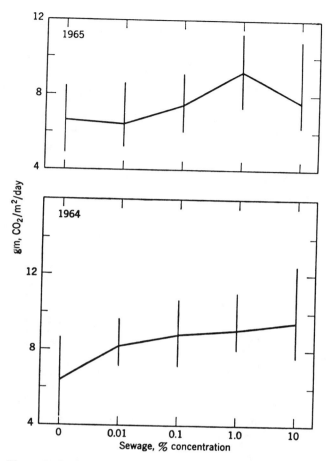

Figure 9.6. Photosynthetic rate as affected by various percentages of sewage in an estuarine microcosm. The vertical lines indicate the range in photosynthetic rate, the curve is drawn through the mean rate of photosynthesis at each percent sewage concentration (Copeland 1967, Pollution and Marine Ecology, T.A. Olson and F.J. Burgess, eds., copyright 1967, reprinted by permission of John Wiley & Sons, Inc.).

over three months. In addition to the *Vallisneria*, the aquaria, contained *Oedogonium*, *Physa*, *Sutroa*, nematodes, protozoa, rotifers, copepods, and flat worms. Water, sediment, and biota had originally been obtained from San Marcos Spring, Texas. At the time of herbivore introduction, the net photosynthesis and nighttime respiration were approximately equal (P/R = 1.1), indicating that succession had reached a stable plateau. Approximately 20 g of snails (*Marisa*) was added to each of three microcosms. Overgrazing immediately began to take place. The snails laid eggs

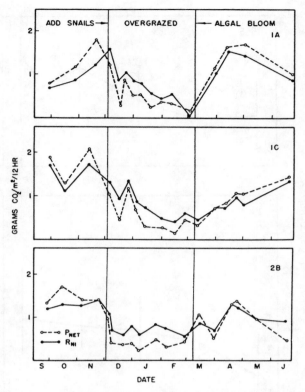

Figure 9.7. Course of nighttime respiration and net photosynthesis of experimental microecosystems, showing time of addition of snails and appearance of unicellular algae (Beyers 1963b).

within two weeks after their introduction into the aquaria, however, most of the young snails disappeared within a month. The adults and young consumed the *Vallisneria* and *Oedogonium* at a rate faster than the plant could replace the biomass with new growth. The result was a decline in macrophyte biomass until all that could be observed in the aquaria was a few snails, very clear water, and some sand. Almost overnight, the water became turbid and green in color. Patches of green unicellular or colonial algae appeared on and in the sand. These green patches eventually spread to the walls of the aquarium. The course of photosynthesis and respiration in these aquaria is illustrated in Figure 9.7. In all three systems the net production exceeded the nighttime respiration prior to the addition of the snails. After the herbivores were introduced, the nighttime respiration immediately exceeded the net production and, subsequently, both dropped to a low level. The initial excess respiration may be attributed to the sudden increase in the herbivore trophic level. In consuming the plants, they destroyed the photosynthetic and respiratory machinery of the primary producers which constituted most of the biota. Therefore, the snails reduced both the nighttime respiration and the net photosynthesis. The conditions were then ideal for the ascendancy of the unicellular photosynthetic forms. The water was high in nutrients, the competition from the rooted plants was eliminated, and the predation pressure from the herbivores was reduced because of the small size of the unicells and their dispersal throughout the water column and the sand. After the algae bloom, the net production again exceed nighttime respiration and both production and respiration returned to a level approximately equal to that attained before the addition of the snails. Neither the drop nor subsequent rise in respiration and photosynthesis could go on indefinitely. The decline was limited by the death and disappearance of most of the organisms present, and the rise in metabolism was limited by the return of the nutrient concentration to its normal low level.

Herbivory in Algal Dominated Aquaria

In a similar set of experiments, Cooper (1973) set up 15 replicate aquaria with 15 liters each of water and 250 g of sediment from a eutrophic farm pond. The microcosms were placed under constant temperature (21°C) and given a 12-h photoperiod. After a 40-day successional period, photosynthesis and respiration came into balance. The dominant primary producers were coccoid green algae and small amounts of *Spirogyra* and bluegreen algae. After climax had been achieved, a native herbivorous fish (*Notropis spilopterus*) was added to ten of the systems. Five were left as controls. Replicate microcosms received 3, 6, 9, 12, and 15 fish. The

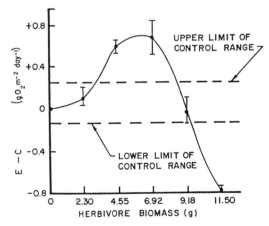

Figure 9.8. Difference in net primary productivity in experimental microcosms and control microcosms vs. grazing biomass of herbivorous fish (Cooper 1973).

fish were allowed to graze for 20 days, and then were removed. Community metabolism was then measured for 3 days using diurnal oxygen techniques. Net primary production was enhanced in those systems having 3, 6, and 9 fish. Those containing 12 and 15 fish exhibited reduction in net primary productivity, probably due to overgrazing (Figure 9.8).

Water Surface Microcosms

The surface of water has specially adapted life which forms a specialized ecosystem of floating plants, animals, microbes, and chemical substances, including many pollutants which tend to concentrate in the surface film. If the large, floating species are omitted (*Sargassum*, water hyacinths, duckweed, *Azolla*, etc.), there remains a microlayer called the neuston. Neuston microcosms have been studied and employed to assay toxic substances. Because these systems have very fast turnover times, self-organization is rapid and responses to foreign elements more rapidly indicated. Riznyk et al. (1987) studied aromatic hydrocarbon effects on the rates of growth of phytoneuston (*Nitzschia*, *Exuviaella*). Inhibition for two days was followed by rapid recovery. This may indicate that physiological adaptation in these species, which are normally required to continually tune adaptation to the harsh circumstances of the water surface, is part of the microecosystem self-organization.

Freshwater Succession with Multiple Seeding

Rees (1979) confirmed the general patterns of the self-organization (Chapter 3) which takes place in small tanks as microcosms develop an ecosystem like those found in shallow freshwater. Populations were followed for 190 days in two frequently multiply seeded tanks containing lake water fertilized with defined nutrients. A carnivorous *Gambusia* and an herbivorous catfish, *Plecostomus*, were added to one tank. Patterns over time are given in Figure 6.4. Algae, cladocera, rotifers, and copepods exhibited prey–predator oscillations. *Tanytarsus* and a rotifer emerged as dominant. There were species substitutions and a gradual establishment of more diversity. Many epiphytic plants and animals were found on the walls. The tank with the fish was not much different from that without. Whether timing differences were due to fish, could not be proven without replications.

Filtered Aquarium Systems

Dating back to the last century, recirculating aquarium systems have been built to exhibit fishes and other life. Particularly for marine organisms away from the ocean, a technology was evolved for maintaining a viable medium by recirculating waters through filters which included both living microorganisms and chemical absorbents, such as charcoal. In a pioneering study, Saeki (1957) investigated the culturing of both freshwater and marine organisms in closed, recirculating systems. He concluded that although dissolved oxygen, carbon dioxide, organic matter, ammonia, and pH were all important in maintaining water quality, controlling the latter two factors (dissolved ammonia and pH) could allow organisms to maintain a healthy condition. He calculated, based on animal biomass, the required size of a sand filter necessary for oxidizing ammonia to nitrate, and recommended frequent adjustments of the pH of recirculating aquarium water.

Microcosms as Lake Simulators

Butler (1964) compared aquarium microcosms with the ecosystems in Oklahoma farm ponds where production is by small plankton algae. He used 64 aquarium microcosms in growth chamber to study the productivity and respiratory metabolism as a function of light (400 and 800 foot candles), temperature (11 and 23°C), turbid suspended solids (0–75 ppm), and dissolved substances (75–600 ppm). Containers with pond water were seeded from 6 ponds and allowed to organize for 3 weeks before measurements. Light intensity levels were those characteristic of turbid ponds in Oklahoma. Gross productivity varied from 0.1 to 0.9 millimoles carbon dioxide per liter per day. Ratio of gross production to total respiration was about 1.0. More production was found at the higher temperature, at the higher light intensity, and at the least turbidity. Productivity was greatest with intermediate concentrations of dissolved solids (300 ppm). Species diversity was between 10 and 21 species found per 1,000 individuals counted when dissolved solids were 300 ppm or less. Species diversity at 600 ppm dissolved solids was about half. Chlorophyll ranged 0.35 to 3.6 mg/liter. Assimilation number was 0.4 to 4.0 mg/mg/hr (photosynthesis per unit chlorophyll A). Ratio of oxygen produced to carbon-dioxide used was between 0.6 and 1.9. For most aspects, the microcosms were similar to the ponds.

In an investigation to determine the effectiveness of still microcosms as simulators of the water column in freshwater lakes, Harte et al. (1980) Levy et al. (1985) conducted a series of experiments with 700-liter systems which included no benthic sediments. They monitored temperature, oxygen, pH, inorganic carbon, organic carbon, ammonia, nitrite, nitrate, inorganic phosphorus, total phosphorus, volume and numbers of phytoplankton and zooplankton, and bacterial plate counts. These experiments were of 3 and 6 months duration. The massive amount of data accumulated led them to the following conclusions:

> The behavior of microcosms, like natural systems, is dynamic, exhibiting frequent and large changes in biotic and abiotic parameters [Figure 9.9].

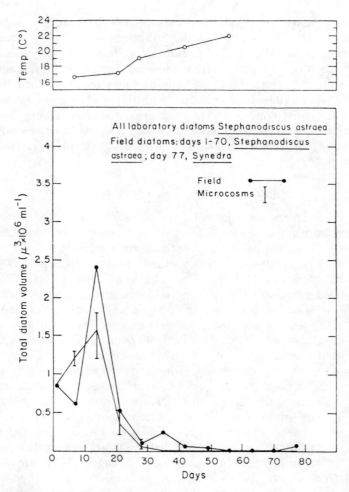

Figure 9.9. The dominant phytoplankton in a microcosm and a reservoir under similar light and temperature regimes (Levy et al. 1985, copyright ASTM, reprinted with permission).

The pelagic biota does not become conditioned by being in the laboratory.
The dynamic behavior of the microcosms often resembles natural systems.
Inclusion of hydraulic flow through a thermocline and benthic substrate does not necessarily increase realism and generally is not warranted in lake-like microcosms.
Inclusion of macrofauna in microcosms reduces rather than increases resemblances to natural systems.
Biological control of side growth is not feasible.

Periodic pouring of microcosms into clean containers eliminates side growth.

The pouring, and consequent mixing, changes microcosm behavior.

Microcosms not subject to surface growth control, exhibit significant size dependent differences. Those subject to periodic pouring to prevent surface growth exhibit fewer size dependent differences.

Good replicability can be achieved if care in initiation is taken.

Mineral Mobilization from Sediments in Microcosms

Taking advantage of the safety of microcosms for radiochemical work, Giesy and Geiger (1980) investigated the role of macrophytes in the mobilization of trace elements from sediments into a water column. They used a specially constructed series of still microcosms which permitted sampling of both the open and interstitial water (Figure 9.10). Sediment and water were collected from South Carolina surface sources. The water

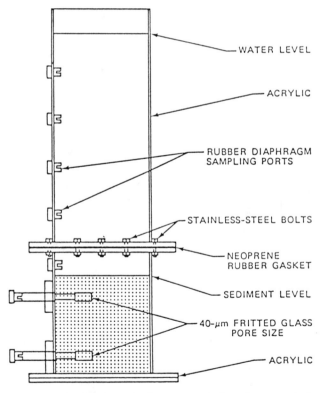

Figure 9.10. Schematic drawing of microcosms designed to sample free and interstitial water. Sediment depth, 15.25 cm, volume 2.6 liters. Water depth 38.1 cm, and water volume, 6.4 liters (Giesy and Geiger 1980).

was low in pH and soft. The systems were aerated by bubbling, and planted with *Juncus* and other plants. The controls did not contain macrophytes. The sediments were mixed with ^{242}americium, ^{144}cerium, and ^{65}zinc, and succession was allowed to take place. The experiment lasted 207 days. At the end of that time there was a significantly greater total number of radioactive atoms in the components above the sediment in microcosms containing macrophytes. This demonstrated that the macrophytes play a role in moving metal ions from the sediments to the water column.

A Microcosm Apparatus for Working with Dangerous Chemicals

In experiments such as the one mentioned above, when an investigator is working with radioactive, toxic, or otherwise dangerous substances, care must be taken to avoid release to the environment. Huckins and Petty (1984) devised an apparatus (Figure 9.11) to contain lentic microcosms which were to be dosed with such chemicals. Particular care was taken to prevent out-gassed chemicals from escaping to the atmosphere or being absorbed by filters, rather than being collected. Experiments on the efficiency of their apparatus during microbial mineralization studies resulted in 97.3% recovery of ^{14}C containing CO_2. Similar efficiencies were found for radioactively labeled pesticides.

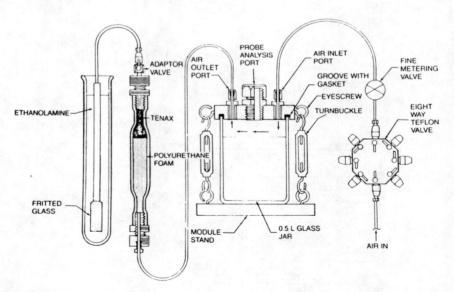

Figure 9.11. Schematic drawing of a 0.5-liter microcosm module with stand and support apparatus (Huckins and Petty, Chemosphere 13(12), copyright 1984, Pergamon Press).

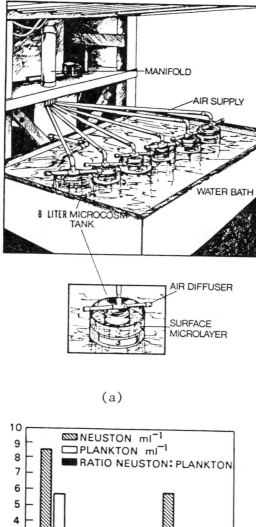

Figure 9.12. Neuston microcosms. (a) Physical setup; (b) densities of bacteria and microalgae (Hardy et al. 1985).

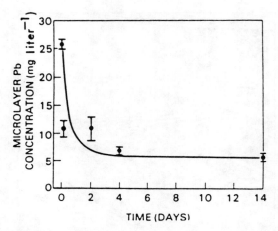

Figure 9.13. Kinetics of lead loss from the surface microlayer under no wind conditions (Hardy et al. 1985).

Neuston Microcosms and Dispersal into Sea Water

The boundary between the physical states of matter in ecological systems is always a place of high organism concentration and activity. Hardy et al. (1985) created a microcosm in order to study the neuston, the community of microorganisms which exists at the boundary between air and sea (Figure 9.12a). These microcosms consisted of cylindrical polycarbonate tanks containing 8 liters of sea water. The tanks were temperature controlled and the surface air flow was regulated. The specific object of the experiment was to study the dispersal of metals from atmospheric particulate matter (dust) into the water column. Greater numbers of bacteria and microalgae were found at the surface layer than in the water (Figure 9.12b). The atmospheric particulate matter was neutron activated in a nuclear reactor so that the dispersal of metals from the neuston to the plankton could be measured (Figure 9.13) under various wind conditions. The studies indicated that particulate metals have residence times in the sea surface microlayer of 1.5–15 h before entering the water column. Microlayer metal concentrations and residence times are decreased in the presence of wind.

Chapter 10
Stream Microcosms

Flowing waters in streams, tidal channels, and underground channels develop intense concentrations of life which make use of the physical stirring energy and draw resources from the moving waters. The living and nonliving components of the climax ecosystem are either continually renewed by the inflow or have means for staying in position. Flowing waters are dominated by eddies and currents, and have a high rate of water turnover. Lotic environments, as they are sometimes called, often have more high-quality energy available than standing waters. Often a flowing ecosystem has a current-dependent reef or encrusting surface subsystem (Chapter 14). Where light is available, characteristic mats of algal cells and associated microzoa develop. These are called periphyton or aufwuchs.

Ecosystem studies of model streams started with Shelford (1929) who studied behavioral responses of aquatic organisms to currents. The investigators in the Office of Naval Research Florida Springs project in 1953–1957 and those studying Silver Springs again in 1980 were stimulated to develop experimental stream microcosms. These investigators included Odum and Hoskin (1957), Whitford (1960) and Whitford and Shumacher (1961), and Knight (1980, 1981). Extensive studies have now been made on stream microcosms, often supported with funds from agencies concerned with stream pollution. Some apparatus (Figure 10.2) was developed with the goal of culturing stream species (Whitford et al. 1964), whereas others were intended for observation of higher levels of ecosystem organization. Some stream microcosms are long, outdoor, concrete structures (Figure 10.1). Others are short, inclined troughs on laboratory benches (Figure 10.2). If water is repeatedly recirculated through a short apparatus, some of the effects are the same as those obtained when waters pass down a

210   10: Stream Microcosms

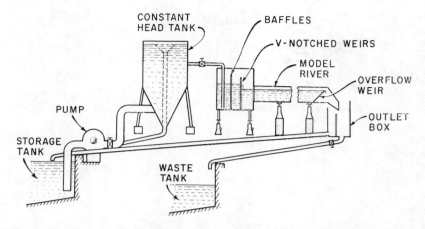

Figure 10.1. Diagram of the Balcones model river (Armstrong and Gloyna 1968).

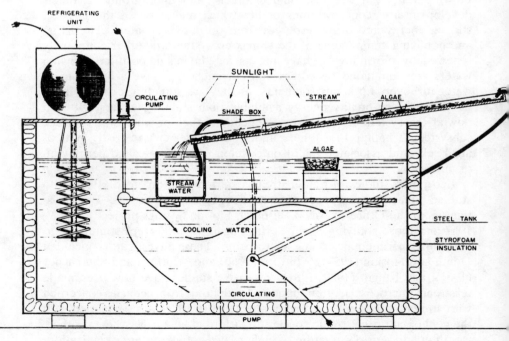

Figure 10.2. Stream microcosm used for algal culture (Whitford et al. 1964).

long stream. The oxygen change over a mile of stream may be simulated by circulating the waters through an apparatus containing a short section of stream ecosystem for the same length of time required for the water to travel the mile of natural stream (Figure 6.5).

Many stream microcosms have been used to study relationships, following the general concept of retaining as much natural complexity as possible while varying factors of interest (Oregon stream microcosms, Figure 10.3). Factors studied include the main external variables: light, circulation rate, nutrients, and organic matter. As with other ecosystems, those with high nutrient inflow develop net organic production. Others, with inflows of organic matter, have net organic consumption. Flowing

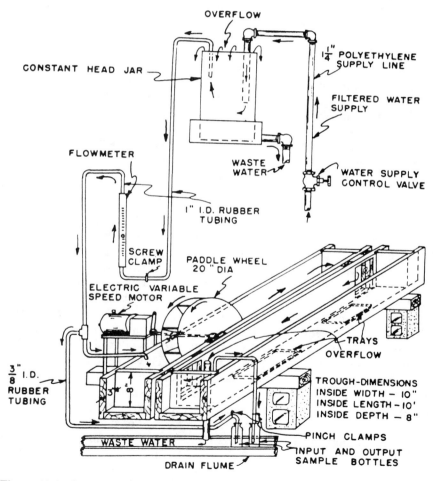

Figure 10.3. Oregon stream microcosm (Warren et al. 1962).

ecosystems depend more on external sources and exchanges than still waters. Rodgers et al. (1980) found metabolic measures to be more uniform (smaller coefficient of variation) than structural measures. This is to be expected if structures are self-organized by stream flows to integrate and maximize performance. Impacts on stream ecosystems have been studied extensively by adding such factors as chemical discharges to ecosystems which have developed some biotic structure and diversity. Rodgers et al. (1980) found photosynthetic metabolism reduced by acute treatment with chlorine, chromium, and copper, whereas heterotrophic metabolism was increased by sucrose addition (Chapter 7). Chemicals may cause change in the organic matter composition of the ecosystem when adaptation requires altered biochemical activity. For example, Clark et al. (1982) found that stream aufwuchs shifted to a higher protein content. Loss of species diversity may be compensated by greater biochemical diversity.

Stream microcosms have also been useful for determining rates of removal of toxic substances from water ecosystems and for determining the short-term toxic effects on components operating under conditions of ecosystem complexity. Pritchard and Bourquin (1984) cite the rapid uptake of 8 ppm P-cresol in 32 h in Monticello riffle-pool recirculating stream microcosms compared with no uptake in the azoic control. Many stream microcosms have been allowed to adapt and self-organize under chronic toxic influences, revealing the long-range capabilities of streams to accept byproducts of our industrial economy (see below and Chapter 7).

Longitudinal Succession

In a flowing stream, changes over time which might take place in one well-defined area of a still water system, become changes that occur through a large, three-dimensional space as the water flows downstream. The changes with distance along a stream are called longitudinal succession. Sequences become series. Many kinds of stream series have been described, such as the "saprobe series" which follows introduction of sewage into a stream (Kolkwitz and Marsson 1908). Another example is the series observed downstream from a clear-water spring, in which the quantity of organic matter and numbers of consumers increase downstream.

Ecosystems along a stream become organized with each other because the outflow from processes in one unit become inflow for the next unit, and there are animals which migrate from one system to others. Whereas small items, such as chemical substances and particles, move generally down stream, many of the animals have life histories that require them to return to the head of the stream for reproduction. Examples are salmon, shad, mussels carried on fish, and eggs from the brief emergent swarms of aquatic insects. Even in small streams, food goes downstream and consumers use some of it to get back upstream (Hall 1972). Also, many stream microcosms provide some information return by means of water recirculation.

Because stream microcosms are short, there is less room to develop long series, yet longitudinal succession is observed in flowing microcosms. For example, in our blue-green algal mat ecosystem (Odum and Hoskin 1957) the first five centimeters at the end where flowing water entered was the brilliant green associated with excess nutrients and net production. The rest of the mat downstream was brilliant orange, showing a low chlorophyll to carotenoid ratio, and higher respiration (possibly photorespiration). The same changes observed in longitudinal space are observed in time with blue-green algal cultures as they go from an early nutrient-abundant, net-production state to a later balance of production and respiration-recycling.

Many streams alternate pools and riffles, and many rivers have been converted into a series of lakes by the building of dams. To some extent, serial microcosms and compound chemostats are their laboratory analogs.

Longitudinal Series and Arrested Succession

The stages at the start of longitudinal succession are analogous to succession that takes place over time, except that the stages are unchanging at steady state, as in the blue-green mat microcosm. In other words, a stage which normally changes with succession becomes arrested as a subclimax zone of longitudinal succession. Some of the low-diversity characteristics of flowing waters, such as cattail marshes, may be a necessary part of the overall diversity of streams and strands, a necessary part of generating diversity downstream. Sampou and Oviatt, 1991 studied an experimental eutrophication gradient which had sulfate reduction and an anaerobic process. Respiration was 10 times higher than at lower nutrient levels.

Nutrient Spiralling and Turnover Length

Radioactive tracers show that in a stream, the nutrients are bound and released and recycled as they would be in a nonflowing system. However, these materials are moved downstream in the process. The cycling and transport is called "nutrient spiralling." What is normally called "turnover time" becomes "turnover length", a useful parameter for representing the dynamics of a chemical substance in a flowing system (Elwood et al. 1981). Rodgers et al. (1980) measured uptake of ^{14}C and ^{35}S isotopes by periphyton covered slides taken from six stream-trough microcosms. The slides were incubated for 4 h in circulating chambers. Photosynthetic carbon fixation was $10-20\,mgC/m^2$ per day. Fixation of radioactive sulfur was used to measure both the autotrophic and heterotrophic metabolism of microbes, but was expressed in carbon units, using the carbon/sulfur ratios of the mat. Then autotrophic carbon fixation rates were subtracted from the sulfur-determined total fixation rates to obtain rates of heterotrophic production.

Tracer Studies in the Balcones Model River

An example of a stream mesocosm which reveals the downstream behavior of metabolic substances was the model river operated at the Balcones Research Laboratory in Austin, Texas (Armstrong and Gloyna 1968). Two parallel channels, each 60 m long, 0.37 m wide and 0.6 m deep (Figure 10.1) contained 16 cm of sediment, *Vallisneria* plants, *Physa* and *Goniabasis* snails, and *Gambusia* fish. Hard ground water was passed through at 0–400 l/min, which resulting in a water turnover time of about 33 min. Studies of the dispersal and uptake of chemical constituents were

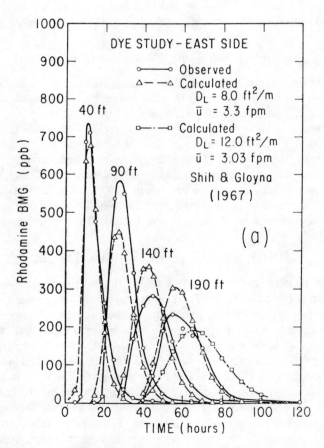

Figure 10.4. Dispersal of dye in the Balcones model river and the curves generated by the dispersion model (Armstrong and Gloyna 1968).

performed with radioactive tracers introduced at the head of the stream. Velocities were 0.5 m/min. Diffusion patterns and coefficients were determined from measurements on dye dispersal after instantaneous releases. As a pulse of some substance entered the flowing systems, its concentration was decreased as it spread over a longer zone as it moved downstream. Figure 10.4 shows how the observed dispersion fitted with the simulation results of the dispersion model. Most of the tracer flushed through immediately, with peak concentrations passing the last sampling station in 2 h. However, the strontium taken up by the plants was held and gradually released, with 10% remaining after 28 days. Figure 10.5 shows the pattern of strontium isotope build-up in plants when the tracer was released continuously. Metabolism was calculated from diurnal oxygen and pH data. In some experiments, sewage waters were included in the inflow.

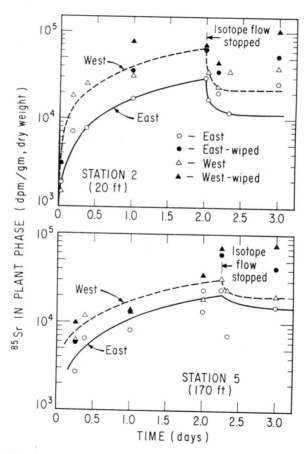

Figure 10.5. Uptake by plants of radioactive strontium continuously introduced at the head of the Balcones model stream (Armstrong and Gloyna 1968).

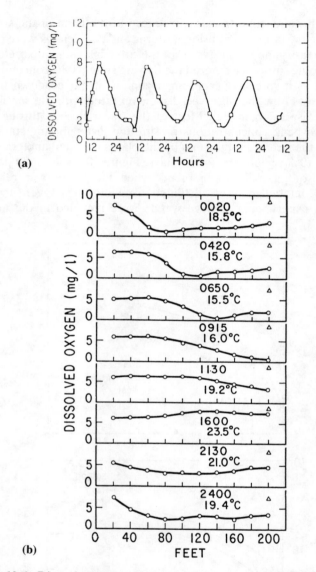

Figure 10.6. Diurnal oxygen variation in the Balcones model river when inflows included sewage (Armstrong and Gloyna 1968). (a) Diurnal variation at one station; (b) oxygen along the channel.

There was a strong diurnal oxygen variation, shown in Figure 10.6. Simultaneous oxygen measurements along the stream showed "oxygen sag curves" dependent on the photosynthesis–respiration balance, according to the time of day.

Self-organization and Current Velocity

For a flowing system, the self-organizational process includes the changes with time and zonal serial organization. The successional process may begin with the immediate colonization of surfaces by microbes, followed by larger algal cells where light permits, and next by small consumers, if there is adequate seeding of species. Steinman et al. (1989), however, found decreasing diversity in stream microcosms with time when the available species were introduced only at the start. The mechanical energy of physical water motion in streams is large, and most of the components in the fully developed ecosystem are adapted to derive benefits from it. Self-organization is dynamic and self-regulating. Cairns (1969) shocked flowing microcosms with high and low pH and temperature change. The greater the shock, the longer the time for repair and return to the previous state.

Although a stream ecosystem in bright sunlight usually receives more solar energy than physical stirring energy, the physical energy is higher in the energy transformation hierarchy. When both kinds of energy are expressed in equivalent units of the same kind of energy (solar Emergy in solar emjoules), the solar Emergy of the physical input is much larger than the direct solar Emergy. Energy analysis studies show that about 25,000 j of solar energy are required to generate 1 j of physical stirring energy. (The solar transformity of physical stream energy is 25,000 solar emj/j. See Table 10.1 and Chapter 15.) Higher quality physical energy benefits production by interacting with the lower quality solar energy as an amplifier to the process. Maximum production occurs when two inter-

Table 10.1. Transformities of Components in Stream Microcosms modified from Knight, 1981a.

Item		Solar Transformity sej/J*
Sources:	Direct sunlight	1
	Physical Stirring energy	25,000
	Cadmium, 10 microgr./liter	840,000,000
Components:	Gross production	4,616
	System respiration	7,720
	Export	12,748
	Detritus-microbes	24,376
	Algae	36,223
	Macrophytes	480,000
	Consumers	5,520,000

*Value of solar emjoules/Joule given by Knight multiplied by 2.4 to include the solar EMERGY in earth geological heat sources and tide.

acting factors have equal marginal effects. If the effects are proportional to Emergy content, then sunlight and physical energy tend to be equally matched when their Emergy contributions are equal. Where physical Emergy is larger than the solar energy, the latter is a limiting factor. The self-organization of stream microcosms results in higher metabolism when currents are present than when they are not.

The Duke Stream Microcosms

A flowing microcosm apparatus was operated at Duke University from 1954 to 1956 (Odum and Hoskin 1957). The apparatus shown in Figure 8.5 had an attached community on a long, glass slide in the inner chamber of an intensely illuminated distilling condenser. The outer jacket was cooled by a once through flow of water. The light reaching the plants was at full daylight intensity and the efficiency of gross photosynthesis was measured as about 3% of visible light received. The inner chamber had recirculating water, with some respiration taking place in the dark recirculating tubes. Oxidation of the tubing plastic contributed some to the respiration until the plastic was replaced with glass. Oxygen changes were used to measure metabolism (Figure 10.7). The recirculating apparatus allowed the experimenters to distinguish between the effect of currents and the effect of physical energy per se in bringing more resources to the system's components. One type of effect of physical energy is the short-term response to a change in velocity when there is no time for the ecosystem to reorganize and adapt. There were increases in production and respiration with an increase in circulation, although results were variable. The next experiment allowed the whole ecosystem to adapt for several weeks to one velocity. Then the velocity was increased and the system was allowed to acclimate again. The faster the current to which the system was adapted, the higher the steady-state metabolism achieved. The ecosystem was using physical energy to maximize production and respiration.

The system was originally started with a long blade of aufwuchs-covered *Sagittaria* from Silver Springs, Florida. The diatom flora were soon replaced with a blue-green algal mat on the walls of the cylinder. During this period there was a decrease in production, respiration, and diversity. This declining succession occurred because the initial ecosystem was too large and diverse to be maintained under the conditions of the microcosm. The diversity of algal cells in the initial periphyton introduced from Silver Springs was 35 species per thousand individuals. In the "declining succession period" the diversity was reduced to 20 species per thousand. A steady state developed rapidly. Nutrients regenerated in the dark tubes caused net production and excess chlorophyll in the upper zone where the nutrients were absorbed. Further down, without the inorganic nutrients, there was a better balance between respiration and photosynthesis, possibly with

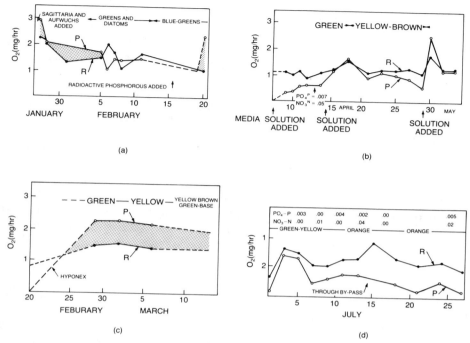

Figure 10.7. Records of gross production and total respiration in the Duke recirculating microcosm. (a) Declining succession during reorganization; (b) homeostatically regulated quasi-steady-state with surge of net production after addition of nutrients; (c) regime of net production when inorganic nutrients were in excess; (d) regime of excess respiration with added organic matter in excess (Odum and Hoskin 1957).

photorespiration contributing to the overall balance of production and respiration. The homeostasis was simulated with the model in Figure 2.3. When new solutions with inorganic nutrients were added, the whole ecosystem mat became green, with metabolic measurements showing a net production until the nutrients were bound into the organic state. The model in Figure 3.13 simulates this response. When the system was restarted with new nutrients and a seeding from an older mat, there was only five species per thousand during early succession, increasing with time to higher steady-state levels. Diversity was greater in the green, net production section of the steady state, compared to the orange section. When organic matter was added, the Duke system shifted metabolically to a larger excess of respiration over photosynthesis. The simulation model of Figure 3.13 predicts this also.

Oregon Stream Microcosms

Laboratory stream microcosms were established by Warren et al. (1962), McIntire and Phinney (1965), and McIntire (1966) for experiments on conditions that maximize production of plants, invertebrates, and fishes. As shown in Figure 10.3, stream microcosms were wooden indoor troughs, seeded with trays of stream-bottom components from outdoors. Water was supplied from a soft ground water source. The water was recirculated through the troughs with a paddle wheel.

Blue-green algae, green algae, diatoms, and iron bacteria formed a periphyton. A variety of insect larvae were present. Successional organization of the periphyton plants was dependent on light intensities (Figure 10.8). Steinman and McIntire (1987) found that different algal taxa developed with different light intensities, after initial similar colonization of periphyton. The same kinds of light–production curves obtained for single plants were obtained for the stream ecosystem. See the example of

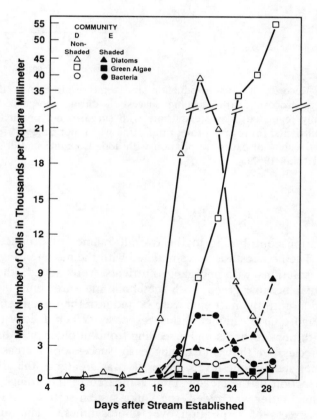

Figure 10.8. Periphyton succession on glass slides (Warren et al. 1962).

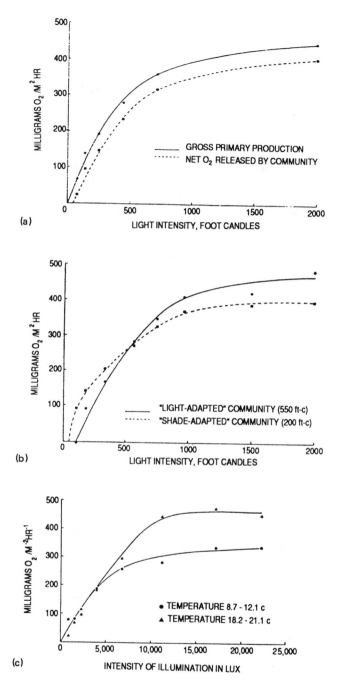

Figure 10.9. Photosynthetic production of periphyton as a function of light intensity (McIntire and Phinney 1965; Phinney and McIntire 1965). (a) Gross and net production; (b) comparison of light and shade adapted communities; (c) production and light intensity at two temperatures.

sun and shade adaptation curves in Figure 10.9. Efficiency of photosynthetic use of light was higher in the shade adapted ecosystem. Algal succession in periphyton was slower in the shade adapted system. A moderately high diversity developed. Measured populations of stonefly naiads, snails, and sculpins were introduced and their growth rates measured as secondary production of biomass. Seasonal differences were observed.

Most of the larger animal components were introduced, rather than allowed to self-organize, although some self-organization is always in process. Then factors were varied to determine whole-ecosystem performances under different conditions. Whereas the smaller components were predominantly self-organizing, their conditions were being controlled by human manipulations, changing animal populations, currents, etc. Stream photosynthesis was a function of light intensity and added carbon dioxide. There was the usual curvilinear graph of diminishing returns, as other factors became limiting when one factor was increased. Such graphs are normal for most production functions in simulation. Figure 10.9 shows the response of production to increases in light.

Gross production increases with diminishing returns when the exchange rate of new water is increased (Figure 10.10). Presumably, this is a

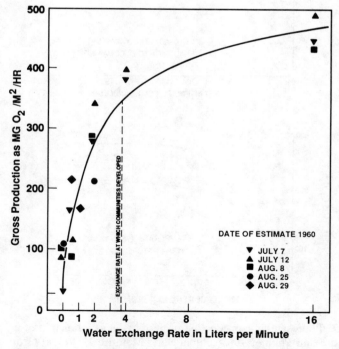

Figure 10.10. Gross production of Oregon stream microcosm as a function of water exchange rate (Warren et al. 1962).

response to limiting nutrients and/or removal of toxins rather than to the current velocity controlled by the paddle wheel. The periphyton part of the ecosystem had twice the respiration at 38 cm/sec velocity compared to the rate at 9 cm/sec (McIntire 1966). Metabolism was slightly higher when periphyton was measured at the current velocity at which it was grown (Figure 10.11). Metabolism was much higher at higher oxygen concentrations, especially at higher temperatures. Higher oxygen concentrations mean more available energy for metabolism.

As illustrated in Figure 10.12, parabolic shaped graphs of biomass production were found for consumers (stoneflies, snails, and sculpins) as a function of biomass introduced (Warren et al. 1962). Such graphs of optimum biomass for maximum production are generated by simulation of autocatalytic consumer growth models, such as the logistic fisheries harvest model applied to tuna. Similar parabolic graphs of production rate as a function of biomass were obtained for steelhead trout in artificial stream microcosms (Hughes and Davis 1986).

Following extensive measurements on stream microcosms, McIntire created stream models by using several approaches. McIntire (1973) simulated periphyton systems, diagraming with Forrester's symbols and methodology, first considering production and biomass of the periphyton as a function of light intensity, current velocity, nutrients, and temperature. Then leaf litter, snails, and recycling were added. All factors were given an asymptotic limit using expressions of the form $kX/(K + X)$. A very curious procedure was employed to account for the fluxes such as res-

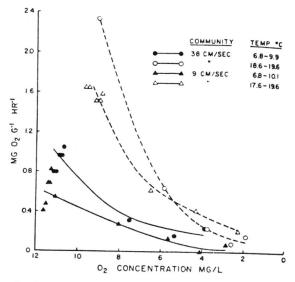

Figure 10.11. Periphyton metabolism as a function of oxygen concentration temperature and current velocity (McIntyre 1966).

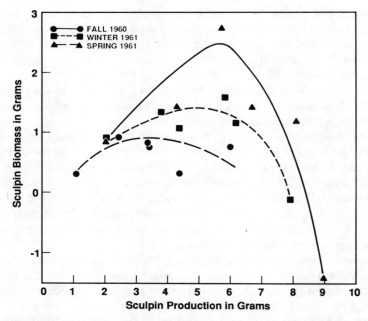

Figure 10.12. Sculpin biomass production as a function of Sculpin stock (Warren et al. 1962).

piration and export. The fluxes were expressed as differential equations with time to a summation. Apparently this total was included in the equation for biomass, and thus was integrated a second time. Forrester (1961) showed how a second integration of a linear effect changes it into a quadratic effect, but the applicability is not clear in this case.

The simulation results as expressed on a phase plane, plotting snails versus periphyton biomass, were stable, perhaps because of the input limiting factor constraints, oscillation occurred only with the applied seasonal variation of light and temperature. The phase plane graph shifted from an initial high periphyton biomass and few snails to a small periphyton biomass and more snails in stable annual oscillation. Rapid turnover of small algal cells allowed support of a larger herbivore biomass.

Laboratory Streams at Michigan State

Kevern and Ball (1965) and Kevern et al. (1966) studied the productivity of algae in two recirculating artificial streams (Figure 10.13). Higher current increased net productivity. High concentrations of a chelating agent (EDTA) lowered productivity, apparently by binding limiting factors. Efficiencies of net production were 3.8–9.8% and efficiencies of gross production were 6.4–16.1%. High efficiencies accompanied lower light

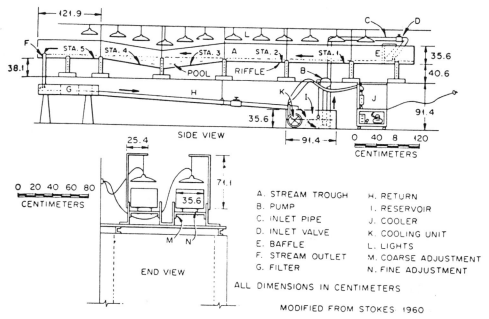

Figure 10.13. Laboratory stream apparatus used by Kevern and Ball (1965).

intensities. Net accumulations of organic matter were much less than the net carbon production calculated from pH, suggesting the release of organic matter into the waters. In later studies, rates of uptake of chemical substances were measured (Pritchard et al. 1984). Periphyton grew at 0.51 dry g/m^2 per day on fresh surfaces, but averaged less than $0.20 \, g/m^2$ per day when there was grazing and sloughing.

Self-organization at Different Temperatures

Sperling and Grunewald (1969) used the model stream apparatus diagramed in Figure 16.3 to measure the effect of current velocity on the net increase of biomass of periphyton on styrofoam. Hot springs algae from Iceland were used to start growths, with the results in Figure 16.4. Waters initially contained Gorham's medium #11 and streams were held at 45°C. Increased current velocity resulted in increased growth. The channel acclimated to the higher velocity had the greater response to current increase.

In the Oregon stream microcosms exposed to short-term temperature changes, Phinney and McIntire (1965) found the respiratory metabolism of periphyton doubled after a 10°C increase (Figure 10.14). Periphyton grown at 13°C had higher metabolism at 23°C and lower metabolism at 3°C (McIntyre and Phinney 1965). At higher temperatures, photosynthetic production at higher light intensities was increased (Figure 10.9c), pre-

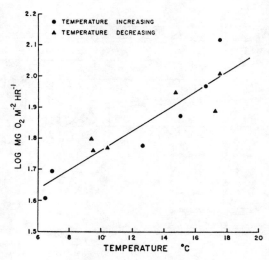

Figure 10.14. Increase of periphyton metabolism with temperature (Phinney and McIntire 1965).

sumably due to the greater nutrient availability from recycling at higher temperatures.

Hughes and Davis (1986) studied the effect of a 5 degree difference in temperature on artificial salmon streams, 10 m long with pools and riffles. The heated stream developed a higher respiration rate, less net production, and fewer steelhead trout and salmon. When food was in excess in hatchery conditions, trout and salmon grew faster at higher temperatures, but in the whole ecosystem, where the base of the food chain was source-limiting, higher temperature merely accelerated the progression to a steady state.

Self-organization and Varying Light Intensity

Well known in many ecosystems and laboratory studies is the increase in the amount of chlorophyll during adaptation to low light intensities. For thermodynamic loading to achieve optimum efficiency and maximum productivity, the receiving system develops less receptor material for photochemical conversion when light intensity is greater. (See discussion in Odum 1983a.) Many examples are known among microcosms. For example, Triska et al. (1981) found chlorophyll quantities to be inversely proportional to light intensity in plexiglass troughs, 10 m long, placed in Lost Man Creek, California under different conditions of canopy shading. Steinman et al. (1988) found different biochemical compositions in the periphyton grown in stream microcosms at different ages and light intensities. With low irradiance, 16:0 carbon fatty acid, glycine and serine were abundant on the 8th day of succession, whereas later, 16:1 and 20:1 fatty acids, aspartic acid, and asparagine acid were more abundant. The nitrogen/carbon ratio was higher in older communities. Steinman and McIntire (1987), using an electron microscope to study periphyton, found more vertical structure and layering at higher light intensities. At low light intensity, Lamberti et al. (1989) found herbivores growing slowly but very efficiently, whereas at higher light intensities, production was greater but less efficient, apparently another example of maximum power priority requiring growth over efficiency when new growth is possible. Succession was pulsing, with periphyton reaching a maximum in about 45 days, after which stocks were grazed or exported. Biomass was larger at higher light intensities even though most of the production was exported. At lower light intensities, biomass was smaller and more of it was retained or grazed.

Stream Hierarchy and Models

Streams are hierarchical, with many small streams converging to form a few larger ones. Biological hierarchy is also arranged in a downstream flow. The energy available to organic productivity from sun and water flowing downhill is transformed successively through food chains going downstream, with more of the larger species which are higher in the food chain, in the larger waters downstream. Even in one section of a stream or microcosm there is visible hierarchy, such as the periphyton communities with numerous broad coverage of diatoms converging into clumps of consumer activity consisting of attached invertebrate animals and their tube structures.

Studies of many kinds of systems show replacement time to be correlated with territory and both to be proportional to their position in the hierarchy. When appropriate turnover times are used to calibrate the components of a simulation model, hierarchy is automatically introduced. The flows of many fast, smaller units are required to support larger, longer-lasting units. The study of streams in the field has allowed investigators to identify some major categories of processes and guilds of species involved. On our energy systems diagrams, hierarchy is represented as a continuum from the small and fast components on the left to those with slow turnover and larger territory on the right. Periphyton microbes and algal cells are at the bottom of the size–time hierarchy, whereas large leaves and large animals are at a higher level. Boling et al. (1974) described the evolution of a stream model as part of the international biological program. The model was changed to recognize the different species' sizes and turnover times, thus beginning to incorporate hierarchy in stream models. They observed a three-fold range in loss rates of riparian leaves. McIntire et al. (1975), McIntire and Colby (1978), and McIntire (1981), using a modeling approach developed by Overton (1977) after Klir (1969), simulated some aspects of streams. The system states at a given time were derived from those at a previous time by state transformation functions. This approach is equivalent to our energy systems diagraming approach, since in both cases, properties at one time are used to determine those at a later time. However, the mode of thinking is different. The state change approach involves more dissection in that it sets up matrices for changes in coefficients and functions in one table, rather than attempting to incorporate and describe the flows and larger mechanisms responsible for change. In this book one equation at a time is written.

The stream consumer model (McIntire 1973) was presented as hierarchical because the compartmental aggregations were divided according to the size of food particles and consumers. Grazers consume algal cells, collectors gather larger particles, and shredders dissect and consume larger organic substances. The state changes were calibrated with rate measurements and in this way, hierarchy was introduced into the simulation. In the simulation runs reported, the predicted turnover times per year were as follows: grazers, 4.27; collectors, 4.80; shredders, 5.25 and predators 0.98. The first three were similar and thus at about the same hierarchical level.

As self-organization and succession develop, the early colonizing by small components is complicated by the development of the components at higher levels in the food chain hierarchies and their feedback control actions. Hershey and Dodson (1985) and Hershey (1987) found sculpins and stonefly predators practicing selective predation in stream microcosms, exerting control actions which were different from those of their role in natural streams. Part of the control was accomplished by means of the tube-forming behavior these preditors induced in the prey organisms.

Eichenberger (1975) studied stream microcosms in Switzerland as they adapted to different sewage rate additions. More light increased the role of the phototrophs and decreased that of the heterotrophs; more organic loading increased the numbers of consumers and, at first, increased the numbers of producers, but these reached a maximum at a sewage addition rate of between 1 and 5%. The greater the amount of sewage organics added, the greater the number of heterotrophs. As the numbers of midges increased, the numbers of filamentous algae decreased. With greater current velocity, uptake interactions were greater but microbiotopes were reduced.

Self Organization of Stream Microcosms with Toxic Substances

As discussed in Chapter 7, toxic substances in microcosms may cause some benefit at very low concentrations, but reorganization and lower diversity result at higher concentrations (mercury and cadmium experiments were discussed in stream Mesocosms, Figures 7.10–7.12). Toxic substances are subject to downstream spiraling but return upstream with animal migrations. Studies of zinc on artificial streams were made by Williams and Mount (1965) where toxicity led to increased organic mats (see Chapter 16). Cushing, Thomas and Eberhart (1970) modeled a stream experiment with zinc uptake (Figure 4.9). Rose and McIntire (1970) studied dieldrin uptake and the effect of toxicity in laboratory streams like that in Figure 10.3.

Self Organization with Heavy Metals in Savannah River Mesocosms

Six concrete channels 91.5 m long were operated as experimental streams at the Department of Energy Savannah River Plant in South Carolina, Giesy, Bowling, Kant, Knight, and Mashburn (1981) and Ferens (1974). As shown in Figure 10.15, each channel was 1.61 m wide and 0.31 m deep. Waters supplied as headwater inflow were soft, acid, low organic (0.5 mg/l) well waters, conditioned somewhat by flowing over limestone rock. Flows were 95 liters per minute, a retention time of two hours and current of 0.013 m/sec. Channels were lined with polyvinyl chloride plastic that could be removed before starting a new kind of study.

The ecosystem that developed was net producing (autotrophic), with its food chains mainly based on photosynthesis of algae. The dominant plants and animals were those that developed during a self-organization period in which many species were seeded. *Juncus diffusissimus* was dominant. The invertebrates that developed were typical of ponds and sluggish streams in southeastern United States.

Figure 10.15. Channel stream mesocosms at the Savannah River Plant, South Carolina.

The channels were built for study of toxic uptake and toxicity. Stream communities were developed with mercury concentrations in 1971–1974: 1.0 microgram per liter in two channels, 5.0 micrograms per liter in two other channels, and two channels were controls. Bacterial populations were low in inflowing waters and increased downstream (Chapter 15). Periphyton biomass was three grams dry weight per square meter, more in some mercury treated channels than in controls. Populations of flowering plants (*Juncus* and *Utricularia*), desmids and filamentous green algae, and blue-green alga *Merismopedia punctata* predominated. Animals included damsel flies, midges, snails, frogs, and the fish *Gambusia*. There was "continual downstream expansion of the macrophytes... rearrangement of insect and algal populations, and a continuous influx of new plant and animal species". Differences developed in concentrations of larger animals that made differences in the lower food chain.

For a year in 1976, four of the channels received steady inflow of cadmium two with five micrograms per liter, two with ten micrograms per liter, and two others serving as control (Giesy 1978; Giesy et al. 1979, 1981). Cadmium levels in organisms and sediments were equilibrated in about 20 days but released soon after the cadmium inflows were stopped. For Gambusia, 180 days were required before concentrations in tissues reached some equilibrated level. At these concentrations the ecosystems were similar to controls, with some periphyton growth and invertebrate populations reduced by cadmium. Controls had higher gross production, net production, and respiration than cadmium treated streams. Decomposition of leaf litter packets placed in the streams was slowed by cadmium. Diversities were less (Chapter 15).

After two years of experimental work studying cadmium in the Savannah River experimental streams, Knight (1980, 1981) developed simulation models for the stream ecosystems that showed the inhibitory effect on the metabolism and main components (Figure 7.10).

Cadmium in Glass Tube Microcosms within Silver Springs, Fl.

Robert Knight (1980, 1981, 1983) also studied a set of glass cylinders arranged in Silver Springs, Florida so the very uniform waters of the Silver river could flow through the cylinders (Figure 10.16). Each cylinder developed heavy growths of periphyton like that found on the *Sagittaria* outside in the main Silver River. Some cylinders were given cadmium released slowly from metal strips, and some were controls. Photosynthetic production and community respiration was determined from changes in oxygen in passing through the tubes (Figure 10.15). Effects of snails and *Gambusia* were also studied. Some positive effects from low concentrations of cadmium were suggested (Figure 10.17), and there were optimal sized

Cadmium in Glass Tube Microcosms within Silver Springs, Fl.   231

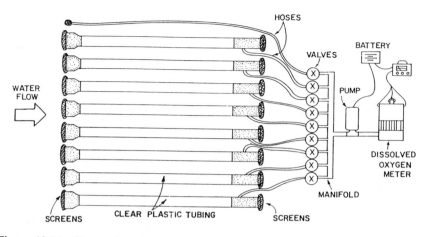

Figure 10.16. Glass tube microcosms within Silver Springs, Florida (Knight 1980, 1981).

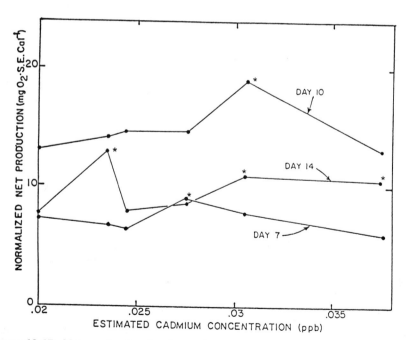

Figure 10.17. Net production in glass tube microcosms in Silver Springs (Figure 10.16) as a function of estimated cadmium concentration (Knight 1980).

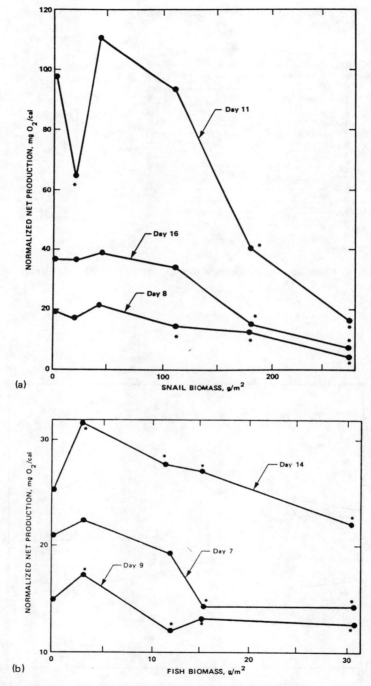

Figure 10.18. Net production in glass tube microcosms in Silver Springs (Figure 10.16) as a function of animal populations (Knight 1980, 1983). (a) Effect of snails, Dec. 1980; (b) effect of *Gambusia*, Dec. 1980.

populations of animals that maximized ecosystem net production (Figure 10.18).

Transformity and Hierarchical Control

Solar transformity, is a measure of position in energy hierarchy. Items at higher levels are those that required more emergy to produce, and by hypothesis these have greater effect because self organization only retains items that have effects commensurate with the high emergy resources used in their making. Many scarce chemical substances have high transformities and by hypothesis have greater effects. Knight (1980, 1981) evaluated the transformity of cadmium and animals in stream microcosms, considering both as important controlling elements (Table 10.1). Apparently, toxic substances are high transformity items capable of large amplifier effect either to increase or decrease ecosystem performance depending on how the system is adapted to use the substances.

Flowing Saltwater Microcosms

Wulff and McIntire (1972) studied assemblages of diatoms in a wood trough 3 m by 0.76 m by 0.8 m receiving flowing saltwater and arranged with varying water level simulating tide. Studies were made on exposure, desiccation, varying light intensity, varying temperature, and varying salinity. Biomass and chlorophyll developed inversely to air exposure, were increased with light intensity, and were reduced by low salinity fluctuations. Heated waters caused low diversity.

Chapter 11
Terraria and Soil Microcosms

Terraria, the terrestrial equivalent of aquaria, are small containers usually containing soil materials, terrestrial plants, animals, and air. Terraria are found by the millions, self-organizing in schoolrooms, offices, and homes. Often open to external air exchange, these microcosms may dry out and become water-limited. Oxygen and carbon dioxide are maintained at external levels by air exchange. Humans occasionally add water. The simulation in Figure 2.11 represents conditions typical of those found in houses and schoolrooms. However, serious studies of typical terraria are scarce. On the other hand, considerable scientific literature exists on transplanted and recently assembled terrestrial microcosms. Gillet and Witt (1978) reviewed earlier work, especially that concerned with study of the way such terrestrial ecosystems process chemical substances.

Minimodels of Terrestrial Microcosms

The main components of terrestrial microcosms are included in the minimodel in Figure 11.1, which describes microcosms with aerobic soil (soil spaces containing air). Compared to aquatic microcosms, a different set of metabolic requirements is limiting and controlling. Water and carbon dioxide are more important in terrestrial systems than oxygen. With terrestrial plants, much water is processed as part of transpiration, due to the Gibbs free energy differential between the drying power of the air and the water concentrations in the soil.

Terrestrial ecosystems over the earth have a global column of air above them which has a large oxygen store (21%) not rapidly depleted by the consumption processes at the ground. The replacement turnover time of the oxygen in the earth's atmosphere for the ordinary metabolism of

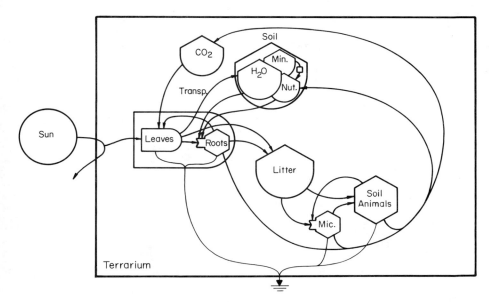

Figure 11.1. Model of main features of an air-filled, closed terrarium microcosm. Min., mineral crystals; Nut., available nutrients.

terrestrial ecosystems is a million years. Carbon dioxide, however, with only about 330 ppm, may turn over in a year. Terrestrial microcosms, having a much smaller air column relative to the living surface of organisms and soils, have even shorter turnover times for carbon dioxide, days or less. The very simple simulation model in Figure 2.10 is mathematically a Michaelis–Menten system since there is closed circulation of a material which stabilizes the system's response to input changes. It represents diurnal processes in overview.

Microbial Decomposition Ecosystem

Ingham et al. (1985) combined a species of bacteria, a fungus, a bacteria-eating nematode, a fungus-eating nematode, and blue grama grass from pure cultures, with Colorado plains soil media to form species-defined microbial ecosystems (gnotobiotic microcosms). Natural soil organic matter was enriched with a chitinous nitrogen source. From rates of growth in different combinations, described in this and other gnotobiotic studies (Anderson et al. 1983; Moore et al. 1985), a model of system relationships was suggested (Figure 11.2). Evidence was obtained that a more complex network increased mutual performance and mineral cycling availability through regulatory actions.

To represent the same model in energy language, relative hierarchical position is inferred from size and turnover time. The small and fast

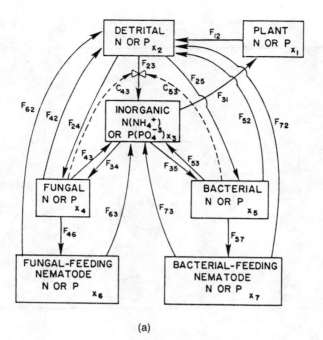

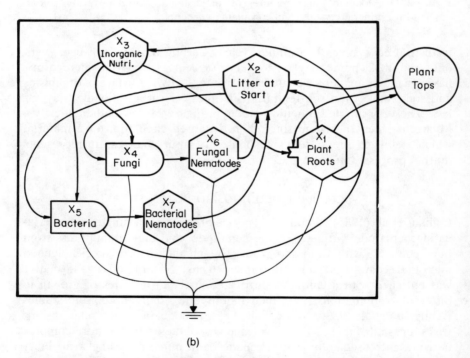

Figure 11.2. Model of microbial soil ecosystem organization (Ingham et al. 1985). (a) Components and pathways studied; (b) energy language representation arranged left to right in order of turnover time of individuals.

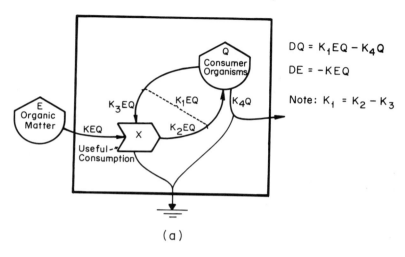

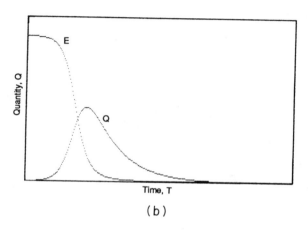

Figure 11.3. Simulation minimodel of a decomposition system based on an initial organic matter stock. (a) Systems diagram and equations; (b) typical simulation results (Program: NONRENEW).

components are on the left and the larger and slower items on the right (Figure 11.3b). Here the nutrient cycle pathways of Figure 11.3a are included with the accompanying energy flows. Consistent with the maximum power principle, ecosystem development reinforces the pathways that are loop-organized, as in Figure 11.2, collectively increasing useful consumption. The minimodel in Figure 11.3 simulates the overall behavior, using one autocatalytic loop to represent the many autocatalytic loops of the real system.

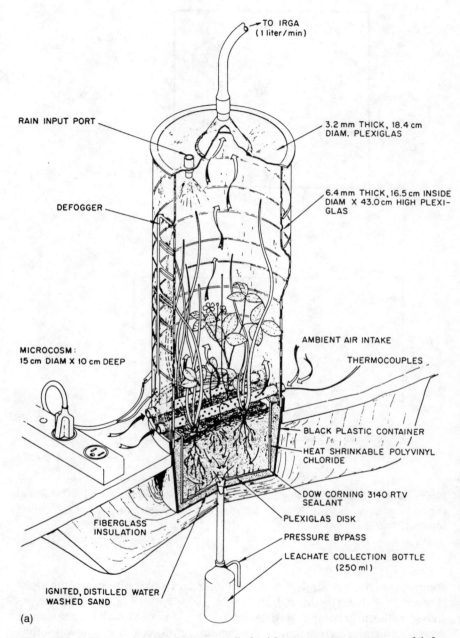

Figure 11.4. Terrestrial microcosms studied with temperature sensors and infrared gas analyzer measuring carbon dioxide metabolism (Van Voris et al. 1980). (a) Microcosm unit; (b) temperature regime; (c) carbon dioxide regime.

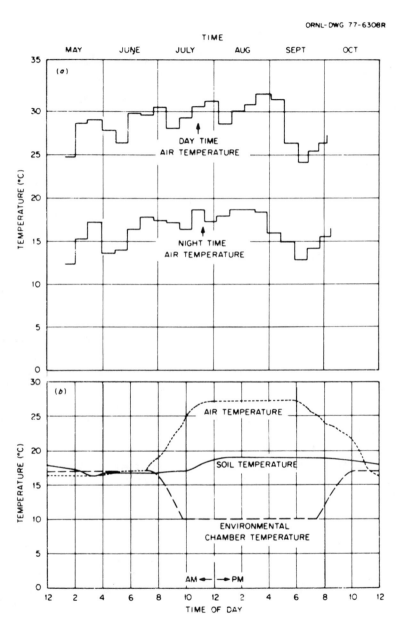

Figure 11.4. *Continued*

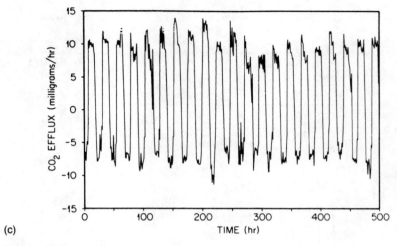

(c)

Figure 11.4. *Continued*

Scale and Hierarchy

In general, terrestrial ecosystems are higher on the scale of size, territory, and turnover time of their components than many aquatic ecosystems (Steele 1985). Consequently, it is more difficult to represent whole terrestrial ecosystems in microcosms. Often, only a part of the terrestrial ecosystem is represented, such as the litter or soil. Because many terrestrial plants are bulky, only a few individuals of a species may be present. Therefore, the dominant organisms in many terraria are not represented by enough individuals to operate as a self-reproducing population. Earthworms are important in terrestrial soils. In microcosms, disproportionate-sized earthworms may develop. Where large species are included, replication is difficult. Terraria started out similarly, often become rapidly different as they become controlled by different dominant plants and animals.

Van Voris et al. (1978, 1980), measuring carbon dioxide metabolism of terrestrial microcosms as shown in Figure 11.4, found that some microcosms had more low frequency variation than others. The low frequency variation represents larger scale components and processes. The microcosms were destabilized by treatment with cadmium to a level of mild toxicity. The microcosms with larger scale metabolic fluctuations were more resistant.

Self-organization, Succession, and Adaptation

With large members having long turnover time, long periods may be required before self-organization can develop new ecosystems or reorganize and adapt to new conditions. Trees and soil profiles take years to develop.

Whereas aquatic microcosms can go through the necessary succession and adapt to a set of conditions in a few years, months, or days, terrestrial microcosms take longer. Consequently, most of the terrestrial microcosm research has been on transplanted soil cores or newly combined associations of organisms. Agricultural microcosms include treatments of recently

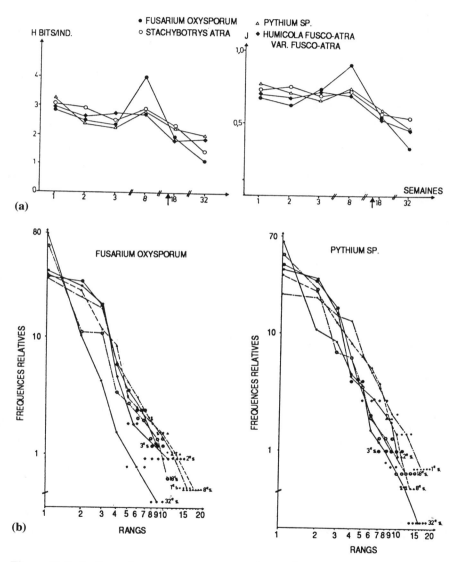

Figure 11.5. Indices of protozoan diversity and organization in microcosms based on fungal food chain (Couteaux and Devaux 1983). (a) Species information; (b) rank order graphs.

rooted plants. As with intensive agriculture, such microcosms are operated without much self-organization. Tolle et al. (1985) compared traditional pots, microcosms, and field tests for assessing waste effects on crops. Microcosms duplicated field plots better than pots (Chapter 17).

As we discussed in Chapter 3, studies of transient, unorganized ecosystems do not reveal full ecosystem functions. The short-term effects and chemical fates of ecosystems not adapted to the stresses they encounter may not be a good indication of long-term impacts or environmental potentials. On the other hand, the diversity of protozoa resulting from self-organization was observed by Couteaux and Devaux (1983) in microcosms (Figure 11.5). Diversities and rank-order graphs were similar to those generated by natural ecosystems with larger components.

The creative results of self-organization are sometimes observed in terrestrial microcosms adapting to special enclosure conditions. When shade-adapted rain forest microcosms were put in growth chamber conditions at higher light intensities (1,000 ft-candles), plant growth filled the small chambers, crowding into all the available space, forming a round, "cabbage-like" leaf mass controlled by the chamber shape.

Homeostasis of Terrestrial Microcosms

The homeostasis of systems with closed circulation of materials is known in the popular press as the GAIA theory. As we discussed in Chapter 2, sealed microcosms are strongly self-regulating of their gaseous contents. Terrestrial microcosms allowed to acclimate in the shade of the rainforest, showed the ups and downs of carbon dioxide fluctuating around a stable mean value (Figure 2.10). These variations were duplicated by a simple simulation model. The simulation model of a terrestrial microcosm, where carbon dioxide is a main variable, is given in Figure 2.10. The daily alternating periods of excess photosynthesis and respiration alternately decrease and increase gaseous carbon dioxide, as observed in these terrestrial microcosms. Figure 11.4 shows another example of steady diurnal rhythms in terrestrial microcosms (Van Voris et al. 1978).

Although temperature changes altered the rates of the chemical processes in the microcosms, the homeostasis was so well regulated that overall processes were little changed by the change in temperature. Faster rates in one part of the system were compensated for by increases in rates elsewhere in the system. The occasional ecological independence of ecosystems from temperature change, seen in aquatic microcosms in Figure 2.18, was found for some terrestrial microcosms also. See the rate of change graphs in Figure 11.6 and compare them with the model-generated rate of change graphs in Figure 2.4b.

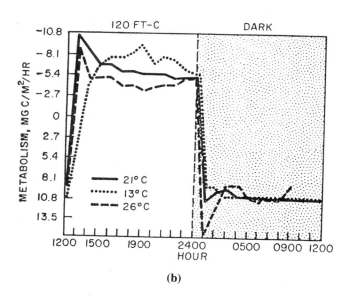

Figure 11.6. Similar rates of metabolism at different temperatures in transplanted terrestrial microcosms from a rainforest in Puerto Rico (Cumming and Beyers 1970). (a) View of microcosms; (b) graph of metabolism.

Terrestrial Microcosms for Global Perspectives

Microcosm homeostasis has global implications for spaceship earth, a much larger closed system also driven by solar photosynthesis. Different composition of living organisms has caused different average carbon dioxide levels about which homeostasis has been regulated. This following quotation from Odum and Lugo (1970) is relevant to the GAIA theory and global warming concerns now becoming popular.

> Since the ratio of litter and consumers to herbs varied from one system to another, it is perhaps not surprising that the carbon dioxide levels achieved at gaseous balance after one day of closure varied markedly. This may be an important demonstration of the control of the atmosphere of the planet by the biotic components existing in the system. The physical properties of the atmosphere of the earth are a result of biological evolution as much as vice versa. Since very large changes in the carbon dioxide level at balance may occur and since carbon dioxide is implicated in the thermal radiation balance on the earth in relation to ice cap maintenance, it is not unreasonable to suspect that ice ages may be caused by the relative evolution of plants and animals and their excesses or deficits in organic matter production.

Arctic Tundra Microcosms and Carbon Dioxide

In order to understand the effects of increasing atmospheric carbon dioxide and climate change on arctic tundra, Billings et al. (1982, 1983, 1984) studied carbon dioxide exchange from frozen tundra cores in their "Phytotron" (Figure 11.7a). Increased temperature and lowered water levels increased carbon dioxide release from microbial decomposition of peaty material. Adding nitrogen enhanced net carbon dioxide uptake by the microcosms. There was little direct effect of doubling the carbon dioxide concentrations in the air. Warming, however, changed the tundra from a carbon dioxide sink to a source. Whereas these studies measured carbon dioxide in air flowing over the soil surface in a natural manner, many other studies of terrestrial microcosms were designed to attempt to measure carbon dioxide metabolism with an alkali cup absorbing carbon dioxide by diffusion in still air. Quiet air microcosms yield an order of magnitude less carbon dioxide than the same soil would yield with normal air movement.

Ecosystem Organization and Chemical Cycling

The components and food chain organization of an ecosystem control the circulation of chemical elements. Witkamp and Frank (1967, 1970) and Witkamp (1972) studied the movement of radioactive cesium in a terrestrial microcosm of mineral soil, plants, leaf litter, microbes, and animals. Adding millipedes accelerated recycling from litter to soil, increasing

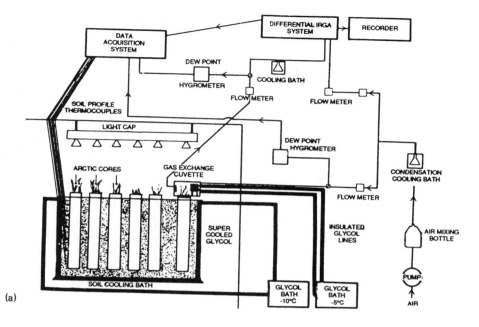

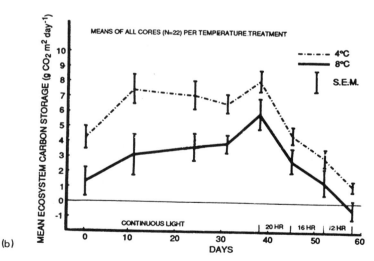

Figure 11.7. Arctic tundra microcosms studied in Phytotron (Billings 1982). (a) Apparatus; (b) decreased net fixing of carbon dioxide with increased temperature.

plant growth. Snails reduced plant growth. When components were added successively, greater complexity increased cycling. Millipedes were more important at lower temperatures. When carbon was lost through metabolism, magnesium was immobilized in humus, cesium in sand, and some potassium was leached out.

A simple simulation model of cesium transfer among the main components, with linear intercompartmental pathways, generated uptake curves not unlike those observed by Patten and Witkamp (1967), Witkamp and Frank (1967) and Witkamp (1972). The model in Figure 11.8a is rewritten in energy language form in Figure 11.8b, with simulation results in Figure 11.8c. When the rate of isotope uptake is high and changes in main system components are slow, uptake is a linear catch-up process. Thus the simulation model was relevant for the short-term tracer behavior. Klopatek, Klopatek, and Debano (1990) studied fire microcosms where were started with forest floor litter, duff and soil. Nitrogen was lost from litter but not from the soil; ammonia was present at the surface. Recovery was underway in 90 days.

Rock Outcrop Microcosms

Where geological history has exposed massive outcrops of granite or other rock, terrestrial ecosystems of small stature develop, and these are readily studied whole in microcosms. Such ecosystems, depending on the

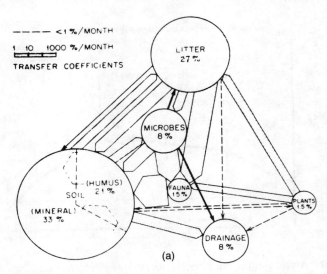

Figure 11.8. Radioactive cesium in litter moving into other components of a leaf-litter microcosm. (a) Model by Witcamp (1972); (b) energy systems diagram of model used for simulation; (c) graphs of observed cesium movement and computer simulation results (Witcamp 1972).

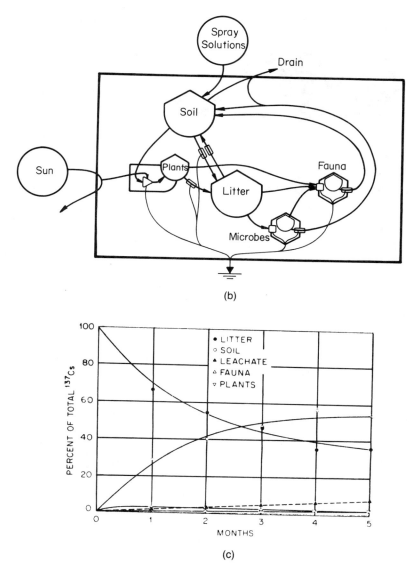

Figure 11.8. *Continued*

available water, may be encrusting lichens, a mosaic of alpine herbs and flowers, or patches of perched wetland marsh. Study of small ecosystem patches in their natural setting is not very different from study of patches transplanted to laboratory and garden situations. Platt, McCormick, Lugo, Murphy, and others studied succession in artificial outcrop communities seeded with transplanted soil and organisms (Figure 11.9). The trans-

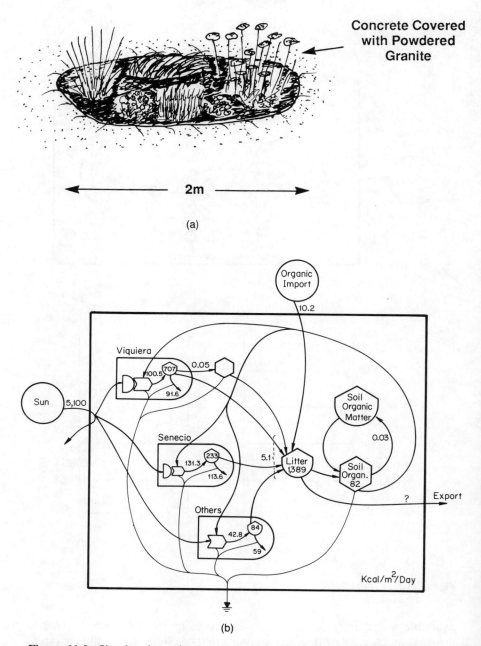

Figure 11.9. Simulated granite outcrop terrestrial microcosms at N.C. Botanical Garden in 1967 by McCormick and others. (a) Sketch of soil and vegetated depression; (b) energy budget in June 1968 (from Lugo 1969).

planting included moving a veneer of the native rock as the substrate for the microcosm experiments (Figure 11.9a) (Platt and McCormick 1964). Effects of radioactive tracers and externally imposed gamma irradiation were also studied (McCormick and Platt 1962).

Rock outcrop (soil depression) microcosms developed by Frank McCormick at the University of North Carolina botanical garden were

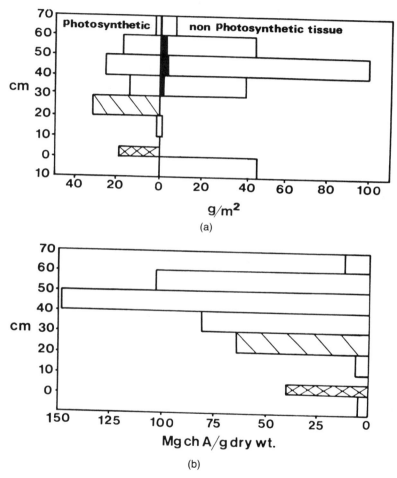

Figure 11.10. Measurements on the granite outcrop microcosm in Figure 11.9 (Lugo 1969); (a) Organic matter distribution Aug 28, 1967 (clear, *Viguiera*; hatched, *Senecia tomentosus*; crisscross, *Polytrichum commune*); (b) Chlorophyll *a* distribution; (c) humidity and temperature, Sept. 12–13, 1967; (d) evapotranspiration and insolation Sept. 12–13, 1967; (e) metabolism and insolation, May 22–23, 1968 when net photosynthesis, dominated by *Viguiera porteri*, exceeded soil and plant night respiration; (f) soil respiration, May 17–18, 1968; (g) soil respiration in 1967 with each point a mean of two 24 hour runs.

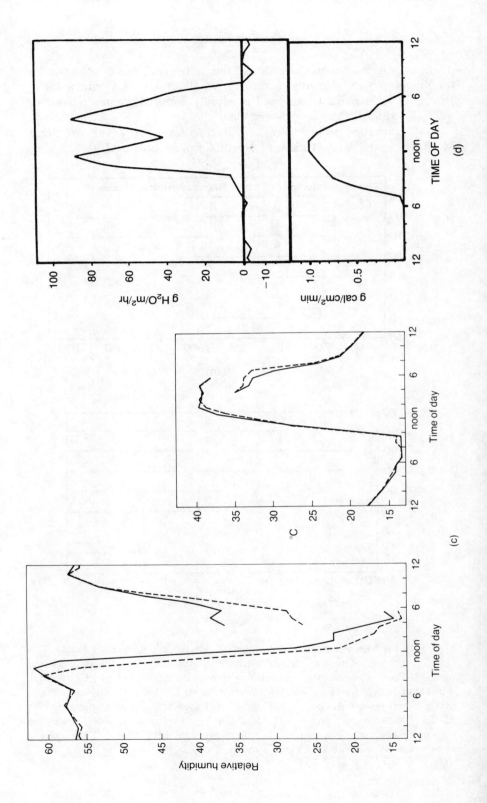

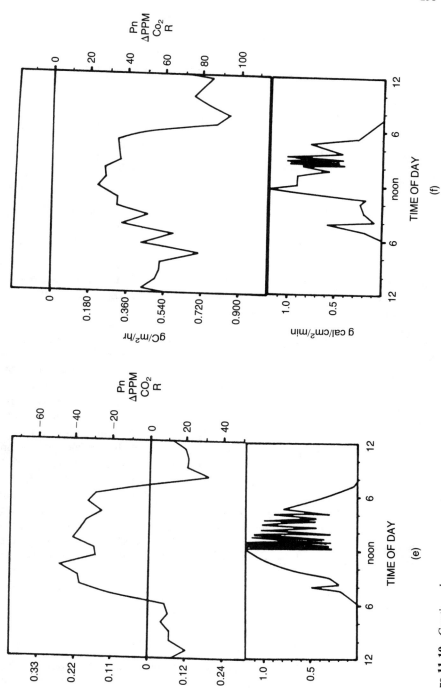

Figure 11.10. *Continued*

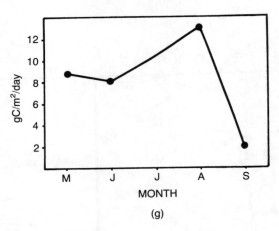

Figure 11.10. *Continued*

described by Cumming (1969). Biomass, chlorophyll, transpiration, photosynthesis, respiration, water, carbon, and energy budgets were measured by Lugo (1969). Systems diagrams with summaries of energy flows are given in Figure 11.9. These were based on gas-flow measurements of metabolism like those in Figure 11.10. Most of the gross production is immediately used in respiration in order to maintain the structure of the system. There were wide temperature ranges (Figure 11.10c), but the annual growth cycle developed a mature ecosystem in the outdoor microcosms by the end of the growing season, with balance of photosynthesis and total respiration, more dead organic matter than live photosynthetic tissues (Figure 11.10a and b), and high carotenoid/chlorophyll ratios of 1.8–10.9 (Margalef's ratio of optical density of the extract at 430 nm divided by optical density of the solutions at 665 nm) (Margalef 1963). Bomb calorimeter measurements of live plant tissues ranged from 3.77 to 4.7 kcal/dry g in whole plants, 4.6 to 4.9 kcal/dry g in dead litter, and 5.1 and 5.8 kcal/dry g in inflorescence. Soil respiration varied diurnally and seasonally (Figure 11.10f and g), with community respiration often exceeding photosynthesis (Figure 11.10e). Growth was limited by water availability from recent rains, which was soon lost with plant transpiration (Figure 11.10d). Most of the main structural and metablic properties found in larger forest ecosystems were illustrated in miniature, especially self-organization resulting in adaptation to somewhat extreme conditions.

Transplanted and Reassembled Soil Columns

Soil column studies have a long history in soils science. Whereas some soil columns are studies of percolation in homogenized, packed tubes, others are sections of living soil removed intact for study. Some have

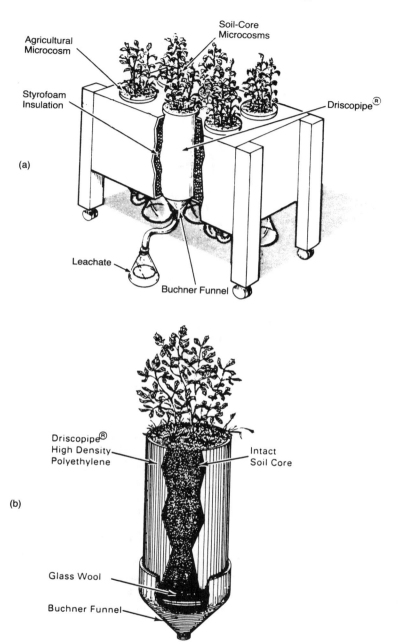

Figure 11.11. Set of terrestrial microcosm tubes containing transplanted soil cores for studying the effect of substances added on top with simulated rain (Van Voris et al. 1983). (a) Apparatus with tubes; (b) one tube unit showing outlet below for collecting leachates; (c) device for collecting soil cores in the field.

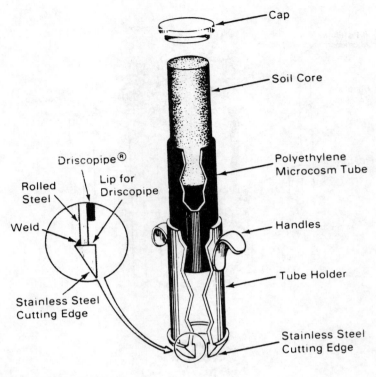

(c)

Figure 11.11. *Continued*

vegetation on top. Transplanted or reassembled columns are used to represent and replicate outside conditions in an indoor environment over a relatively short time, maintaining as much as possible the inflows and outflows of the original source sites (Figures 11.11c, 11.7, and 12.4 are examples). Some are aerobic and some anaerobic.

Effects of Acid on Terrestrial Microcosms

With the increase of acid rain and air pollution effects on forests and other land ecosystems, more and more terrestrial microcosms have been studied in an attempt to understand and measure these effects under experimentally controlled conditions. An important parameter is pH, which is affected by carbon dioxide concentrations, sulfur redox processes, and the inflows of acids or bases. Different species are adapted to different pH conditions. Microcosms provide insights on the self-organization that is now adapting new species to new pH conditions in many terrestrial ecosystems of the world. Cronan (1980) found acid leaching of soil nutrients from terrestrial microcosms to be faster at higher temperatures and where

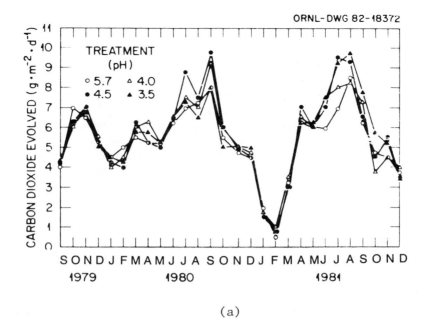

(a)

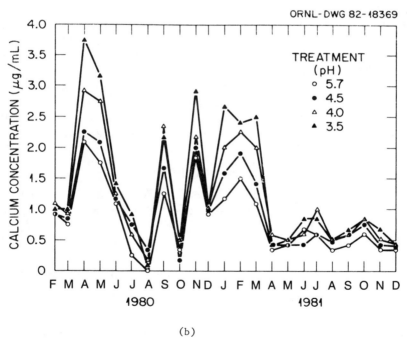

(b)

Figure 11.12. Effects of simulated acid rain on forest seedling microcosms. (a) Effect on litter decomposition (Kelly and Strickland 1984); (b) effect on nutrients of the throughfall off tulip tree and pine seedling canopy (Kelly and Strickland 1986).

green plants were absent. There was little effect on respiration in soils from pine and hardwood ecosystems (Cronan 1985).

Kelly and Strickland (1984, 1986, 1987) studied the effect of simulated acid rain on the decomposition of litter added to soil-box microcosms with homogenized soil and tree seedlings. A full canopy developed after three years. Figure 11.12a shows that there was little effect of the acid on decomposition, as measured by carbon dioxide emission. Low pH in simulated rain increased throughfall of all nutrients, being more efficient in washing nutrients from leaf surfaces (Figure 11.12b). The acid rain treatments leached calcium, sulfate, and other nutrients out of the upper soil horizons in proportion to the acidity.

Toxic Elements in Terrestrial Microcosms

Nickel

DeCatanzaro and Hutchinson (1985) set up 30 forest floor microcosms, each 36 cm wide and 30 cm tall, containing a cylinder of surface Canadian podsol including, litter, ground vegetation, and one Jack pine seedling in each. Soil plugs rested on gravel with a container drain so that the leachate from 52 in of rain per year could be collected for analysis. After a month in the greenhouse, nickel–copper solutions at pH 2 were added, simulating the effect of emissions from a nickel smelter. Some microcosms were given stack dust. Most of the copper and some of the nickel were retained in the soil and plants, the rest appeared rapidly in the leachates. The treatments interfered with the normal, tight recycling and conservation of nutrients, causing loss of nitrogen and some plant mortality. Similar studies were made of soil columns packed with homogenized soil.

Cadmium and Zinc

Gas flow measurements of carbon dioxide released from soil were used by Chaney et al. (1978) to measure the effect of cadmium and zinc on 105 soil microcosms. The chemical additions to soil without heavy metals reduced respiratory metabolism in such a way that it resembled the metabolism of soils with heavy metals collected in a city environment.

Terrestrial microcosms receiving fly ash are considered in Chapter 18.

Organic Chemicals in Terrestrial Microcosms

Some of the major questions facing those with responsibility for environmental management concern the ability of terrestrial and wetland ecosystems to receive pesticides and other organic chemical substance, denature or vaporize the agents, and repair losses due to toxicity. Study of transplanted or recently established terrestrial ecosystems can reveal the abilities of existing ecosystems to accomodate isolated stresses. Studies of recently synthesized (newly combined mimic ecosystems) microcosms involve more ecological interactions than single organism toxicity testing, but still can not provide a realistic representation of the performance of an organized ecosystem.

A more difficult question is how useful and resistant are the new ecosystems that self-organize under the chronic influence of organic toxic chemicals. Learning how adapted systems perform will ultimately determine what kinds and concentrations of chemicals can be released into the environment. Unfortunately, few studies have investigated ecosystem adaptation to chronic chemical influence. Gile et al. (1980, 1982) applied organic chemicals, tagged with ^{14}C as a tracer, to terrestrial ecosystems in the chamber shown in Figure 17.2. There was an arrangement for measuring the radioactive carbon dioxide released into the air, thus measuring the decomposition and vaporization of the compounds (Figure 11.13a). The terrestrial ecosystems had been recently synthesized by combining soil materials and nutrients, seedlings, and other living components. These papers describe the studies of the mass balance of the added agents as well as their toxic effects compared to controls. Plants were added first, with animals added later. Adaptation time was two months or less.

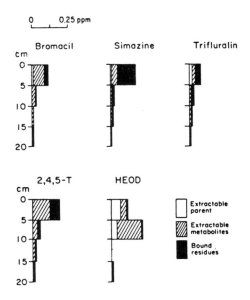

(a)

Figure 11.13. Results of monitoring chemical substances in a simulated forest floor microcosm (Gile et al. 1980, reprinted with permission from Environ. Sci. Technol. 14(9), copyright 1980, American Chemical Society). (a) Distribution of substances in vertical soil profiles; (b) network diagram showing percent of added chemical ion TMC in various parts of the system at the end of three week experiments.

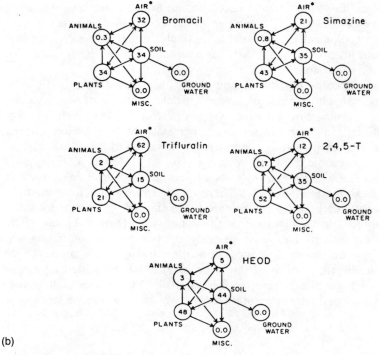

(b)

Figure 11.13. *Continued*

Figure 11.13 shows the results of pesticides added to a microecosystem mimic of the forest floor ecosystems of Oregon, including Douglas fir, red alder, rye grass, earthworms, pillbugs, mealworms, crickets, snails, and a vole. Gile et al. (1980) also studied the effect of chemical pressure-treated wood placed in similar synthetic microcosms (without tree seedlings). Very little of the preservative reached soil and living components, and none reached simulated ground water. Gillett et al. (1983) examined cricket mortality from creosote, dieldrin, and other substances within a terrestrial microcosm before and after adding a vole predator. Predator–prey relationships were disturbed in several ways as chemicals increased cricket mortality and toxic effects decreased vole predation efficiency.

Need for Terrestrial Adaptive Microcosm Studies

Unfortunately, most of the literature on terrestrial microcosms reviewed here deals with the short-range effects of changed inputs rather than with ecosystem organization, adaptation, and homeostasis. Since terrestrial ecosystems have longer turnover times for their components and structures,

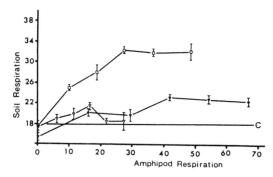

Figure 11.14. Effect of animal activity on soil function (Richardson and Morton 1986, Soil Biol. Biochem., Copyright 1986, Pergamon Press).

it is difficult to address the more important ecosystem processes and changes in research projects of a few years' duration. The short-term studies will be highly misleading if they are used as the basis for judging what happens to terrestrial ecosystems as they reorganize, change species, and establish new mechanisms for maximizing functions in new conditions. Learning how acid rain is damaging to an unadapted forest is not very useful for learning about the new ecosystems which will be adapted to acid conditions in the future. Multiply seeded, self-organizing microcosm studies, like those done with faster-responding aquatic media, are needed, even though more years will be required for results. For example, to study acid adaptation, multiple seeding of species should include species from the environs of active volcanos, where longer times for species selection have already occurred.

Holistic systems ecology developed first with plankton. This provided early investigators with a window in the fast part of ecosystem hierarchies so they could observe collective properties which developed rapidly, measure total metabolism, diversity, and track nutrient cycles. Aquatic systems microcosmology may be well ahead of terrestrial microcosmology. The window of time and space for terrestrial microcosms is somewhat longer, requiring time for the animals to organize. The control and stimulus to total performance which one expects from the maximum power principle is often similar to what is shown in the soil metabolism graph in Figure 11.14 (Richardson and Morton 1986).

Chapter 12
Wetland Microcosms

Wetlands are ecosystems with water-saturated soils and rooted plants which are sometimes exposed to air. If the pore spaces of a microcosm soil become occupied by water, the gaseous reservoir is excluded and wetland conditions prevail, especially low oxygen availability to roots and soil animals. With a high proportion of organic matter favoring respiration, oxygen is quickly depleted and the system becomes anaerobic, a typical wetland condition.

Sulfates and nitrates are used by microorganisms as a source of oxidized input, and are reduced by microbial metabolism. When nitrate reduction (denitrification) occurs, nitrogen fixing is often associated. When sulfate reduction occurs and hydrogen sulfide is formed, pH is increased. This is opposite to the usual effect of respiratory consumption which reduces pH as a result of the release of carbon dioxide. Rooted plants adapted to anaerobic soils are specialized for anaerobic conditions. Figure 12.1 is a minimodel of a wetland microcosm with some of the main processes that predominate in these conditions.

Succession in a Miniature Bog

Bellamy and Riley (1967) studied a small hummock (56 cm high, 70 cm across). Growth of sphagnum on top of a previously high-nutrient fen resulted in elevation and isolation of the surface, generating conditions for succession to a low-nutrient bog in seven years.

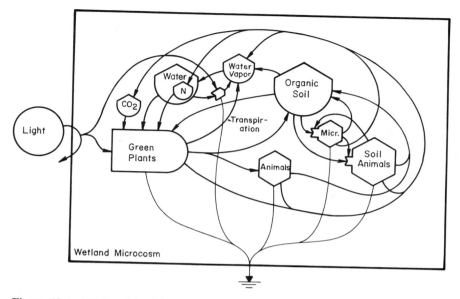

Figure 12.1. Minimodel of the main components of a wetland microcosm.

Wetland Decomposition

A wetland microcosm study by Day (1983) found faster decomposition in laboratory microcosms kept at a higher uniform temperature than that prevailing in field conditions. However, Kemp et al. (1985) found slower decomposition of leaf litter (within litter bags) in 84 3-liter bottle microcosms than in field conditions, possibly due to the absence of animal shredders. Four conditions of atmospheric interface were tested: argon, pure oxygen, air in microcosms and air in litter bags in the field. Nutrient retention was controlled by available oxygen and its control of redox potentials (Figure 12.2). Day et al. (1989) developed the apparatus shown in Figure 12.3 for decomposition studies.

Metabolism of Transplanted Salt Marsh

Gallagher and Daiber (1973) measured photosynthesis and respiration of transplanted salt marsh cores in microcosms (Figures 4.6 and 12.4). Respiration in the dark was steady, but photosynthesis in constant light showed a rhythm. The system depicted in Figure 4.6 had higher gross photosynthesis at higher temperature without diminished metabolism at high light intensity.

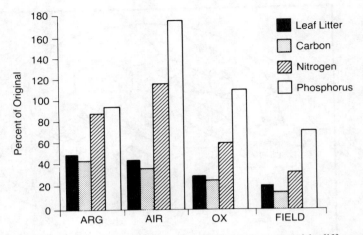

Figure 12.2. Decomposition of litter in wetland microcosms with different gases in air space (Kemp et al. 1985).

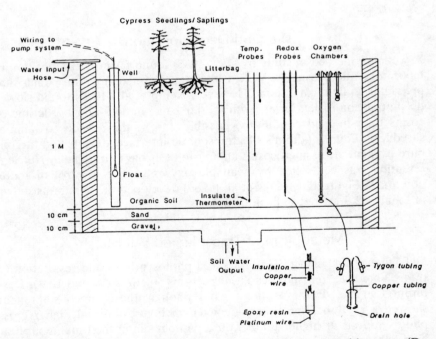

Figure 12.3. Cypress wetland mesocosm used to study decomposition rates (Day et al. 1989).

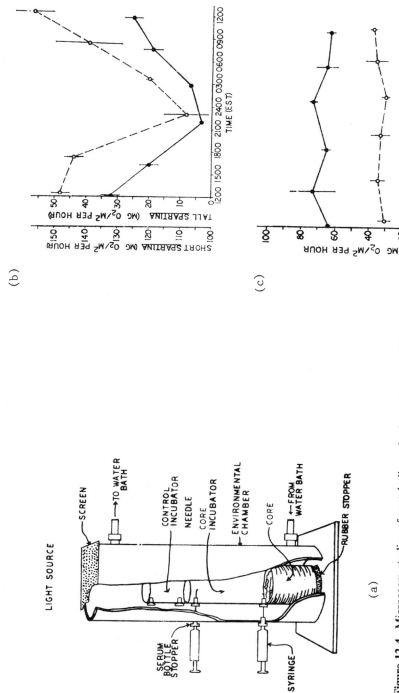

Figure 12.4. Microcosm studies of metabolism of saltmarsh cores (Gallagher and Daiber 1973). See also Figure 4.6. (a) Apparatus; (b) diurnal rhythm in photosynthesis at constant light; (c) steady respiration at constant darkness. Solid circles, tall *Spartina*; open circles, short *Spartina*.

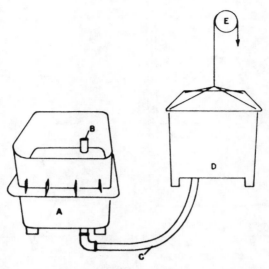

Figure 12.5. Salt marsh microcosm with artificial tide (Everest and Davis 1979, reproduced from the Journal of Environmental Quality: 8(4). 465–468 by permission of the American Society of Agronomy, Inc.; Crop Science Society of America, Inc.; and Soil Science Society of America, Inc.).

Nutrient Flows and Metabolism of a Salt Marsh Tidal Microscosm

As shown in Figure 12.5 a wetland microcosm for a coastal marsh can be given a tide by regularly raising and lowering an adjacent water reservoir.

Everest and Davis (1979) studied phosphorus movements by means of a radioactive phosphorus tracer placed in the sediment 17 cm below the surface. It appeared at the mud surface due to crab movements starting the 1st day, and was found in *Spartina* leaves in 4 days. Kitchens et al. (1979) compared marsh transplanted to microcosms with field control sites. Transplanted marsh filled four 6-m^2 tidal tanks which were fitted to study production and respiration with gas metabolism measurements taken from closed air circulation. Air conditioners were used to maintain normal temperature regimes. Nutrient fluxes were measured in the ebb and flow of the artificial tide. Typical curves of carbon dioxide uptake and release are shown in Figure 12.6. After four and eight months, animal species were similar in the microcosms and the marsh, although some species developed increased numbers. Dominants were fiddler crabs, polychaetes (*Heteromastus* and *Nereis*). Meiofaunal populations were statistically similar. Reichgott and Stevenson (1978) compared ATP, carbon, and bacterial counts for the four marsh tanks with corresponding data from the natural marsh. Mean values for bacterial numbers and ATP were about twice as high in the microcosms as in the outdoor system. Microcosms were somewhat depleted of clay and some differences had developed among the microcosms.

Burney et al. (1981) studied salt marsh microcosms and documented a diel rise and fall of dissolved sugars and the bacteria that consume them. Concentrations were 300–800 ppb with a diel range of 20–30 ppb.

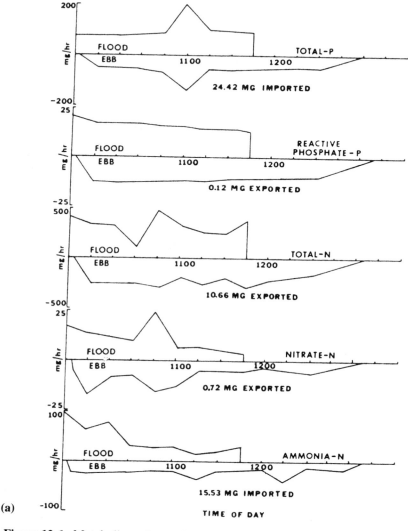

Figure 12.6. Metabolic exchange in a tidal, saltmarsh microcosm (Kitchens et al. 1979). (a) Inflows and outflows of nutrients; (b) carbon dioxide release with night respiration (R) and uptake with daytime net production (P).

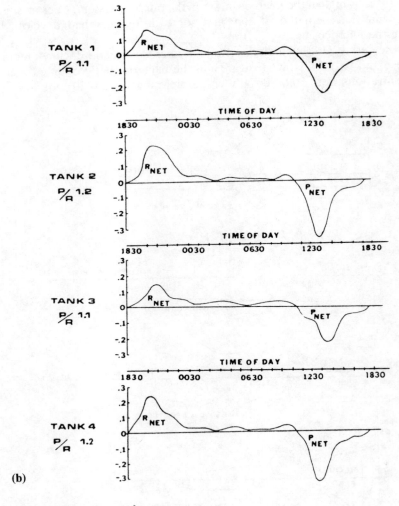

Figure 12.6. *Continued*

Artificial Marshes

Among the systems of aquatic plants used by Wolverton (1982) for waste treatment studies were marsh microcosms and mesocosms. The system in Figure 12.7 has anaerobic waters from primary sewage treatment percolating through gravels in which reeds (*Phragmites communis*) are growing. With a residence time of 24 h, the presence of the reeds decreased BOD to 3 ppm and caused absorption of ammonia to increase to 0.6 ppm. Kjeldahl nitrogen went to 2.9 ppm, and phosphorus to 2 ppm. See Chapter 18.

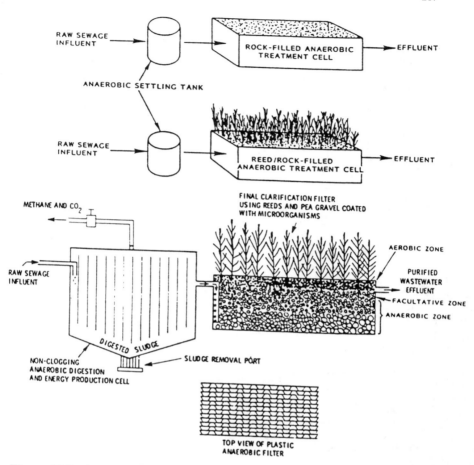

Figure 12.7. Artificial *Phragmites* marsh used by Wolverton (1982) for waste treatment.

Heavy Metals in Wetland Microcosms

Sanders and Osman (1985) added arsenic to salt marsh core microcosms imbedded in salt marsh so as to receive a 2-h flooding in each tidal cycle. Analyses of components were made after 9 days. Elevated concentrations of arsenic were found in *Spartina* blades and detritus.

Pesticides in Wetland Microcosms

The microecosystem chamber in Figure 12.8 was used by Isensee et al. (1982) to study chemical persistence in an agricultural wetland condition (rice paddy). The herbicide propanil produced aniline-based metabolites

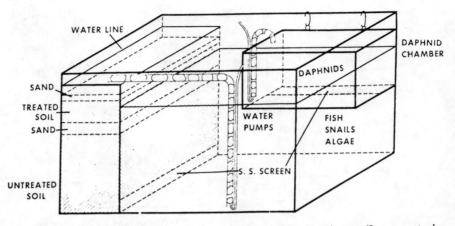

Figure 12.8. Microecosystem chamber for rice paddy experiment (Isensee et al. 1982).

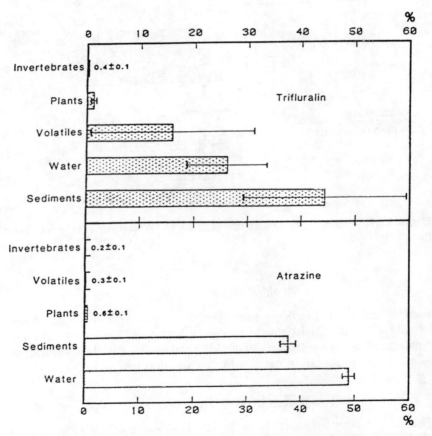

Figure 12.9. Fate of pesticides in wet prairie pot hole microcosms after six weeks (Huckins, Petty, and England, Chemosphere 15, Copyright 1986 Pergamon Press).

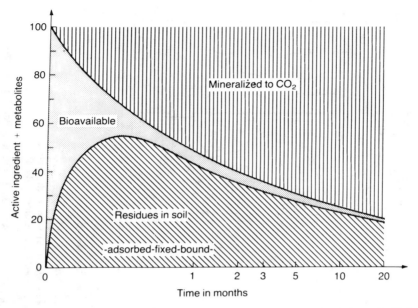

Figure 12.10. Summary diagram of the fate of organic chemicals in soil (Fuhr 1987).

which were adsorbed into the soil with and without flooding. Johnson (1986) performed short-run studies of prairie pothole microcosms, including water fleas and midges. He tested the effects of pesticides in fresh runoff from agriculture, and later, after microcosm interactions. The microbial activities examined were little affected. Atrazine inhibited plant productivity. The toxicity of carbofuran, fonofos, phorate, and triallate in microcosms disappeared rapidly. Huckins et al. (1984) studied the fate of pesticides in prairie pothole microcosms after 6 weeks (Figure 12.9). The main accumulations were found in sediments, apparently in humus compounds not readily available for further interactions. Wolverton and McDonald-McCaleb (1986), using the artificial marsh system in Figure 12.7, obtained 93% and 99% removal of aromatic and aliphatic hydrocarbon pollutants. Fuhr (1987) summarizes the fate of organic chemicals which become unavailable in soil (Figure 12.10). Part of the ability of ecosystems to neutralize toxicities may be in their tight binding mechanisms, developed long before human pollution. If this general ability has been reinforced in past evolution, the mechanisms seem also to be preadapted to protect the biosphere against toxicity generated by the human economy.

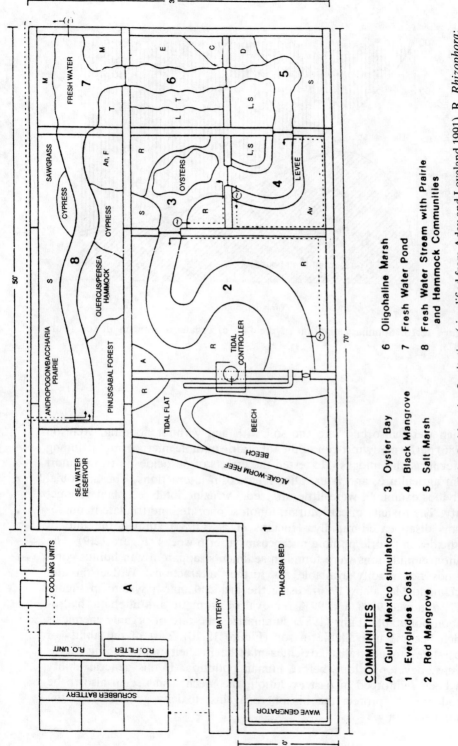

Figure 12.11. Layout of Everglades mesocosm of the Smithsonian Institution (simplified from Adey and Loveland 1991). R, *Rhizophora*; Av, *Avicenia*; S, *Spartina*; D, *Distichlis*; F, *Fraxinus*; C, *Conocarpus*; M, *Ludwigia, Pontederia, Pistia, Ceratophylum, Azolla*; E, *Eleocharis*; T, *Typha*; L, *Laguncularia*; S, *Salicornia*; An, *Annona*.

Smithsonian Everglades Mesocosm

Through initiatives of W.H. Adey and others, a series of microcosms and mesocosms has been developed at the Smithsonian Institution, both as a tool for research into understanding ecosystems and to develop functional ecosystems for public display. Figure 12.11 depicts the plan for the system operating in Washington, D.C. Display microcosms are an ecotecnological system with part of the organization left to the organisms, and essentials, such as filtering and conditioning of the water, being aided by pumping and filtering components. For more on these systems see Adey and Loveland (1991).

Chapter 13
Ponds and Pools

Ponds are small bodies of water, usually much wider than they are deep, and usually containing self-organizing ecosystems. Ponds have been called microcosms by many writers in literature, such as Thoreau in his description of Walden pond in Massachusetts. Forbes (1887) and Hutchinson (1964) called ponds microcosms in the scientific sense of being a miniature demonstration of the processes occurring on a larger scale in lakes and oceans. In our terminology, most ponds are actually mesocosms, being larger than the small experimental laboratory microcosms.

It is beyond the scope of this small volume to review the entire literature on ponds, which is a good part of the limnology literature. However, we have included studies where pools and small natural and artificial ponds were used as experimental microcosms. That is, they were used to help elucidate basic processes of self-organization, system characteristics, and ecosystem responses to experimental treatments, including replications.

Eutrophication Experiments in Ponds

In the earlier days of limnology and oceanography, increasing productivity through fertilization was a major goal and was extensively practiced with freshwater and marine fish ponds. Many times there were batch treatments which produced a transient pulse of change. There was a return to the earlier condition after the pulse of fertilizer had been dispersed (Raymont and Miller 1962). Smaller components responded quickly, self-organizing in accordance with the new circumstances, and changing again with the return to a lower nutrient level. In other experiments, fertilization was continued at regular intervals so that more extensive self-organization of large components could occur. Some experimental studies were made with replicate ponds and control ponds, at Auburn, Alabama (Swingle and Smith 1939). Higher fish production was achieved with fertilization. Intense green blooms were encouraged, and pond sides were made steep to help shade out rooted plants, the lack of which channeled plant production into zooplankton and then into sunfish and bass.

Plankton Ecosystem in Fertilized Ponds at Cornell

In 1968–1969, eight ponds, previously built in 1964 for the Agronomy Department at Cornell University in Ithaca, NY, were fertilized with three levels of nutrient concentrations three times a week during the summers (O'Brien and deNouvellos 1974). Ponds were 0.1 acre each and 1.37 m deep. No fishes were present. The study involved tracking nutrients, phytoplankton, and zooplankton, mainly during the summer season. The levels of chlorophyll and plankton increased. As shown in Figure 13.1, prey–predator oscillations were generated among the component populations. The effects of these oscillations were reflected in the nutrient levels.

Over Consumption by an Uncontrolled Fish Consumer

With containers large enough for one species of fish carnivore, the large-scale roles of the components with a longer turnover time become apparent. Hurlbert et al. (1972a) studied the effects of the presence of the mosquito fish, *Gambusia affinis*, in plastic pool mesocosms. With the fish, intermediate consumers were reduced and stocks of algae increased, increasing solar heat absorption and the binding of nutrients. Species substitutions and replacements took place. These experiments illustrate a principle pertaining to microcosms. Whatever is the largest component is the species that is without its normal consumer–regulator and likely to overgrow the optimum level for maximum ecosystem production. In the long run, these largest members may tend to be eliminated by their own extreme oscillation. A similar phenomenon occurred in open aquarium microcosms (Beyers 1963) when larvae of the herbivorous snail, *Marisa*, were introduced and overgrazed the standing crop of rooted *Vallisneria* (see Chapter 7).

Pool Microcosms Relating Fish Growth to Nutrient Concentration

Goodyear et al. (1972) set up 15 plastic pools 2.4 m in diameter and 37 cm deep, multiple seeded them with local aquatic organisms and *Gambusia affinis*. The pools were placed outdoors under natural conditions of weather and light. A general plant fertilizer was added in several concentrations to

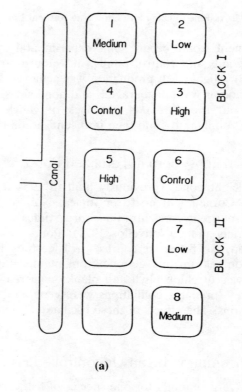

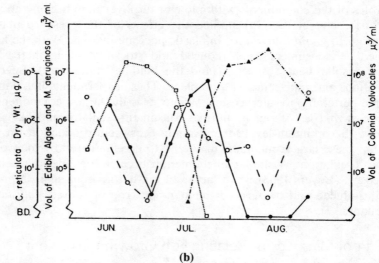

Figure 13.1. Plankton oscillation in Cornell ponds. (a) Pond layout (b) plankton: open circles with dashed line, edible phytoplankton; solid triangles with dot-dash line, *Microcystis*; dotted line with open squares, colonial Volvocales; solid line with solid circles, zooplankter, *Ceriodaphnia* dry weight (O'Brien and deNovelles 1974, Reprinted by permission of Kluwer Academic Publishers).

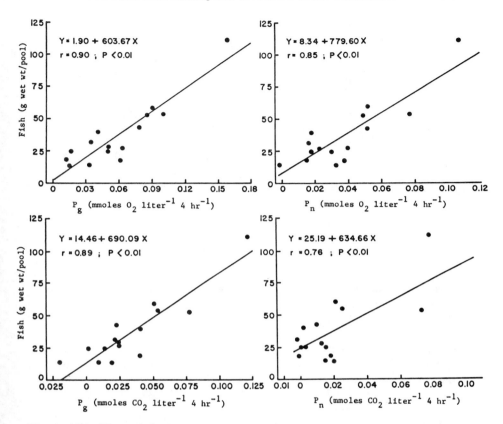

Figure 13.2. Plots of the average gross productivity P_g and net productivity P_N against final mosquitofish biomass in 15 pools. Each productivity measurement is the mean of determinations made on seven dates (Goodyear et al. 1972).

form an experimental series. Total metabolism was measured by means of oxygen and carbon dioxide diurnal curve methods. After four months, growth of *Gambusia* was related to production and to the fertilization. There was excellent correlation between average gross and net photosynthesis, and final fish biomas (Figure 13.2). In a similar experiment, McConnell (1965) reared the fry of *Tilapia mossambica* in 310- and 2,840-liter microcosms constructed from galvanized steel. Inorganic phosphates and nitrates were added in varying amounts to establish a gradient of rates of photosynthesis in the experimental systems. The experiments lasted 74–100 days. Again, there was an excellent correlation between gross photosynthesis, as measured by diurnal oxygen methods and fish growth (Figure 13.3).

In the microcosm of Goodyear et al. (1972), no attempt was made to limit the fish population, and reproduction took place. In the

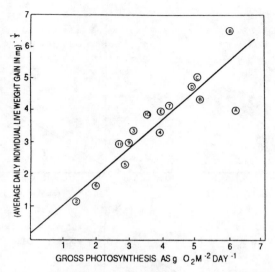

Figure 13.3. Relation of *T. mosambica* growth rate to the gross productivity in pond type microcosms (McConnell 1965).

McConnell experiments, the fry were below reproductive age and the population was artificially reduced when the projected rate of fish biomass increase suggested an impending effect on their growth from accumulating metabolites. However, even under these different conditions, there was still a correlation between fish growth and photosynthesis.

In yet a third similar experiment, Hurlbert et al. (1972) constructed 8 plastic pools, 2 m in diameter and 30 cm deep. The outdoor pools were filled with tap water and inoculated with biota from a nearby lake and a laboratory colony of *Daphnia pulex*. Fifty *Gambusia* were added to each of 3 experimental pools. The *Gambusia* greatly reduced the rotifer, crustacean, and insect populations, and thus permitted extraordinary development of the phytoplankton populations. Other effects included decreased optical transmissivity and increased temperature of the water, decreased amounts of dissolved inorganic phosphorus and increased amounts of dissolved organic phosphorus. Inhibition of the macroalga, *Spirogyra*, and the replacement of one species of annelid worm with another were observed.

Microcosms and Pond Mesocosms Treated with Organic Compounds

Giddings et al. (1985) and Hook et al. (1986) compared the response of a set of laboratory microcosms and a set of outside pond mesocosms, both sets treated with oil which contained toxic phenols (Figure 13.4a–c). The

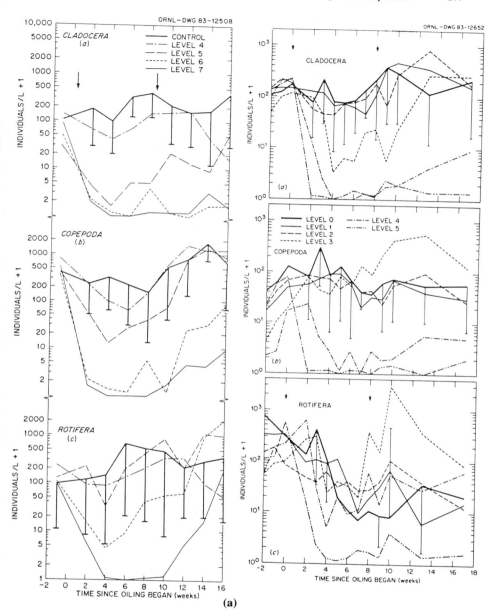

Figure 13.4. Response of aquatic microcosms and mesocosms to various concentrations of oil containing phenols (Giddings et al. 1985). Arrows indicate first and last days of oil addition. (a) Animal populations; (b) total dissolved phenolics in pond water, measured by HPLC; (c) total dissolved phenolic concentrations. W, weekly dosed microcosms; D, daily dosed microcosms.

278 13: Ponds and Pools

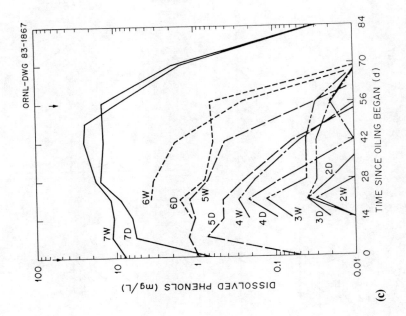

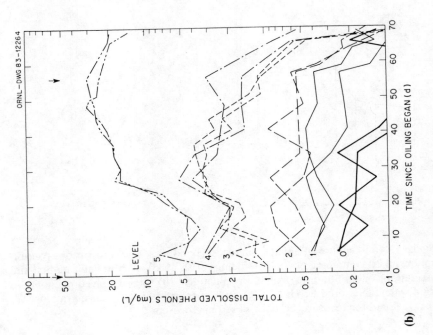

Figure 13.4. Continued

12 experimental ponds had a volume of approximately 15 m^3 each. The synthetic coal oil contained 13.4% water soluble compounds, 95% of which were phenolic. The communities were established with sediment and biota from local ponds near Oak Ridge, Tennessee. All were seeded with *Elodea* and miscellaneous water, detritus, plankton, etc. The larger mesocosms contained *Gambusia*, insects, and a variety of other organisms. The ponds were treated with oil in six different dosages ranging from 0 to 16 ml/m^3 per day. Water quality parameters, chlorophyll, phytoplankton productivity, phenol concentration, temperature, dissolved oxygen, fish populations, emerging insects, bacteria, and periphyton were measured. Zooplankton species, numbers, and biomass were also determined. The microcosms and mesocosms receiving low phenol concentrations were able to metabolize the toxic substances, although specialized, low-diversity biota became dominant. Bacterial populations and total respiration increased. In the microcosms, nutrients generated from decomposition caused large chlorophyll-bearing blooms. The systems receiving higher concentrations lost most of their biota, and the toxic substances accumulated without much decomposition. Results for microcosms and mesocosms were very similar, suggesting that the microcosms equalled the ponds for purposes of testing for phenol–oil toxicities. Species sensitivities in the microcosms and mesocosms were similar to those observed among the same species isolated from their ecosystems. Community similarity coefficients were derived using the BIOSIM program and converted to Z Scores, a method of describing distance from the mean. Response was generally proportional to dose (Figure 13.5) with the larger doses deviating more greatly from the control mean. The authors concluded that ponds and microcosms were suitable for the study of ecosystem-level phenomena associated with oil spills.

Larsen et al. (1986) compared single species algal assays, a Taub microcosm, and 470-m^3 experimental pond mesocosms exposed to similar concentrations of atrazine. They compared concentrations of atrazine that reduced algal activity to 50% of the control values (EC50). The EC50 values, based on radiocarbon uptake, were: single algae, 37–308 µg/l; Taub microcosms, 103–159 µg/l; and ponds, 100 µg/l. They postulated that the basic similarities among EC50 values suggested that the microcosm method would be adequate to provide a reasonable estimate of the concentration of atrazine that would produce a similar effect on natural communities.

Hurlbert et al. (1972a) treated 4 experimental pond mesocosms with Dursban, an insecticide, at the rate of 0.028 kg/ha and 4 at 0.28 kg/ha. Four of the 8 by 17 by 0.24 m ponds were kept as controls. After the treatment, predacious insects showed a greater reduction in numbers than herbivorous insects. Predacious insect populations generally recovered more slowly to control-pond levels than herbivorous populations. Five weeks after treatment, the numbers of predacious insects in the high-dose

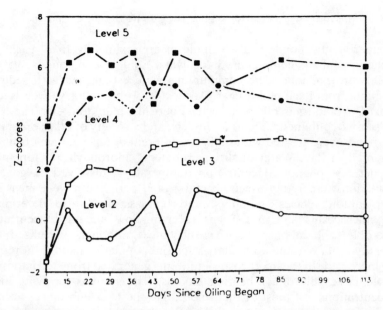

Figure 13.5. Zooplankton community similarity values (BIOSIM) converted to Z scores plotted against days since oiling began for various doses of synthetic oil in freshwater ponds (Giddings et al. 1985).

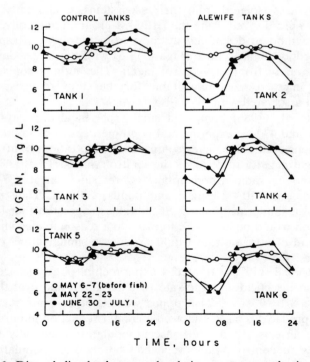

Figure 13.6. Diurnal dissolved oxygen levels in concrete pond microcosms with and without decomposing alewives (Durbin et al. 1979).

ponds averaged only 9% of the numbers in the controls; in the low-dose mesocosms they averaged 45% of controls. Simultaneously, the comparable figures for the herbivorous insects were 206% and 108%, respectively. Herbivorous crustacean zooplankters suffered high mortality due to the treatment and required 3–6 weeks to recover. Herbivorous rotifers increased dramatically in the treated ponds, from 5- to 25-fold. A predacious rotifer was 35 times more abundant in the treated ponds than in the control ponds. It was feeding on the herbivorous rotifers and other herbivores. The reduction of the herbivorous zooplankton by the Dursban and the restraint of the herbivorous rotifers by the predaceous rotifer permitted the rapid increase of phytoplankton populations in the treated ponds. Six weeks after treatment, the phytoplankton was 2 to 16 times more abundant in the low dose and high dose ponds, respectively, than in the control ponds. The results of this study illustrate the complexity of natural food webs and the sometimes unexpected results that occur when the population of one trophic level is differentially removed from an ecosystem.

Aquatic Leaf Litter Microcosms and Effects of Dead Fish Migrants

Where forest leaves accumulate in small woodland streams and pools, there is a high carbon/nitrogen ratio. Decomposition is slow due to a lack of inorganic nutrients for bacterial action. Durbin et al. (1979) set up six 7.6 by 2.4 by 0.55 m concrete pond microcosms with a 5 cm layer of leaf litter. They were studing the effect of dead alewives which had migrated to spawn in the head waters of fresh water streams. Fish were stocked in 3 of the ponds at a rate similar to the weight of fish that would occur naturally in a nearby spawning ground. The added nutrients from the fish initiated a short-lived plankton bloom, and increased community respiration through the decomposition of the leaf litter. The heightened quantities of inorganic nutrients stimulated photosynthesis in the detrital layer (Figure 13.6). As with salmon, the alewife life cycle couples freshwater and marine areas, resulting in useful exchange. The authors speculated that the most important ecological result of the alewife spawning migration may be its ability to release the vast supply of energy stored in leaf litter by providing the nitrogen and phosphorus that stimulate microbial activity.

Ecological Succession in Natural Vernal Pools

As part of a massive study of the vernal pools of southern California, Zedler (1987) presented data from the temporary pools of the Kearny Mesa in San Diego County. The occurrence of species of vegetation growing on the pool bottoms was recorded during the 1979 wet season in

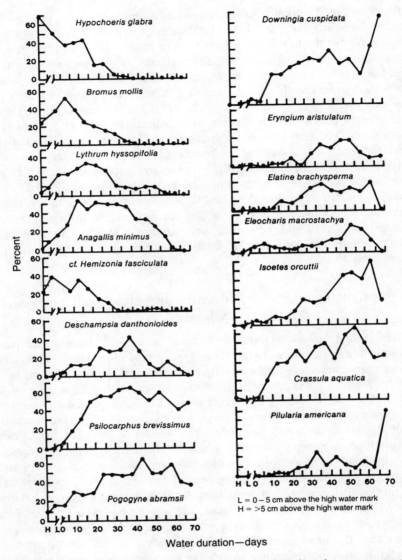

Figure 13.7. The relative occurrence of species plotted against the water retention time for pools on Kearny Mesa, San Diego County (Zedler 1987).

decimeter square quadrats placed along transects in 26 pools. The water levels in the pools were simultaneously monitored. With these data it was possible to express the occurrence of plant species relative to the length of time that a quadrat was under water during the study year (Figure 13.7). His data demonstrate that the percentage of occurrence of each species peaks at a different time, indicating the same type of replacement that occurs in other ecological successions (Chapter 4). Using 30 square meter pond mesocosm, Walton, Tietze, and Mulla (1990) found flooding reduced mosquito larvae as more invertebrate and small fish predators were increased.

Simulation of Marine Coastal Environments

Odum et al. (1963a) simulated Texas coastal ecosystems in outdoor concrete ponds. Conditions in the pond systems were devised to simulate three important bay system types. The first type was a low-salinity, turbid bay containing oyster reefs (see Chapter 14). The second type was a shallow, grassy bottom bay at about 30 ppt salinity. The third was a high-salinity blue-green algal mat community. The ponds were seeded with biota from appropriate natural communities in the Texas bays.

The grass pond developed animal and plant populations similar to the bay prototype. *Thalassia testudinum*, *Diplanthera wrightii*, and *Ruppia maritima* were the dominant rooted plants. Benthic diatom populations were observed in the sediments and water. The copepod, *Acartia*, exhibited a typical vertical migration. Typical animal populations were present, including the brown shrimp, *Bairdiella*, and the fishes *Cyprinodon*, *Fundulus*, and *Menidia*. Barnacles and oysters occupied the solid substrates. A dense population of bottom foraminifera was observed. Typical diurnal measurements for the grass pond are shown in Figure 13.8. The grass pond resembled its prototype ecosystem in many population and metabolic parameters. Still, as in all microcosm studies, the correlation was not perfect.

In a week after seeding, the blue-green algal mat pond developed new, fresh, green mats over the bare sand between the transplanted mats so there was a continuous carpet. Metabolic curves showed an anaerobic nighttime phase with low redox potentials, just as in the environmental prototype (Figure 13.9). The mats acquired the typical vertical structure of filaments and black organic ooze reported for algal mat communities throughout the world. Again the simulation was not perfect. A heavy population of corixid water bugs developed in these microcosms and persisted over time. These organisms are not typical of the parent type of community, such as the Laguna Madre of Texas.

Community Metabolism in a "Natural" Microcosm

The microcosm technique would be impossible were it not for the fact that Nature herself has created natural situations that have many properties of microcosms. Such a situation was exploited by Ganning and Wulff (1969, 1970) in eight brackish water rock pools on the island of Vrangskar in the Baltic Sea. These pools normally contain between 1,000

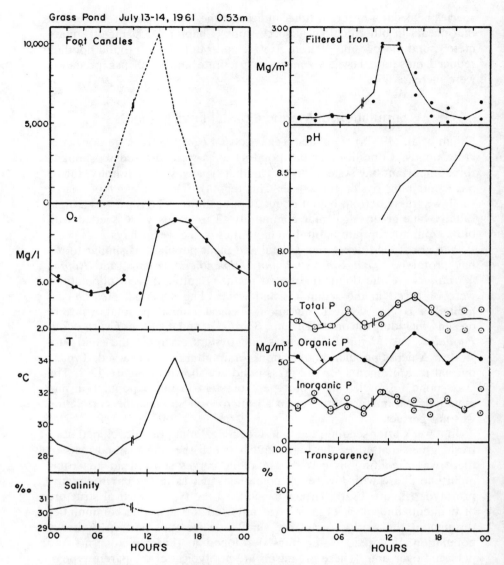

Figure 13.8. Diurnal record of incident ft-candles, oxygen concentration, temperature, salinity, iron in millipore filtrates, pH, phosphorus, and bottom transparency in the grass pond (Odum et al. 1963b).

and 2,000 liters of low-salinity water. Estimates of community metabolism were made using the diurnal oxygen technique (Figure 13.10). Gross photosynthesis values were between 2.7 and 21.8 g O_2/m^2 per day and community respiration values were between 2.7 and 18.2 g O_2/m^2 per

Figure 13.9. Diurnal record of pH, oxygen concentration, temperature, and carbon and oxygen metabolic rates in the blue-green mat pond (Odum et al. 1963b).

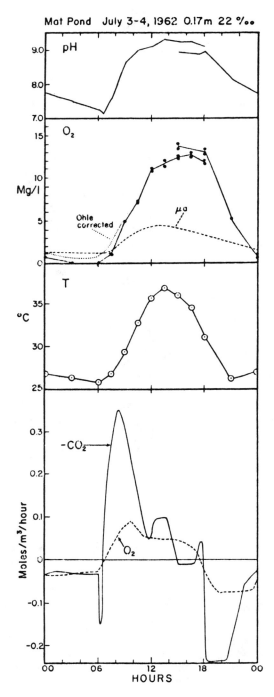

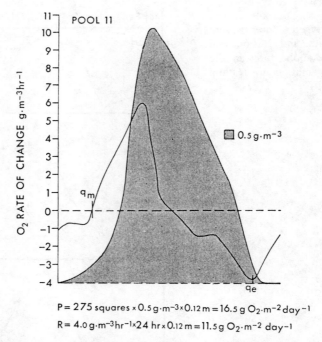

Figure 13.10. Oxygen rate of change curve for a brackish water rockpool (Ganning and Wulff 1970).

day. Photosynthesis exceeded respiration in all of the pools and these autotrophic conditions were generally found throughout the year, as the pools were fertilized by sea-bird droppings. The shape of the diurnal curve (with maximum photosynthesis in the morning) resembled those reported by many other authors.

The "Natural" Microcosm as a Teaching Tool

Bovbjerg and Glynn (1960) used a small, semiisolated and seemingly barren permanent rock pool near Pacific Grove, California as a marine microcosm to be studied as a class project. Their objectives were to describe the environment, measure environmental factors daily, and to determine the species present. The class identified 45 species of macroscopic animals and two algae. Population sizes and vertical and horizontal patterns of distribution were determined. A systematic study of food habits was done by field observation, laboratory feeding, and by stomach analysis. The local food web was delineated and standing crops measured. Finally, community metabolism and turnover rates were investigated. The authors listed the advantages of a microcosm approach for the students as follows:

1. Untrained students were exposed to many techniques in a meaningful field context.
2. Many basic ecological concepts were corroborated.
3. Field and laboratory approaches were seen to be necessarily complementary.
4. Accuracy and replication of measurement were stressed, as were quiet and prolonged observation.
5. The class was continuously reminded of the complexities of ecological relationships and the necessity of year-round studies.
6. The class experienced actual research.
7. The greatest value derived from the microcosm approach was the holistic view, the ability of the student to have a picture of the whole at all times and therefore a more meaningful appreciation of the structure and forces maintaining that whole.

These advantages are usually recognized by instructors who use some form of microcosm technique, either natural or artificial.

Estuarine Ponds Receiving Treated Domestic Sewage

Beginning in 1968, a large group of investigators, headed by H. T. Odum, conducted multifacted studies on control and experimental ponds designed to simulate estuarine conditions amended with treated sewage (Odum 1985, 1989). Located near Morehead City, North Carolina, the ponds were approximately 18.3 m by 24.4 m and averaged 0.5 m deep (Figure 13.11). The bottom was covered with an impervious layer of red clay, topped with a layer of black marsh mud. The control ponds received a mixture of tap water and pumped estuarine water. The experimental ponds received secondarily treated domestic sewage and estuarine water. Salinities were maintained in the 15–25 ppt range. The pumping inputs continued through 1971.

Initially, there was a period of build-up of larger organisms, during which the smaller ones were variable and changing. A heavy fringe of *Spartina* grass, an enormous winter stock of *Monodus*, and a substantial population of killifish developed in the waste ponds. A cover of *Ruppia* and a larger diversity of fishes developed in the control ponds. Figure 13.12 is a graphic summary of the seasonal sequence that emerged in the waste ponds after initial succession and was repeated for three years. A *Monodus* build-up occured in the fall, resulting in a huge chlorophyll and plankton biomass with high pH and oxygen values peaking in the early spring. This was followed by a crash in early May. The cells sedimented out, accompanied by a burst of respiration with a concomitant release of nutrients. This is the aquatic equivalent of a terrestrial forest fire. Photosynthesis then gradually recovered in parallel with a rise in the diversity of phytoplankton, microbial activity, animal biomass, and temperature.

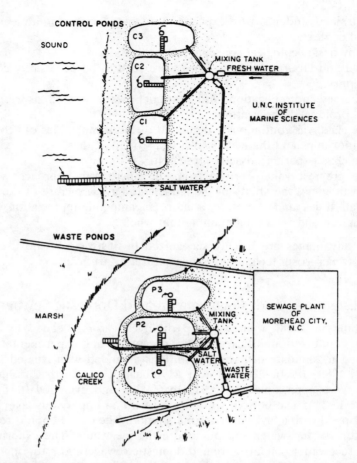

Figure 13.11. Experimental mesocosm ponds at Morehead City, N.C. used to study self-organization with and without flux of treated sewage waters (Odum 1985).

Finally, organism stocks and metabolic processes diminished with the decline of sunlight and temperature in the fall. Some ecological observations that grew out of this study were:

1. The ponds could develop functional ecosystems in three years.
2. Self-organization was accompanied by a gradual build-up of storages of detritus, animal and plant biomass and increases in productivity, respiration, and diversity.
3. Diversities were lower in the experimental ponds than in the controls, and the diversities in all the ponds were lower than in surrounding, comparable estuaries.

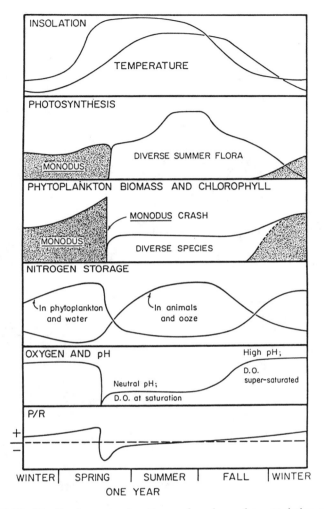

Figure 13.12. Idealized seasonal pattern of various characteristics of ponds receiving treated domestic sewage (Odum 1985).

4. The experimental ponds had the typical characteristics of cultural eutrophy (algal blooms, low diversity, extremes in oxygen concentration).
5. Self-organization of new ecosystems for new conditions occurs rapidly, using available species.

As described in Chapter 2, an energy network model was constructed for the pond systems (Figure 13.13) and a microcomputer simulation was run for both the high and low nutrient states (Figure 13.14). To be successful, the overview was required to generate behavior that simulated

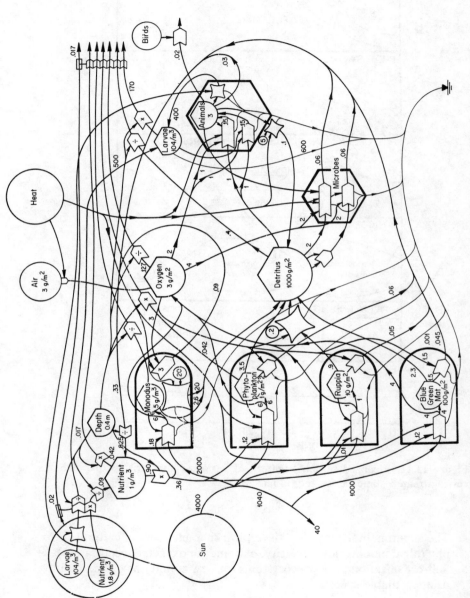

Figure 13.13. Simulation model of successional and seasonal changes in estuarine ponds. Numbers represent values of storages and pathways (Odum 1985).

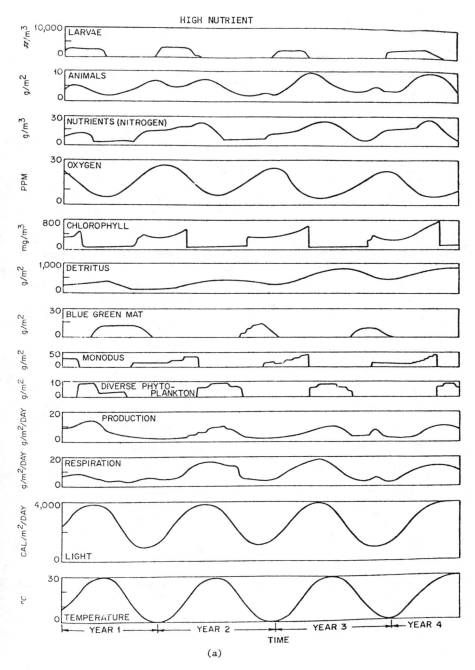

Figure 13.14. Results of computer simulation run of the successional and seasonal model of the ponds in Figure 13.11. (a) Simulation results with high nutrient levels; (b) simulation results with low nutrient levels (Odum 1985).

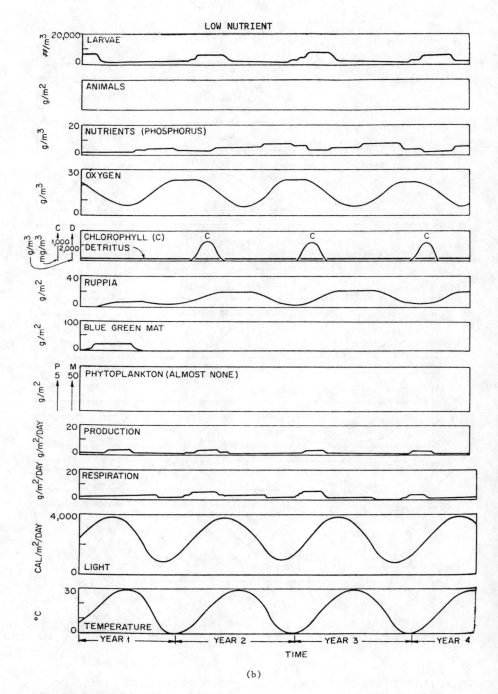

Figure 13.14. *Continued*

both the control and experimental ponds with different levels of nutrients. Therefore, the model was supplied with pathways for both regimes. This is a method for representing the multiple seeding available to both types of ponds. It also introduces self-organization as observed in nature, where some pathways are reinforced and prevail over others, depending upon environmental conditions. The values of storages (state variables) and flows are written on the diagram (Figure 13.13) By inspection, a reader may compare the flow in or out of a storage tank with the storage number in the tank to calculate a turnover time. The output of the simulation run closely emulates observed phenomena in the ponds (Figure 13.12).

Chapter 14
Reefs and Benthic Microcosms

The ecosystems of underwater surfaces are important, often conspicuous, intensively metabolic, and some are the most organized and diverse in the biosphere. Water motions relative to the fixed surface help to maintain the attached organisms and high rates of metabolism. We have already considered stream microcosms where the water flow is unidirectional. This chapter concerns microcosms which mimic the reefs, bottom communities, and encrusting ecosystems in lakes and seas where water motions are multidirectional.

In some places the accumulations of skeletons of encrusting organisms build reefs and deposits of geologic importance. The ecosystems of surfaces occur as sub-systems within most other ecosystems. Microcosm studies of surface-dominant ecosystems are relevant to small patches within lakes and estuaries as well as to large areas of reef or bottom. Microcosms with container walls readily develop attached organisms if water-stirring energy sources are available. Where there is a sedimentary bottom, spaces among the particles are small, water-filled, and if shaded from light, usually anaerobic, whereas the water above may be oxygenated. Like the real systems which they mimic, aquatic sedimentary microcosms include the many organisms and processes that use the aerobic–anaerobic interface.

Where currents exist, many bottom fauna build their own structures for reaching above the bottom to utilize the passing waters for obtaining food and oxygen and for the dispersal of waste products. Sometimes called epifauna, these organisms and their structures are "microreefs" in their way of life. Benthic plants, including larger algae and vascular plants, are similarly adapted to utilize the resources in moving water.

Resuspension of bottom sediments is a major part of the way physical energies of water motions periodically affect the bottom, an influence that adapted communities can utilize for positive gains. It is appropriate here to include microcosms designed for evaluating the effects of turbidity and sediment additions.

Reefs

Where currents and waves are strong, life which can hold its position can utilize the physical energies to do productive work such as bringing food, removing wastes, aiding reproduction, helping transport, etc. Photosynthetic reefs have a food chain based more on the sun. Consumer reefs have a food chain based on an inflow of organic matter. Chemosynthetic reefs are based on inflowing inorganic chemical potential energies. One of the main characteristics of reefs is the autocatalytic building of a reef structure that increases the ability of the system to capture more nutrients, food, and mechanical energy. Some of the most concentrated and highly metabolic ecosystems are reefs. See the general models of reefs in Figure 14.1.

The solid structures made by living organisms are ecosystem skeletons, often the cemented union of the skeletons of the component living organisms plus matter of nonliving origin. Reef structure, like organic biomass, is structure that provides means for capturing more resources. Where rocks are found in the beach zone of the western Gulf of Mexico, reefs of cemented sand are found giving tunicates and worm tubes additional armor against breaking waves.

The life of some reefs is based on filtered organic matter (Figure 14.1a). Others are based on photosynthesis within skeletal deposition plants, such as the red and green calcareous algae. The most complex of all reefs are the hermatypic coral reefs where food is largely produced by zooxanthellae which live as symbionts within animal tissues (Figure 14.1b). Where the special conditions of water motion that generate reefs are reproduced in the laboratory, successful reef microcosms of each type have been developed.

Consumer Reef Microcosms

Oyster reefs and other reefs of consumer animals depend on physical currents to make oxygen and food readily available for the attached animals so that a high population density is possible. Perpetual water motions insure removal of wastes (Figure 14.1a). Odum et al. (1963b) studied artificial reef microcosms (Figure 14.2) which were multiply seeded from nearby estuaries. A pump maintained a recirculating current over a biota initially transferred from an oyster reef in a Texas bay. This initial seeding included 123 *Crassostrea virginica*, 410 *Brachidontes exusta*, 36

296   14: Reefs and Benthic Microcosms

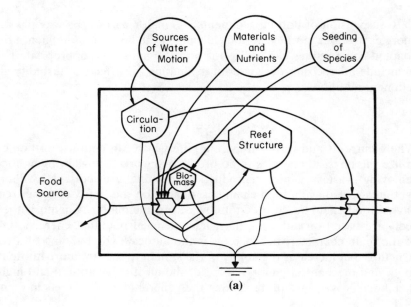

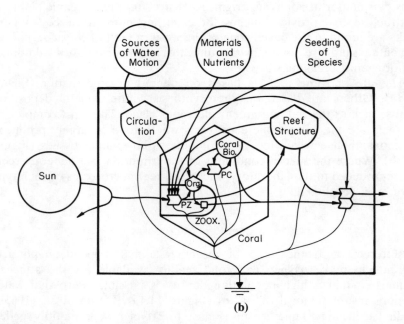

Figure 14.1. Aggregated overview model of reef systems and the autocatalytic role of reef structure. (a) Consumer reef plan; (b) photosynthetic producer reef plan. PZ, production by zooxanthellae; PC, animal biomass production by corals.

small mud crabs, and many *Balanus eburnius* and bryozoa. The area of sunlight-receiving waters was 20 times larger than the reef. The diurnal oxygen curves showed a coupling of production to consumption (Figure 14.2b). Early experiments maintained heavy diatom populations (1,300 *Pleurosigma* and *Amphora* per mm^2 of bottom ooze) and healthy oysters.

In later experiments, although seeding was similar, the water was fertilized. The phytoplankton in this case were small green nanoplankters and the zooplankton were rotifers. The oysters died, and the reef developed ivory barnacles, mud crabs, and sea anemones. The three replicate ponds had similar metabolism (Figure 14.2c).

Fouling Surfaces

Because of the importance of encrusting growths on boat hulls, many studies have been made of "fouling surfaces." These encrusting growths resemble microreefs. These small ecosystems are microcosms of larger reef systems and often are part of larger reef assemblages. Bacteria cover surfaces in the first 48 hours, followed by encrusting diatoms over the next few weeks, and complex associations of many attached plants and animals in a few months. See Chapter 10.

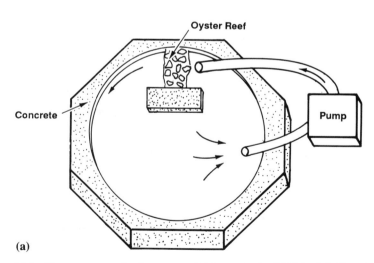

Figure 14.2. Consumer reef microcosm (Odum et al. 1963b). (a) Structure of microcosm and pumped circulation; (b) diurnal curves at low nutrient levels with diatom populations and oysters; Salinities are in parts per thousand (‰); (c) metabolism during eutrophic period with nanoplankton and barnacles. 1, 2, and 3 are replicate ponds.

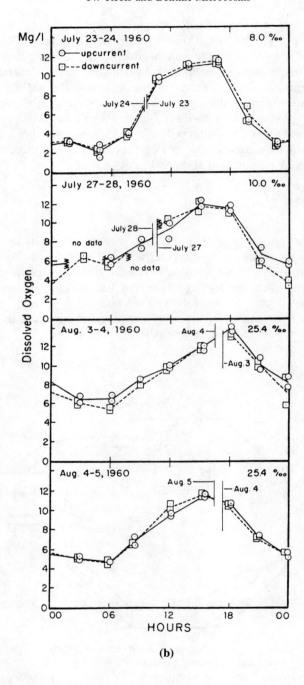

Figure 14.2. *Continued*

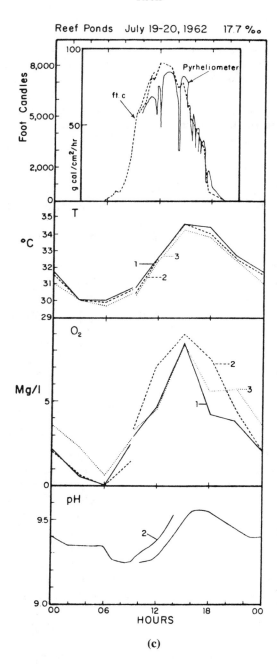

Figure 14.2. *Continued*

Artificial Substrates for Water Quality Studies

Considerable research on water quality and toxicity has been done by studying what grows on an artificial substrate such as glass slides, plastic, ceramic plates, or suspended trays, etc. For example, diversity measurements are used to infer the level of health of running waters. The observed results are partly a function of the seeding of available propagules and partly a function of the resources and stresses in the circulating waters. See the review volume edited by Cairns (1982).

Artificial Reefs

Artificial reefs have been widely used to provide better sport fishing. Old cars, tires, and other strong, bulky structures have been placed on lake or coastal bottoms, where their surfaces soon develop encrusting growths which are mostly dependent on the foods and physical stirring work of the passing waters. Soon there is a complex food chain supporting larger fishes. Artificial reefs of broken concrete blocks were introduced into the marine pond mesocosms at Morehead City, North Carolina (Figure 12.11) (Odum et al. 1985). In low-nutrient ponds, xanthid crabs were abundant, as were snapping shrimp (*Alpheus heterochaelos*). Blue-green algae coated the reefs and fouling surfaces in the sewage fertilized ponds. Wood blocks and ceramic fouling plates were also suspended in the water. The attached flora and fauna on these substrates increased the diversity of the estuarine ecosystems which developed.

Ecological Engineering of Coral Reef Microcosms

Culturing of live coral reef fauna in recirculating sea water systems using filters and other technology was only partly successful until Adey (1983) realized the role of algae in nutrient regulation. He designed an algal subsystem called an "algal turf scrubber," which he patented for laboratory and aquarium display mesocosms. Adey (1983) described the operation of a small unit for five years under natural light. This greenhouse experiment led to a larger one that has been on display indoors under artificial light at the Smithsonian Institution since 1983. Over 300 species are present. More details are given in the book by Adey and Loveland (1991). A larger version was assembled for the Biosphere 2 (Chapter 19; Sagan 1990).

In Adey's "algal turf scrubber," water passing through and over the microcosm reef is circulated through a trough which receives a high level of light and contains a plastic mesh with attached filamentous algae. The water containing the wastes of the reef stimulates the growth of heavy beds of algae. These are self-organizing populations that utilize all the nutrient resources. The water flowing out of the scrubber accumulates in a V-shaped plastic tank supported only at its ends. When the tank fills, its

Table 14.1. Characteristics of Smithsonian Caribbean Reef Lagoon Microcosm*
See Figure 14.3.

Operation 1980–1990
System water volume 15.3 m^3
Reef surface area 5 m^2
Lagoon area 1.5 m^2
Refugium tanks 0.6 m^2
Algal turf scrubbers 7 m^2
Reef lighting
    Five multivapor and five mercury vapor metal halide lamps, 1,000 watts each
        Summer regime: 16 h on 8 h off
        Winter regime: 14 h on 10 h off
Scrubbers
    Lighting 16 400 watt metal halide lamps for 14 h
    Production 5 to 20 g dry organic matter per m^2 per day
Temperature 28.5 for summer; 25.0 for winter
Oxygen 7.5 ppm day; 5.0 ppm night
pH 8.3 in day; 8.0 at night
Salinity 35.5 to 36.5
Nitrate and nitrite nitrogen 1.4 to 14 ppb
Phosphate phosphorus 0.3 to 3 ppb

---

*Courtesy of W. Adey and M. Finn.

weight causes it to rotate and empty its contents of clean water onto the reef. This creates a wave and at the same time aerates the water. This mix of technology and self-organization is an example of ecological engineering, a new field of study that fits technology and nature together. Mesocosm characteristics are listed in Table 14.1.

Tropical Reef Microcosms Based on Symbiont Photosynthesis

Shown in Figure 14.3 is a microcosm operated by the Smithsonian (Williams and Adey 1983) which provides circulating water, waves, and a supply of plankton to supplement photosynthetic nutrition. There is a second tank that receives the output from the reef, simulating the quieter water in the lagoon behind coral reefs. A wide diversity of coral reef components is maintained. The main difficulty has been temperature changes in the building which cause mortality of the corals. Brawley and Adey (1981) found grazing amphipods able to keep filamentous and soft bodies algae from overgrowing the reef structure. With moderate crustacean populations, the effect was to increase the diversity of algae (Figure 14.4). The microcosms had more coral species diversity than is found in nature, as is appropriate for public display. Fish diversity was necessarily less in the confined volume of the microcosm, but was within the range observed in nature.

Hierarchical Control of Coral Reef Ecosystems

The interplay of macrophytic algae and corals occurs in nature as in the microcosms. Some of the relationships are shown in Figure 14.5. The algae can overgrow the corals, but not if the algal consumers, sea urchins, and other herbivores, are present. When nutrients are low, the corals, with their tight internal nutrient cycles, easily prevail. When nutrients are high, the algae act as scrubbers, keeping the system clear. McClanahan (1990) found that in over-fished reefs, sea urchin populations overgrazed, damaging reef substrates and reducing productivity. Contrast the concepts in Figure 14.1a and 14.1b. Autocatalytic loops of structural support and nutrient recycling cause parallel units (algae and corals) to be cooperative rather than competitive. If one isolates parallel species from these organizational controls, a mutualistic network becomes a competitive one (Figure 14.5).

Adey found the algal scrubber–coral microcosms helpful in maintaining a higher ratio of large consumer fishes and other animals to plants

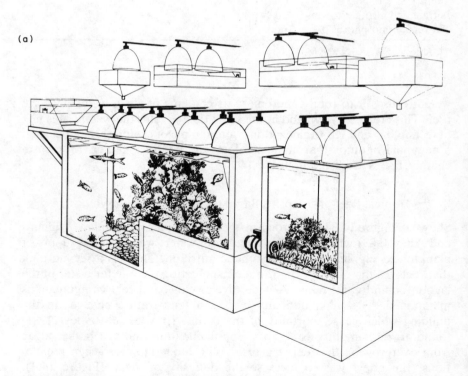

Figure 14.3. Smithsonian Tropical Recirculating Coral Reef Microcosm (Adey 1983; Williams and Adey 1983). (a) Structure; (b) comparison of diurnal oxygen curves in the microcosm with those on a reef in the Virgin Islands.

Comparisons between microcosm and St. Croix natural reef — yearly mean or range.

	Microcosm	St. Croix Reef
O_2 concentration	5.5–6.2 mg/l 7.5–8.3 mg/l	(min.) 5.2–6.6 (max.) 7.4–9.3
GPP	32.5 g O_2/m^2/day (13 g C/m^2/day)	20–80 g O_2 (back) 18–20 g O_2 (fore)
Respiration	0.9 g O_2/m^2/day	2.6 g O_2 (back) 0.7 g O_2 (fore)
pH	8.2–8.3	No data
Nutrients (mg-at/l) (typical lower values)	NO_2^- 0.3 NO_3^2 0.9 PO_4 0.2 Ammonia 0.2	$NO_2^- = 0.1$ $NO_3^2 = 0.0$ } outflowing $PO_4^2 = 0.09$ } water — —
Light. reef flat ($\mu E/m^2$/day)	35.3 10^6 (14 h)	39.6 10^6

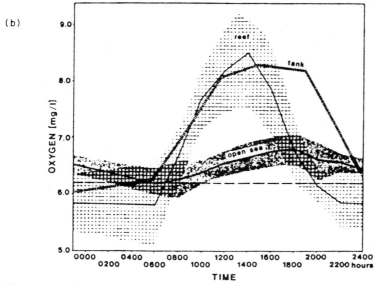

Figure 14.3. *Continued*

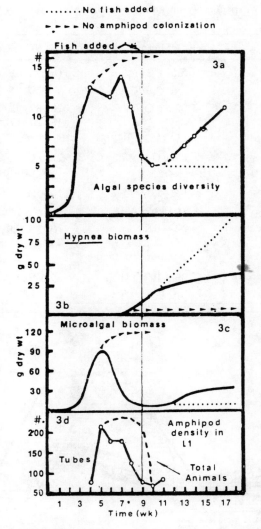

Figure 14.4. Effects of amphipods and fish on algal diversity in coral reef microcosms (Brawley and Adey 1981).

than could normally be supported in a microcosm. Supplemental feeding helps increase the number of fish to levels higher than in nature. This is desirable for public displays where people tend to be organismally interested. However, supplemental feeding results in net nutrient increase. Algal scrubbers are able to absorb the nutrients, converting the eutrophic tendencies into generally usable organic detritus which is directed into growth. In the Smithsonian's display microcosms, an organic-matter

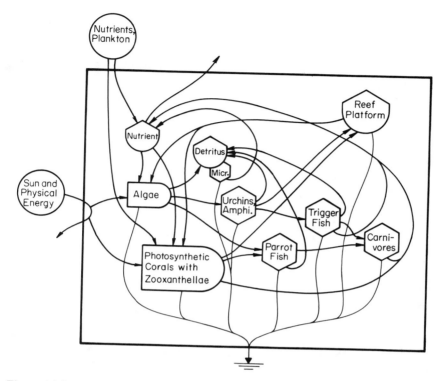

Figure 14.5. Aggregated overview model of coral reef ecosystem for interpretation of exclusion microcosms.

steady-state was maintained by cleaning out the algae in the scrubber every one or two weeks.

Corals and Bacteria

Moriarty et al. (1985) reported on some experiments with components of coral reefs combined in isolated chambers. They were able to study bacterial production in coral reef waters somewhat separate from bacterial consumption. Coral mucous and high concentrations of meiofauna increased bacterial populations. The low concentrations of bacteria normally observed are apparently the result of a broad diversity of effective consumers of bacteria and bacterial detritus.

Rocky Shore Mesocosms

Where rocky shores and bottoms receive strong wave and current energy, attached plants and animals form rich ecosystems, some intertidal. Bakke (1990) reviews experimental mesocosms set up to represent hard bottom, attached ecosystems. Notini et al. (1977) with seeding from the Baltic Sea studied beds of the brown alga *Fucus vesiculosus* grown on rocks with propeller driven current at Karlskrona, Sweden. Diurnal variations in oxygen were observed. The species composition of the transplanted community did not change.

With four experimental mesocosms at Solbergstrand, Norway, Bakke (1986), Gray (1987) and Pedersen (1987) used wave generators to study the impact of oil on an attached community of *Fucus* and *Laminaria*, 1 meter deep, bathed with running sea water. There was a gradual increase in biomass and diversity over a 3 year period. Mussels were followed by starfish and crabs. Oil caused mussel mortality and interfered with predation.

Benthic Sediment Microcosms

Turtle Grass–Coral Microcosms

Turtle grass and corals grow together in many patch reefs in the Caribbean. Williams and Adey (1983) studied the survival and growth of turtle grass seedlings collected in the Bahamas and placed in the "lagoon" tank. Survival was 25% after eight month, but growth of surviving blades was 2 cm per month, values as high as those observed in grass flats in the Bahamas.

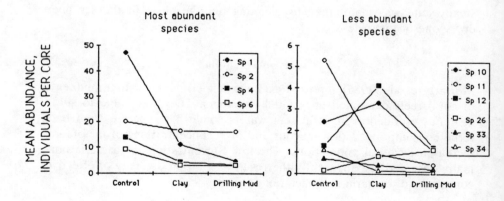

Figure 14.6. Changes in dominant species observed after sea grass cores were transferred to flowing microcosms for treatments with drilling mud and clay (Kelly et al. 1987).

Oil-drilling Mud and Turtle Grass Benthic Core Microcosms

Price et al. (1986), Morton et al. (1986), and Kelly et al. (1987) used sea grass microcosms to study the effects of oil-drilling muds (high density mud containing barium sulfate) and suspended clays. Cores from the shallow Gulf of Mexico grass flats were transplanted to start flowing benthic microcosms which were supplied with Gulf water pumped through screens. After 10-day and 6-week tests, many more species were lost from the drilling mud microcosms than from controls. Controls still retained most of the species found in the cores initially. Decomposition rates of turtle grass blades with drilling mud was half those with clay (controls). The 10-day studies showed toxic effects before much reorganization (succession) occurred in the microcosms. However, the data (Figure 14.6) showed population substitutions beginning as part of ecosystem adaptation. After 6 weeks, new population dominants were maintaining normal functions. It may be argued that an appropriate test of a new substance in the environment is not the short-term change immediately after exposure, but the longer term response after adaptation and reorganization.

Sediment Control of Water Chemistry in Microcosms

Microcosms that generate ecosystems like those of lake and oceanic sea bottoms develop in containers with sediment, especially if provided with physical stirring energy. The bottom sediment organisms of the tall, cylindrical MERL tanks in Rhode Island have been extensively studied. Hunt (1983) put a smaller chamber over the bottom sediments within the microcosm to study the flux of benthic manganese emerging from the bottom. He found it was proportional to the productivity of the main microcosm in the previous weeks. These studies reveal the coupling of top and bottom subsystems (see Chapter 15). Similarly, Kelly (1984) found that isolated sediment core microcosms lost nitrogen, and did not maintain their profiles of pore water ammonia when there was no influx of organic matter.

Kimmel and Lind (1970) found that phosphate phosphorus was removed from waters in microcosms with Texas lake sediments, primarily by inorganic absorption not involving iron. This may occur through exchange with calcareous particles. Federle et al. (1982), studying sediment microcosms fed with plant litter and Alaskan lake waters, found changes in bacterial species to be dependent on physical factors. Electron micrographs showed greater numbers of filamentous bacteria when conditions included stirring and darkness, although levels of ATP were little changed. Higher temperature increased decomposition rates. Baker et al. (1985) observed that sulfuric acids added to lake bottom microcosms were neutralized through sulfate reduction and cation exchange with calcium and magnesium.

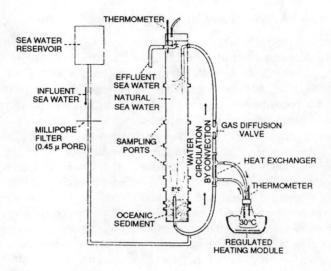

(a)

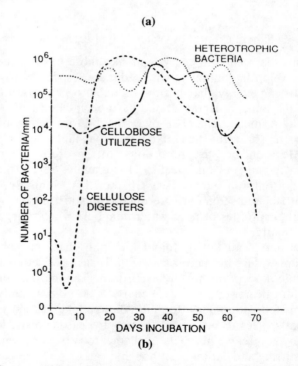

(b)

Figure 14.7. Decomposition observed in an ocean floor microcosm (Liston et al. 1963). (a) Apparatus; (b) bacterial populations after introduction of strips of cellulose.

Decomposition in an Ocean Floor Microbial Microcosm

The study of cellulose decomposition in an ocean floor microcosm (Figure 14.7) by Liston et al. (1963) showed growth of large populations of cellulose-adapted bacteria in 10 days, even though the temperature was 2°C.

Floating Field Microcosms for Benthic Communities

Thorp and Cothran (1982) devised a raft of floating tanks (Figure 14.8) so that a set of benthic sediment microcosms could be exposed to normal interactions with sun, temperature, aerial borne eggs, microbes, and especially dragon flies, but with nekton excluded. There was continual recruitment of aquatic insects and the microcosm populations were very similar to those obtained in transect Ekman grab-sampling of the lake

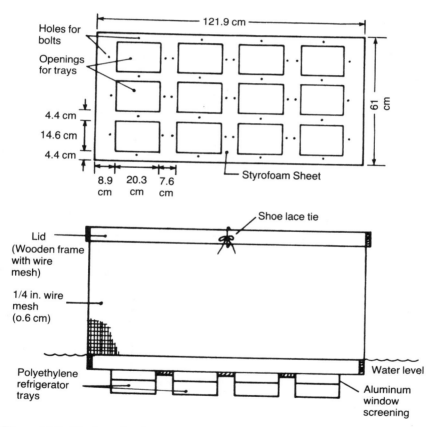

Figure 14.8. Platform for floating, open to air, benthic sediment microcosms used by Thorp and Cothran (1982) for studying the role of bottom insects.

310 14: Reefs and Benthic Microcosms

bottom. Fewer species of chironomids were in the microcosms. Amphipods were more abundant in the microcosms, and oligochaetes more abundant in the lake bottom.

Estuarine Bottom Communities Microcosm with Salinity Gradient

Copeland et al. (1972) studied a set of 5 microcosms containing sediment and bottom animals, including clams, through which an exchange of fresh and salt waters was passed to represent a salinity gradient in tidal exchange (Figure 14.9). Diurnal measurements of oxygen were used to estimate production and consumption. As in the reference estuary, consumption exceeding production was supported by organic matter from the freshwater input.

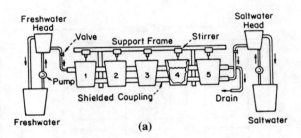

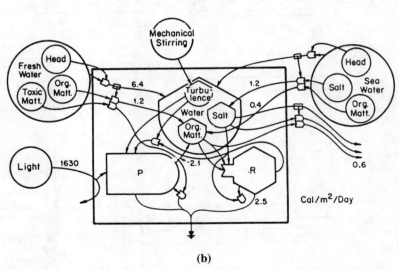

Figure 14.9. Estuarine microecosystems (Copeland et al. 1972). (a) Apparatus; (b) evaluated model of one cell.

Tubificid Control of Benthic Pore Space pH in Microcosms

Fisher and Matisoff (1981) studied the vertical changes in pH through the bottom sediment of microcosms with and without sewage and tubificids, and compared these results with lake bottom cores. Without tubificids, pH was lower. With added sewage sludge, pH dropped as low as the pH in a zone 4 cm below the sediment surface. With tubificids, the usual pH was around 7.

Capitellid Tube Control of Microbes in Microcosms

Alongi (1985) discovered that microbial production and meiofaunal abundances in marine sediment microcosms were influenced by the presence of the small polychaete tube worms *Capitella capitata*. Protozoa grazing the tubes stimulated microbial production.

Anaerobic Soil and Sediment Microcosms

If a microcosm is filled with sediment and the only available oxygen is in pore spaces between particles (10 to 40% by volume), the gas reservoir is small and turnover time rapid. Where the level of organic matter is high and respiration exceeds photosynthesis, oxygen is easily depleted and carbon dioxide concentrations are typically large. If the pore spaces are water-filled, the only oxygen is that dissolved in the pore-space water, and it is used up quickly. Wet soil microcosms without much air space and microcosms of aquatic sediments are typically anaerobic. Such microcosms may represent the conditions usually found in lake and marine sediment bottoms. If the solids have physical properties of weathered rock and windblown particles, such microcosms are like wet soil; if the solids have properties of water-transported particles, they are better named sediment microcosms.

Ramm and Bella (1974), after studying microcosms seeded with tidal flat components, provided a network diagram of their view of the microbiology and sulfur cycle of a tidal flat sediment surface ecosystem (Figure 14.10). The diagram was organized with components and processes in vertical position, according to water, aerobic, and anaerobic zones. Figure 14.11 is a minimodel in energy systems language of these same processes in an anaerobic microcosm with an air interface. This diagram puts items in their energy hierarchical position with small, fast items on the left and slower, larger, controlling processes on the right. Note the autocatalytic reinforcement pathways (pathways from the right circling back to reinforce processes on the left).

312   14: Reefs and Benthic Microcosms

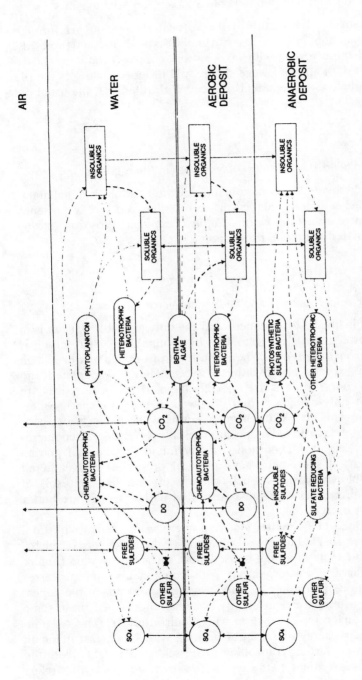

Figure 14.10. Conceptual model of a sediment–water interface from Ramm and Bella (1974).

Benthic Sediment Microcosms 313

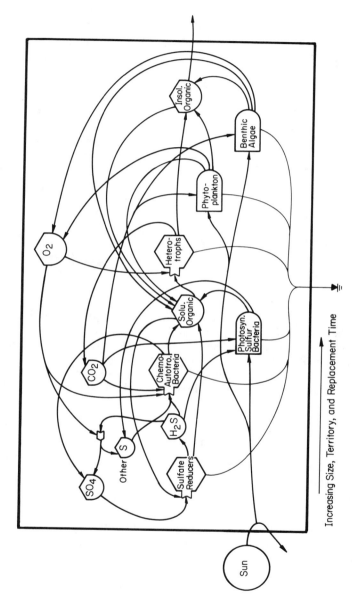

Figure 14.11. Model of the main features of an anaerobic–aerobic interface microcosm, with components from left to right in order of size, turnover time, and hierarchical position.

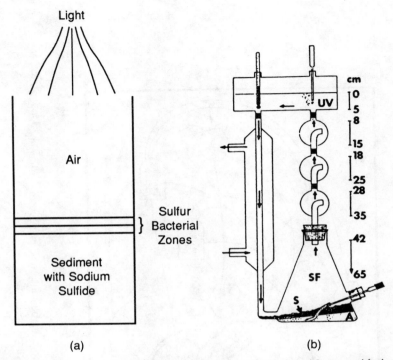

Figure 14.12. Anaerobic microcosm for sulfur bacteria with an oxidation–reduction gradient maintained by sedimentary sulfide. (a) Original concept after Winogradsky (1949); (b) complex apparatus (Hirsch 1988. Reprinted with permission from Handbook of Lab Model Systems for Microbial Ecosystems, Vol. II, copyright CRC Press, Inc.) with water drawn upward by suction, developing anaerobic sediment below and with oxidized globes above, representing plankton levels of a pond.

Winogradsky Columns

Important to the history of microbiology are the microcosms of Winogradsky (1949) in which anaerobic sedimentary conditions were created in microcosm by adding a reservoir of anaerobic "fuel," sodium sulfide. The interface between anaerobic sediment and oxygen-containing air produced the ecosystem conditions appropriate for sulfur-utilizing bacteria. These include the white sulfur bacteria which oxidize sulfides as a source of energy. See Figure 14.12.

Beach Sand Microcosms and Grain Size

Hockin (1982), with the apparatus shown in Figure 14.13, studied the interstitial fauna in microcosms with different sized soda-glass beads, and thus different pore space dimensions. Bead diameters were 0.147 mm,

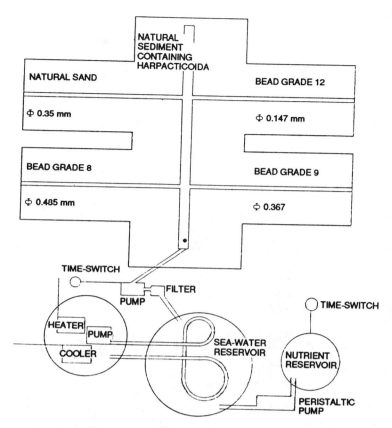

Figure 14.13. Apparatus for study of glass bead "sand" microcosms (Hockin 1982).

0.267 mm, 0.367 mm, and 0.485 mm. Microcosms were provided with fluorescent lights and a bacterial nutrient supply mixture of bacto-casamino acid, yeast extract, and protease-peptone and agar (16:2.1:1). EDTA was added to chelate potentially toxic substances from the vessels. Species from the environment were seeded into the central chamber from which the species could move out into microcosms of different grain size. Different species developed in different grain-size realms.

Recirculating Sea Water Through a Sand Bed

Plessis (1956) described a marine aquarium system in which the water was circulated up through the sand sediment bottom, assisting in the development of a diversity of organisms which helped maintain appropriate conditions for larger animals in the open waters above (Figure 14.14).

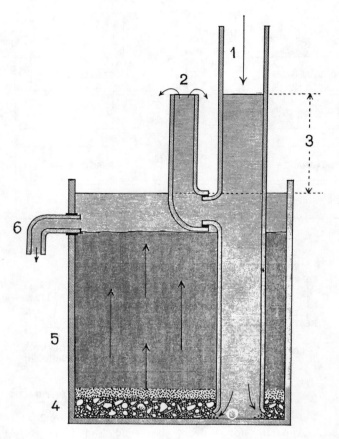

Figure 14.14. Recirculating sea water system with updraft filtration ecosystem in a sandbed (Plessis 1961).

Twenty-eight species of larger invertebrates were kept for a year or more. See also Figure 18.14.

Microcosms for Predicting the Role of Sediment in New Lakes

Craft (1983a, 1983b), in an interesting attempt to use microcosms to predict the limnological consequences of impoundment, studied the sediment–water interactions in microcosms derived from Provo River water and river basin soils. The area is the site of the then-proposed Jordanelle Reservoir, near Heber City, Utah. One of the soil sources (Site 4) was an abandoned acid mine-tailings pond and another (Site 1) was an area near the future dam. Representative soil samples of 40 g were place in 500 ml Erlenmeyer flasks and shaken with 500 ml of river water. One set of

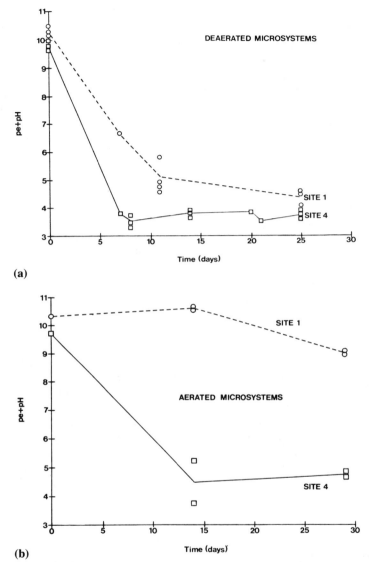

Figure 14.15. Redox versus time for soil-water microcosms. Site 1 is normal soil. Site 4 is soil from a mine tailing-pond. (a) Anaerobic system; (b) aerobic system (after Craft 1983b).

microcosms was to be analyzed under aerobic conditions. Other flasks were degassed and sealed. Both types were held for a 27-day incubation period at 10°C. Microcosms were sacrificed and analyzed every 2 days throughout the incubation period. Figure 14.15 shows a plot of redox (pE

+ pH) over time for the aerated and deaerated microcosms. The anaerobic systems showed an immediate drop in redox for both soil types, while only the mine-tailing microcosms showed a drop in redox in the aerated systems. The systems were analyzed for the presence of various forms of nitrogen and phosphorus as well as cyanide and a number of trace metals.

As a result of these studies, the author predicted that the reservoir, when built, would probably exhibit high initial productivity with large nitrogen and phosphorus releases from the sediments. In the deeper portions of the river bed, some problems with high iron, manganese, and zinc concentrations could be expected. The tailings-pond area would represent the most serious water quality problem for the reservoir, with 6 heavy metal concentrations exceeding Utah State aquatic wildlife standards.

Effects of Macroinvertebrates on Heavy Metal Partitioning

Krantzbert and Stokes (1985), went one step further than Craft. They added chironomids and chaoborids to sieved lake sediment and water microcosms from three different Canadian lakes (Chub, Lohi, and Port Credit). Two, Chub and Lohi, were low pH (5.0 to 5.8) lakes and Port Credit was alkaline (8.1 to 8.3). The control and experimental systems, 8 liters in volume, were held in the dark at temperatures encountered at lake bottom. One set of microcosms (from Lohi), was spiked with additional copper in order to simulate highly polluted environments. The proportion of cation exchangeable and specifically adsorbed copper observed in the Chub Lake microcosms, colonized by chironomids and chaoborids, was greater than that for uncolonized sediment. The same relationship held for copper in the Lohi microcosms supporting a similar community, and for copper and zinc in the Port Credit sediment which was inhabited by tubificids. The authors concluded that macroinvertebrate communities have the potential to increase metal concentrations in the water column, especially during periods of high burrowing activity.

Effects of Macroinvertebrates on Pollutant Transport

Studying another aspect of macroinvertebrates and pollutants in sediments, Karickhoff and Morris (1985) evaluated the role of tubificid oligochaetes on the transport of hexachlorobenzene, pentachlorobenzene, and trifluralin. Using sediments and tubificids from a north Georgia stream, they placed approximately 1 kg of sediment, which had been previously equilibrated with test chemical, in 4 1-liter glass bottles. Distilled water and 100–5,000 worms were added. Worms were omitted from the controls. Air was released through fritted glass to aerate the microcosms and to carry the released organics to a tenax resin column for collection and measurement (Figure 14.16).

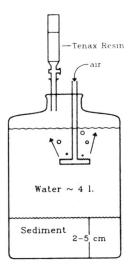

Figure 14.16. Microcosm designed to study the effects of tubificid worms on the release of organic pollutants from benthic sediments (Karickhoff and Morris 1985. Reprinted with permission from Environ. Sci. Technol. 19, copyright 1985, American Chemical Society).

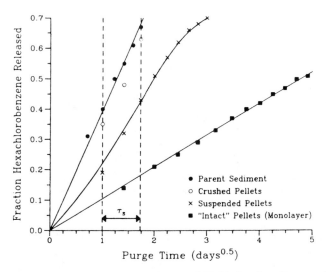

Figure 14.17. Release of chemicals from tubificid fecal pellets under various conditions (Karickhoff and Morris 1985. Reprinted with permission from Environ. Sci. Technol. 19, copyright 1985, American Chemical Society).

More than 90% of the organic chemicals contained in the sediments were transported to the sediment surface during a 30- to 50-day period. However, pollutant release into the water column was not comparably enhanced. The presence of the worms caused only a four- to six-fold increase. This may be explained by the fact that pollutant release from intact fecal pellets was highly retarded by sorption. Less than 20% of the pollutant contained in the pellet was released while the pellet was on the surface. The parent sediment released more chemical than the crushed pellets which, in turn, released more than suspended pellets or a monolayer of pellets (Figure 14.17).

Microbial Degradation of Xenobiotics

Although macroinvertebrates play a role in the fate of organic pollutants in aquatic sediments, the major controlling factor is microbial metabolism. This fact has been amply illustrated by a series of papers from the

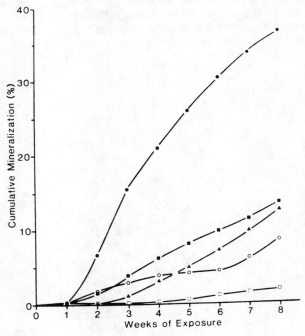

Figure 14.18. Mineralization of BPDP in sediment–water microcosms from five diverse ecosystems with varying histories of exposure to xenobiotics. Solid circles, Late Chicot; solid squares, Little Dixie Reservoir; solid triangles, Redfish Bay; open circles, Arkansas River; open squares, DeGray Reservoir (Heitkamp and Cerniglia 1986. Toxicity Assessment, copyright 1986, reprinted with permission of John Wiley & Sons, Inc.).

National Center for Toxicological Research (Huckins and Petty 1984; Heitkamp et al. 1984; Johnson et al. 1984; Heitkamp et al. 1986, 1987; Heitkamp and Cerniglia 1986, 1987).

Typical of this series is the 1986 paper by Heitkamp and Cerniglia, who constructed microcosms containing sediment and water from lacustrine, riverine, and estuarine ecosystems in the apparatus pictured in Figure 9.11. The history of previous exposure to chemical contaminants of these ecosystems varied from extreme (Lake Chicot) to pristine (Degray Reservoir) through intermediate (Little Dixie Reservoir, Redfish Bay, Arkansas River). The microcosms were spiked with radiocarbon-labeled t-butylphenyl diphenyl phosphate (BPDP). Mineralization was followed by measuring of the labeled carbon residues released by the microcosms. Over one-third of the BPDP was mineralized in 8 weeks in microcosms containing components from a eutrophic ecosystem which had a high organic carbon content in the sediment and a history of chronic exposure to xenobiotics (Lake Chicot). In contrast, only 1.7% of the BPDP was mineralized after identical incubation in sediment–water microcosms collected from a noncontaminated mesotrophic ecosystem with a low level of organic carbon in the sediment (DeGray Reservoir) (Figure 14.18). These results illustrate the ecological principle that previous exposure to xenobiotics frequently induces the enzymatic ability of the decomposers to handle these substances and to utilize them as energy sources.

Chapter 15
Plankton Columns

Taller microcosms are relevant to plankton ecosystems. Many are large enough to be called mesocosms. The pond mesocosms covered in Chapter 13 are characteristically broad. The systems considered in this chapter are proportionately deeper, often constructed so as to have dominant plankton zones, and represent deeper waters. The ratio of benthic surface to lateral surface is 1,000 for a pond, 0.25 for a cubic enclosure, and 0.05 for a tall cylinder.

Because mesocosms are larger, more of the larger scale processes of nature may be included, such as populations of larger animals, larger fluid eddies, and smaller ratios of container surface to volume. Tall cylindrical mesocosms operated to simulate plankton communities differ from nature by having a large bacterial–microzoal wall surface. The surface processes of self-organization can be used to improve total metabolism. Figures 15.1–15.5 depict the many types of containers referred to in this chapter, including experimental bags, tanks, and cylinders (Figures 15.1–15.5). Rectangular, prism-shaped plastic enclosures have been called limnocorrals (Figure 15.6; Kaushik et al. 1986).

To get closer to natural plankton conditions, to reduce the surface to volume ratio, to permit larger species to be included, and to reduce the stress due to sampling, large microcosms, called mesocosms have been popular, but are expensive. Mesocosms have volume-to-surface ratios of 2 to $4\,m^3/m^2$. Without the side walls, the surface-to-volume ratio for the same depth may be 0.1 or less.

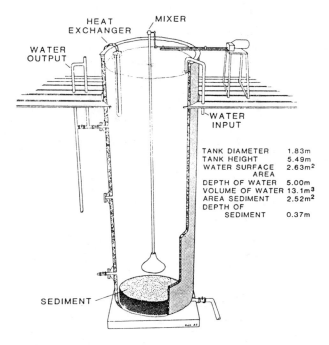

Figure 15.1. Cylindrical Microcosm Estuarine Research Laboratory (MERL) Microcosm, Narragansett, Rhode Island (Pilson et al. 1979).

Time Scales and Sizes

One method of using enclosures is to suddenly contain the typical components of an ecosystem. This is done to observe the immediate effects before much self-organization can take place. Studies where nature is enclosed for only a few hours for purposes of measuring processes as they occur are not really employing microcosms. Examples of such enclosures are the light and dark bottle measurements.

Starting just after enclosure, there is a long period of transition as self-organization responds to the new set of conditions. There are many studies of larger containers where measurements were made for periods of a few weeks to a few months. No matter the original purpose of these containers, these data are valuable records of the self-organization process. Enclosed plankton columns (Figure 3.8) in lakes in one year generate ecosystems that are very different from those of the lake in which they are suspended (Lund 1978). The time required for transition depends on the size of the system (and thus the size of the larger components). Mesocosms, being larger, require two or three years for self-organization to form a new ecosystem. Once formed, this system can be reseeded into

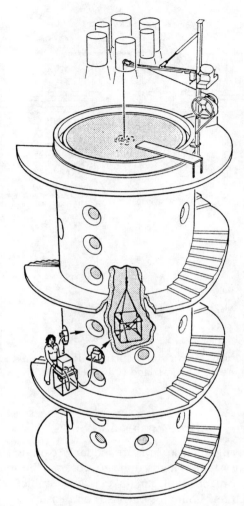

Figure 15.2. Aquatron marine mesocosm, Halifax, Nova Scotia (Price 1989).

a another microcosm and produce the same conditions in less time. Small microcosms, because of their size, can maintain only small components, and thus all their processes have fast turnover times. Processes of self-organization, transition, and adaptation are rapid, and some kind of steady state may be reached quickly. With mesocosms, however, large dimensions mean larger species and larger controlling turnover times. Longer times are required for self-organization and adaptation to new conditions. Figure 15.1 shows an example of a cylindrical mesocosm which was studied for three years and developed a repeating pattern, indicating that some quasi-steady state had been achieved (Figure 15.8).

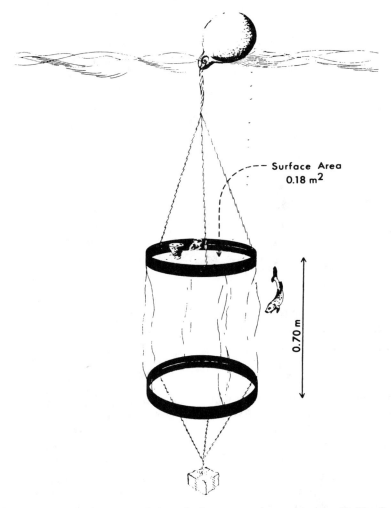

Figure 15.3. Plastic bag used for plankton experiments in the Baltic Sea by McKellar and Hobro (1976).

Although a mesocosm may be in a self-organizing transition, experiments can be conducted, comparing controls with treatments. Kaushik et al. (1986) utilized three sizes of prismatic plastic enclosures in Lake George, Ontario (Figure 15.6) in their study of the effects of the pesticide methoxychlor. For these effects see Chapter 17. In the controls, the dominant cladocera and rotifers decreased by an order of magnitude. Calanoid copepods, after initial decline, increased again, and cyclopoids increased.

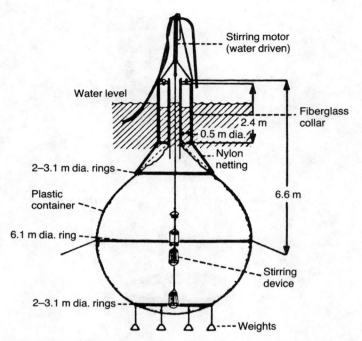

Figure 15.4. Submerged spherical container, 6 m in diameter, used by Strickland and Terhune (1961) in Departure Bay, B.C. Canada.

Bags Suspended in Waters

Suspended bags have often been employed for the development of plankton-type ecosystems. By suspending the enclosure in the lake or ocean, use is made of natural fields of light, temperature, and some of the water's turbulence. One of the first enclosures was the bag used by Strickland and Terhune (1961), shown in Figure 15.4. Goldman (1962) found temperature stratification in a 10-m long plastic bag (held open by aluminum hoops) in a lake to be close to that found in normal surroundings, but the thermocline in the bag was somewhat shorter. A large cylinder used in lakes is shown in Figure 3.10 (Lund 1975). Being vulnerable to storms, the suspended plastic containers have not often been operated for periods longer than several months. McKellar and Hobro (1976) suspended 100-liter bags (Figure 15.3), in which they varied proportions of zooplankton, in the Baltic Sea. Their data showed that zooplankton increased algal production through better recycling. One is reminded of the New Zealand sheep husbandry adage: "You have to eat grass to grow grass."

Early microcosm studies at the Institute of Marine Sciences (1957–1967), of The University of Texas at Port Aransas included plastic and fiberglass field enclosures of the marine environment. With this background, Pat Parker, on leave as program manager of the Office for the

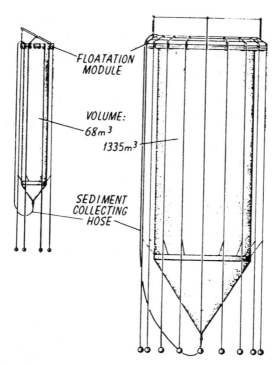

Figure 15.5. Plastic cylinder suspended in water in CEPEX studies (Grice and Reeve 1982).

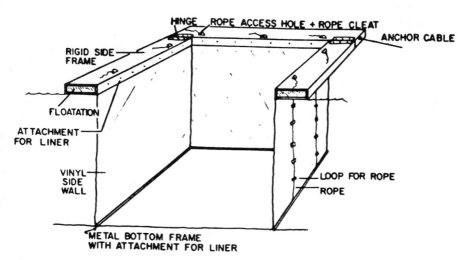

Figure 15.6. Floating plastic tank (limnocorral) used by Kaushik, Solomon, Stephenson, and Day (1986, copyright ASTM, reprinted with permission).

International Decade of Ocean exploration, National Science Foundation, in 1972 coordinated an International effort to study marine enclosures. The CEPEX (Controlled ecosystem pollution experiments) mesocosm studies consisted of plankton waters in large plastic cylinders suspended in the deep sounds of coastal Washington State (Figure 15.5). With runs of three months or less, there were extensive examinations of a system during early succession (self-organization). Many chemical studies were conducted. The investigation of processes within the self-organizing ecosystem was very appropriate in that it could provide a basis for the study of many pollution situations where the system is reorganizing due to the new waste inflow. Data on short term transitions that occur after enclosure are included in Figure 3.11.

Long, Rigid Cylinders

Illustrating the idea of a plankton dominated microcosm is a small, plankton column microcosm studied by Patricia R. Roberts in Odum's limnology course at Duke in 1955. Her report describes a 1.1-m glass tube, 10 cm in diameter, seeded with lake water, plankton, and lake peat on the bottom (Figure 1.1). The upper half was lighted and heated by a floodlight. Thermal stratification developed in the upper 50 cm with epilimnion at 27°C and hypolimnion at 20°C. After two weeks, the epilimnion was green with *Gleocystis* algae on the glass surface and 252,000 cells per ml in the water. The lower half was wrapped with cheesecloth. Oxygen, depleted due to the bottom organic matter, was recorded as 1 ppm in the epilimnion and zero in the bottom water. Green sulfur bacteria developed at the thermocline. *Daphnia* populations remained in the middle and lower zones.

Margalef (1963) studied turbulence and the rate of descent of solid particles in 2-m vertical tubes. They had temperature–salinity controlled pycnoclines, like those in nature. The complexity of the thermocline increased with time (Figure 15.7). Turbulence and sinking rates were highest at the top. The million dollar MERL tanks discussed below are the full flowering of the controlled cylinder method.

Tall Estuarine Cylinders of the MERL Laboratory in Rhode Island

The MERL mesocosms are large vertical tanks which have been operated over longer times, allowing more mature ecosystems to develop. Figure 15.1 shows a saltwater MERL microcosm used to study the conditions of deep water estuaries (Pilson et al. 1979; Nixon et al. 1980; Oviatt et al. 1986a). Initially, nine replications were seeded and maintained so that they would resemble the waters of Naragansett Bay, Rhode Island. Waters from the bay were exchanged with the mesocosms on a turnover

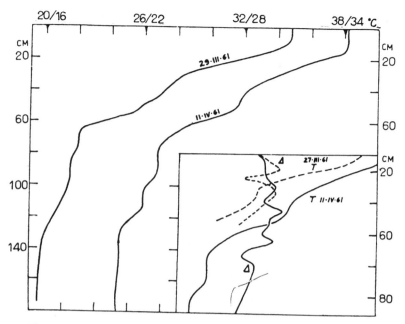

Figure 15.7. Thermoclines developing in microcosm 2 m high. Inset shows thermoclines modified by artificial turbulence (Margalef 1963).

schedule of 27 days. The salinity range was 27–31 ppt and temperature ranged from 0 to 23°C. Plungers introduced physical eddies in the waters. Fiberglass walls were white and reflective, and were brushed twice a week in summer to reduce attached organisms. This resulted in systems more like the typical bay column of plankton and bottom organisms. The rise and fall of oxygen and pH, as in the model in Figure 2.14, was used to calculate metabolic rates. From 1976 to 1988 there were several experimental management regimes. With runs of several years' duration, self-organization produced ecosystems with recurring patterns. See Table 3.1 for a comparison of inputs.

Control Tanks

In many ways, the ecosystems developing in control tanks resembled the bay. See the metabolism graphs in Figure 15.8 (Oviatt et al. 1986). Oviatt et al. (1981a) found daytime net production (276 g C/m^2) and respiration (163 g C/m^2), with some net production for the year (113 g C/m^2), values much like those calculated for Naragansett Bay. The nutrient patterns investigated by Pilson et al. (1980) were found to be similar (Figure 15.9). Small oscillations of plankton were observed by Oviatt et al. (1989), mostly in the summer (Figure 15.10). Donaghay (1985) observed the dom-

330   15: Plankton Columns

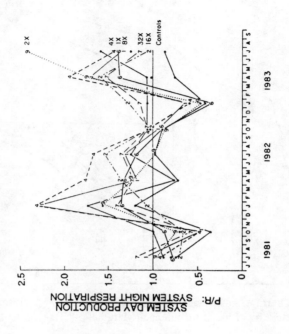

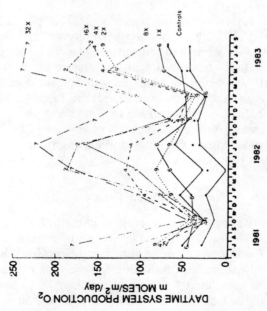

Control Tanks

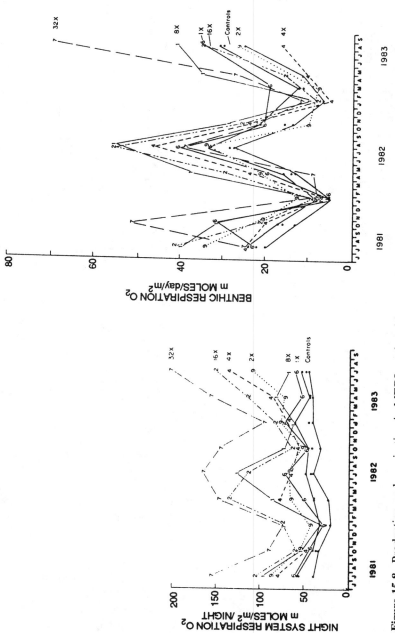

Figure 15.8. Production and respiration in MERL tanks with eutrophication inputs (Oviatt et al. 1986a). 1×, 2×, 4× etc. indicates enrichment factor.

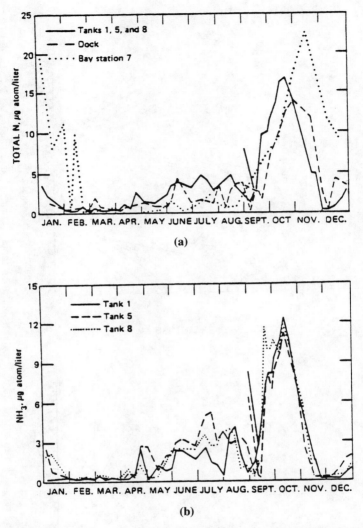

Figure 15.9. Comparison of nutrients in MERL tanks and those in the bay (Pilson et al. 1980). (a) Total nitrogen; (b) ammonia; (c) phosphate; and (d) silica.

inant copepod *Acartia*, finding that the previous feeding history affected the ability of the population to use high-quality food when available. In other words there was a time lag which tended to cause oscillations. Grassle et al. (1986) found similar benthic diversities and chlorophyll in the tanks as compared to the bay (Figure 15.11). Low organic concentrations due to high consumption rates in August caused benthic mortalities, but rapid recovery. Eutrophication produced some population increases.

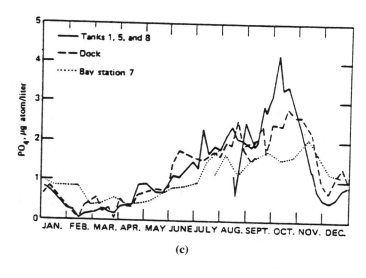

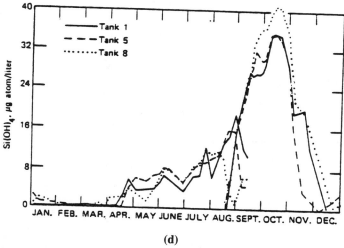

Figure 15.9. *Continued*

Stable isotope ratios of carbon were similar to those in the bay, with lower values for diatoms and diatom consumers and higher values (smaller negative anomaly) in nekton (Gearing et al. 1984).

Effects of Varying Physical Turbulence

Turbulence caused by the plungers in base runs contributed similar amounts of mechanical energy on an area basis as found in the bay (Nixon et al. 1980). Oviatt (1981) studied the effects of varying the plunger energy

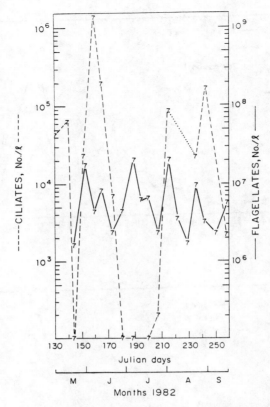

Figure 15.10. Oscillations in plankton populations in MERL tanks (Oviatt et al. 1989).

(Figure 15.12). Without mixing, self-organization produced stratification of chlorophyll and nutrients, with a large biomass of flagellates and rotifers adapted to low physical energy conditions. With high levels of continuous mixing, large quantities of chlorophyll were maintained, with more nutrients kept in a bound state, but the numbers of the zooplankton *Acartia* were diminished. Turbulence had substituted for zooplankton as a recycling mechanism, and plants were maintained at higher stocks with less grazing. Questions of turbulence and scaling are discussed in Chapter 3.

Fuel Oil Experiment, 1977–1979

One of the first series of experiments in the MERL tanks was the chronic addition of fuel oil to 3 tanks (330–600 ppb, containing 0.8 to 11 ppb toxic hydrocarbons), a condition simulating a common problem in navigation estuaries. There were 3 controls. Gearing et al. (1980) reported

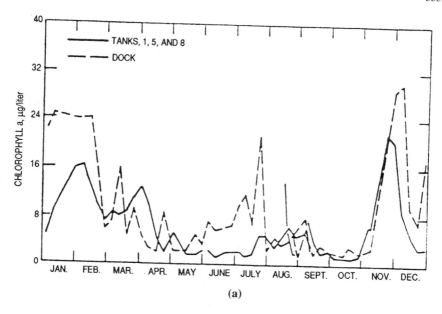

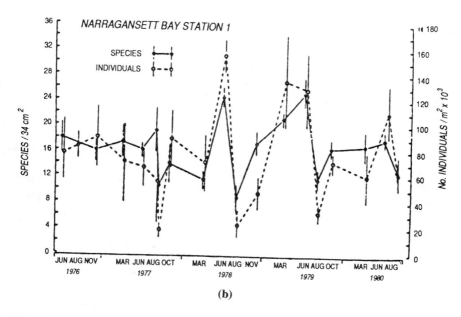

Figure 15.11. Ecosystem characteristics in MERL tanks (Grassle, Grassle, Brown-Leger, Petrecca, and Copley, 1986 Copyright 1986, John Wiley and Sons, Inc.). (a) Chlorophyll; (b) species diversity.

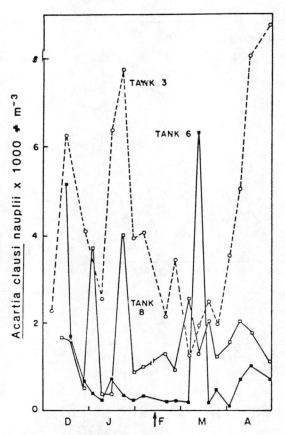

Figure 15.12. Plankton populations in MERL tanks with different levels of physical stirring (Oviatt 1981).

fractionation and sedimentation of half the saturated hydrocarbons, but only 20% of the aromatics and 10% of the aliphatics remained in sediments after a year. Effects were greatest in sediments, reducing the numbers of animals and respiration rates. Later, with oil additions stopped, recovery was accompanied by increased respiration and disappearance of oil and its effects within a year (Oviatt et al. 1982). See Figure 15.13. Frithsen et al. (1985), studying small benthic organisms, found fewer ostracods and harpactacoid copepods, but foraminifera and ciliate populations on the bottom increased. Caron and Seiburth (1981) examined the successional growth of peritrich protozoa on glass slides in the tanks. They normally reached full development in a month, but the oil adsorbing on these surfaces prevented much of the expected development.

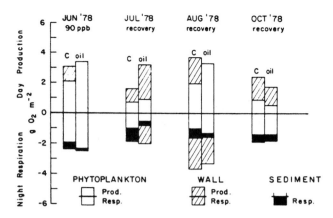

Figure 15.13. Metabolism in tanks receiving low levels of fuel oil (Oviatt et al. 1982).

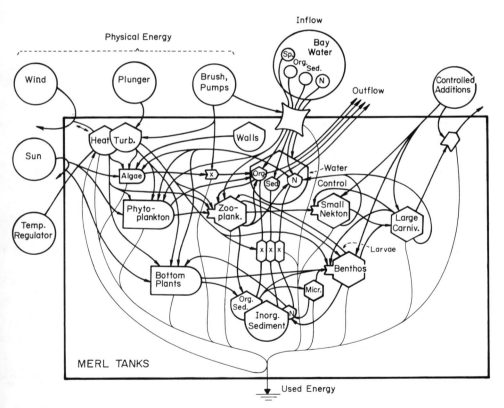

Figure 15.14. Systems diagram of the ecosystem of the MERL microcosm, emphasizing the main external controls and influences; sp, species; sed, inorganic sediments; org, organic detritus; n, one or more nutrients; micr, consumer microorganisms.

Simulated Storm

Oviatt et al. (1981b) simulated a storm by plunger stirring for 14 h in 3 tanks. The treatment resuspended 0.3 cm of sediment, sharply increasing levels of turbidity, nutrients, and metals in the waters. Gearing and Gearing (1983) reported little effect from the storm on the decomposition rate of the oil which was resuspended by the stirring.

Radioactive Tracer Studies of Chemical Cycles

Tracers for iron, manganese, chromium, mercury, selenium, zinc, cesium, cobalt, and sodium were tested in three MERL tanks from 1978-1981. The tanks provided a very sensitive way to follow metal processing by the food chain, the bacteria on the walls, and the benthic community (Amdurer 1982). Iron, chromium, and most other metals were removed by uptake onto particles with subsequent sedimentation in about two months. The rate was faster in summer. Particle transport was also studied by introducing plastic microspheres. The tracking of radioactive sodium revealed sea water exchange with 3 cm of the bottom sediment during the tanks' regime of settling and resuspension. Large fractions of cobalt and manganese were discovered on the walls. Cesium and manganese were rapidly transported to the sediments. Data curves for introduced tracer manganese became congruent with those for the carrier in about 30 days. This was the time required for the tracer to develop a similar specific activity in various phases of the system. Rudnick and Oviatt (1986) followed inorganic radioactive carbon through phytoplankton uptake. Fifteen percent was found in sediments after 6 months. Since no flux model was evaluated, turnover times could not be correctly estimated. Frithsen and Doering (1986) found that filter-feeding bottom-tube annelids were important in removing particulate matter containing radiotracers from the water in 2 days.

Processing of Volatile Organic Compounds

MERL tanks were used by Wakeham et al. (1983) to determine the fate of 18 toxic organic substances. Most of the chlorinated hydrocarbons and alkylbenzene were lost by volatile emission from the surface in a few days. Alkylbenzene and chlorobenzenes were partly metabolized, and aliphatic hydrocarbons were removed to particles and the sediments. One tank, poisoned with mercuric chloride, had much slower loss of organic compounds metabolized.

Eutrophication Series, 1981–1983

The studies in 1981–1983 had 3 control microcosms and 9 microcosms which were given higher levels of nutrients in order to simulate sewage eutrophication. Inorganic chemical nutrients were added in atomic ratios: 12.80 N, 1.00 P, 0.91 Si. As shown in Figure 15.8, the added nutrients increased the magnitudes of production and total respiration. Seasonally, the excess of respiration in early winter provided enough nutrients to stimulate an excess of production in late winter and spring, as a result of the steadily increasing light. Bottom respiration did not increase as much as photosynthetic production (Kelly et al. 1985). At low nutrient levels accompanied by stirring, bottom feeders received a physical energy supplement to their feeding and waste removals. Thus, they could dominate consumption and recycling. At higher organic matter concentrations, recycling by plankton components can prevail. The authors generalized about the predominance of plankton food chains in eutrophic circumstances — an occurrence long known in fish ponds. Our theory (Chapter 3) is that net nutrient influx causes arrested succession and an export of net organic production, as part of a longitudinal succession.

The normally dominant copepods, *Acartia*, were much diminished by the fertilization in which nitrogen was provided as ammonia. Sullivan and Ritacco (1985) confirmed *Acartia* toxicity with tests in- and outside the tanks at concentrations of 10–600 ppb. As reported by Berounsky and Nixon (1985), the MERL tanks which were fertilized with ammonia, developed very high rates of nitrification, apparently due to bacterial conversion of ammonia to nitrates and nitrites. This was a demonstration of self-organizational opportunism in which organization favors organisms capable of using a resource existing in excess, thereby reinforcing the main ecosystem production processes.

Predation

The action of larger consumers was simulated by removing 20–60% of the zooplankton with plankton nets two or three times a week (Oviatt et al. 1979). The rest of the system was little disturbed. Whereas removal simulates one aspect of higher animal activity, it does not provide the managerial actions of higher species. Part of the importance of larger animals is their control actions, often much more complex than simple removal of lower species stocks. Nektonic shrimps and crabs produced large populations in some tanks, thus providing control units with a longer time constant (Nixon 1986). *Mercenaria* clams were added to the tanks, causing an increase in the flux of nitrogen and silicate from the sediments and the consumption of more oxygen (Doering et al. 1987). Plant production was stimulated, not hurt by the increased grazing. As models show, timing of nutrient availability and blooms of fast-turnover phytoplankton may be controlled more by animal recycling than vice versa, because the animals have longer time constants. Clam recycling increased with increasing temperature (Doering et al. 1986).

Stratification Experiments

Donaghay and Klos (1985) operated two MERL tanks to produce a thermally stratified system in which a typical high nutrient hypolimnion develolped below a nutrient-bound epilimnion. After eutrophication, the increased prevalence of plankton and bacterial recycling over that of benthic and larger components was observed. The results were generalized by Nixon et al. (1980). As already suggested in Chapter 3, net nutrient influx tends to produce an arrested stage of net production, with lower diversity which passes its organic matter on to the next stage of longitudinal succession.

Opportunity for Simulation Synthesis

The MERL studies include a splendid series of measurement papers. Each of these reports on the behavior of a different component, and each speculates as to how the observed data resulted from the existing balances of plus and minus processes. Often, inferences were made about rates based on the balance between input and output. These deductions may be incorrect because accurate calculations require values determined by a model. Each paper should be integrated with the others and with a simulation model, so that quantitative backing for the guessing can be shown and each variable can be properly linked to each study. This sort of synthesis is long overdue. Given in Figure 15.14 is an aggregated view of the ecosystem described by the MERL studies.

Would it not be better if all these measurements were consolidated in one major systems understanding, with plausibility of interrelationships proven with simulation models? Something is wrong with our system of funding, promotion, and scientific reporting when it causes the publishing of little pieces of research in short papers in many journals, without systems synthesis.

Chapter 16
Thermal and Brine Microcosms

Hot springs, brines, anaerobic waters, temporary ponds, tide pools, frigid water, and other extreme conditions are recurring in the biosphere. They are accompanied by characteristic ecosystems composed of adapted organisms. Often occurring in small volumes as natural microcosms (Duke 1967; Ganning and Wulff 1970), these ecosystems can be readily duplicated in experimental microcosms.

General Principles

Physiological Adaptation and Low Diversity

In general, the energy of such systems goes into physiological adaptation of a few species rather than into a large diversity and division of labor among species. Nixon (1970) reported a species diversity of less than eight species per thousand in a hypersaline environment, while the New Mexico hot spring studied by Duke (1967) contained only one alga and two bacteria. Brock and Brock (1966) took quantitative cores of the blue-green algal mats occurring along a temperature gradient in a Yellowstone National Park hot spring. The cores were analyzed for DNA, protein, and chlorophyll (Figure 16.1). There was a sharp maximum at 56°C. Since all other factors in the spring environment remained essentially constant, these data demonstrate that the community was best adapted to a temperature generally considered above the optimum for most aquatic organisms.

To obtain insights about the management of the saline Pyramid Lake in Nevada, Galat and McConnell (1981), Galat and Robinson (1983), and Galat et al. (1988) studied the effects of increasing salinity from 3% salt

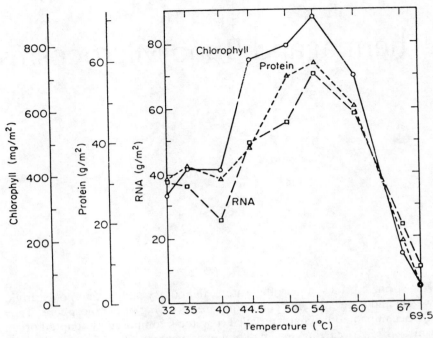

Figure 16.1. DNA, protein, and chlorophyll optima in an algal mat taken from a hot spring in Yellowstone National Park (Brock and Brock 1966. Reprinted by permission from Nature, Vol. 209, pp. 733–734, copyright 1966, Macmillan Magazines, Ltd.).

to 11% salt, using greenhouse mesocosms floored with plastic Astroturf. There was a substitution of zooplankton dominants and a decrease in the levels of benthic invertebrates important to fishes. Short-term bioassays showed less effect than the long-term mesocosms.

Taxonomic Similarities

There are relatively few species adapted to these extreme conditions, so the fine-tuning that takes place through species substitutions is not as prevalent. In surveying the species lists of aquatic organisms from extreme environments on a world-wide basis, one is struck with the similarity of genera, and sometimes even species, one encounters. This was noted by Tuxen (1944) in his classic work on Icelandic hot springs. Considering only the faunal portion of the biota, he states: "... the species occurring in the absolutely hot springs throughout the world originate from some few groups of animals, not from all the animal groups occurring in fresh water." Maintaining the extreme condition is all that is necessary to maintain the integrity of the adapted ecosystems.

Duke noted no changes in the populations over a period of 4 years, and no gross observable seasonal changes in the activities of the ecosystem in her 53–60°C hot spring.

Community Metabolism

Beyers (1965) (Figure 2.5) included four types of highly stressed systems in his comparative study of the diurnal metabolism of aquatic systems. It is interesting that two of the systems were algal mats, dominated by blue-green algae (*Oscillatoria*). They were the algal mat system from the Laguna Madre of Texas, and the system maintained at 51°C, originally derived from Mimbres Hot Springs in New Mexico. Although the temperature and salinity regimes were quite different, the well known ability of blue-greens to adapt to extreme conditions is evident in the fact that the dominant primary producer organisms belonged to the same genus. However, the diurnal metabolic pattern of the two mat systems differed. The "hot" system much more closely resembled the pattern of the brine system, which was planktonic in nature. The thermal and brine systems were more extremely stressed than the other two. The more highly stressed systems showed a reduced species diversity and an extreme "spike" in photosynthesis and respiration at dawn and dusk. One might speculate that the storage compartments in these systems were extremely small in comparison to more normal systems, and thus the reactants of the photosynthetic and respiratory equations were used up more quickly.

Although adapted to extreme conditions, algal mat systems often react metabolically in ways similar to other aquatic systems. Sollins (1970) presents diurnal oxygen rate of change curves taken from algal mat microcosms which resemble those of other aquatic microcosms and natural bodies of water (Figure 16.2).

Adaptation to Drying

The blue-green algal mat community, whether derived from hot springs or hypersaline environments, is remarkably tenacious to life. Two of the hypersaline algal mat microcosms used by Beyers (1965) were still in existence in 1993. When they were established, commercial soluble plant fertilizer was added to one but not the other. The biomass in the fertilized system was several times larger than the unfertilized microecosystem. During most of the 28 years since their inception, they were in a dry state. The appearance was a leathery mass with encrusted salt. Every two to five years, mainly as required for class demonstration purposes, distilled water has been added to restore the original salinity. Within 12 hours the fertilized community becomes bright green, with obvious photosynthetic activity taking place. The non-fertilized system requires a longer period to swing into operation. This serendipitous experiment illustrates that nu-

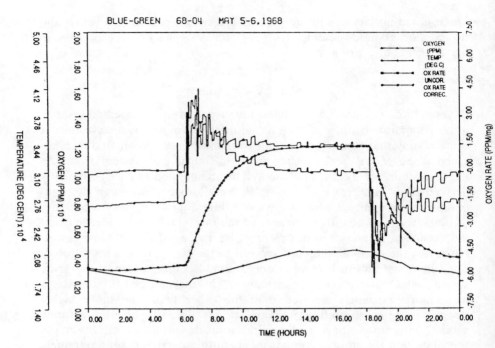

Figure 16.2. Diurnal curves of oxygen concentration in the water above two hypersaline blue-green algal mats. Also shown are rate of change of oxygen concentration and this rate corrected for aeration (Sollins 1970).

trients within a closed ecological system are never lost, but recycled. Sollins (1970) mentions that, as in nature, his algal mat microcosms often dried out completely during periods of inattention. The addition of distilled water to restore the previous salinity resulted in vigorous photosynthesis within hours, and no permanent changes in the mat due to desiccation were ever noted.

Unusual Biogeochemistry

Physiological adaptation to stressed environments is frequently exhibited as unusual biogeochemistry. Sulfate-reducing bacteria are a characteristic feature of algal mats. Paterek (1983) points out that methanogenesis has been reported from several aquatic ecosystems which were under stress from extremes of temperature, pressure, pH, or salinity. Such environments include hot springs, submarine hyperthermal vents, and hypersaline bodies of water. Nixon (1970) presents evidence for photorespiration in hypersaline environments and implicates carotenoid pigments in the process. Childres, Fisher, Favuzzi, Sanders, and Alayse (1991) studied the sulfide and carbon metabolism of hydrothermal vent tubeworm *Riftia pachyptila* Jones in flow-through pressure aquaria.

Examples of Stressed Environments

Iceland Hot Streams

The ability of blue-green algae to adapt to high temperatures was also documented by Sperling and Grunewald (1969). After isolating *Mastigocladus laminosus* from a hot spring in Iceland, they maintained it at 45°C in batch culture. Material from the batch culture was then transferred to a model stream at the same temperature (Figure 16.3). Two streams were used, one flowing at a fast and one at a slow rate. Biomass increase and phosphorus uptake by the algal mat was measured (Figure 16.4).

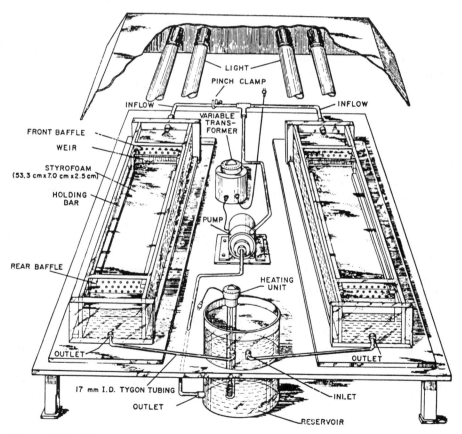

Figure 16.3. Diagrammatic sketch of the two channel laboratory stream model (Sperling and Grunewald 1969).

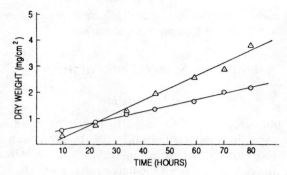

Figure 16.4. Effect of fast and slow currents on biomass increase and phosphorus uptake by *Mastigocladus laminosus* in a model stream at 45°C (Sperling and Grunewald 1969).

As has been demonstrated in aufwuchs communities under more normal temperature regimes, both the biomass increase and phosphorus uptake were enhanced by an increase in stream flow. Thus, the algae taken from a hot spring in nature were acting under elevated temperatures the same way algae from more normal environments respond at more usual temperatures.

Yellowstone Hot Springs

Oscillatoria and *Mastigocladus* also were the dominants found in Yellowstone National Park (Fraleigh 1971; Wiegert and Fraleigh 1972; Wiegert and Mitchell 1973; Fraleigh and Wiegert 1975). These workers constructed wooden-substrate outdoor microcosms, called "boards," which were fed by a flow of water from a hot spring at 40–58°C. The flow rates on the boards were 30 and 15 l/min. Each was 24.4 m in length and 1.2 m wide. After sampling at various intervals down the length of the boards, like Sperling and Grunewald, they found that algal mat biomass increased with flow (Figure 16.5).

Gross productivity and changes in the standing crop of both biomass and chlorophyll *a* were measured during ecological succession in the thermal blue-green algal community. The rate of gross production was directly related to chlorophyll *a* and independent of light intensity. There was a log–log relationship of chlorophyll *a* to biomass, and the respiration rate of the algal mat was directly proportional to biomass. These relationships were used to construct a nonlinear model, and algal growth during succession was predicted, based on algal density, free carbon dioxide concentration in the water, and day length. The model predictions agreed closely with the field data on measured biomass increase during succession (Figure 16.6).

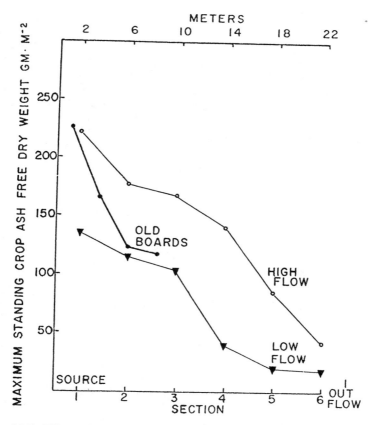

Figure 16.5. Effect of fast and slow currents on maximum standing crop of algal mat in a wooden substrate microcosm (Fraleigh 1971).

Mimbres Hot Springs

Duke (1967) studied one of the 25 hot springs located in Cold Spring Canyon near Mimbres, New Mexico. The mean temperature of the water was 56°C and the mean flow 10.5 l/min. A man-made basin, 56 cm deep and containing 550 liters of water, surrounded the spring and its contained ecosystem was the object of study. The biota consisted mainly of the blue-green alga, *Phormidium*, and two genera of bacteria. Oxygen was monitored with oxygen electrodes (Figure 16.7) and diurnal community metabolism calculated. Water entering the bottom of the spring basin was oxygen free, and despite photosynthesis and atmospheric diffusion, never reached the half saturation point. Dissolved oxygen varied between a nighttime minimum of 0.25 mg/l and a daytime maximum of 2.05 mg/l. The gross photosynthesis of the thermal spring was between 3.96 and

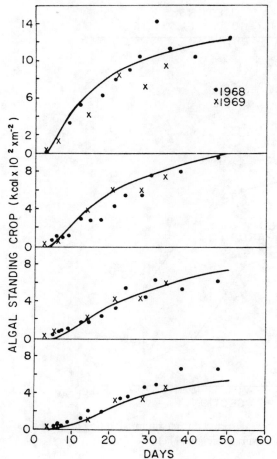

Figure 16.6. Increase in algal standing crop over succession in experimental boards in Yellowstone Hot Springs. Lines are from the predictive model. Crosses and circles represent points from different experiments (Fraleigh and Wiegert 1975).

7.56 g O_2/m^2 per day, and the community respiration was between 3.58 and 7.18 g O_2/m^2. The P/R ratio varied between 1.11 and 1.0. Once again, although existing in an extreme environment, this community produced data well within the range of data from other, more normal aquatic ecosystems.

Hypersaline Algal Mats

An algal mat usually consists of three major layers. These are a photosynthetic layer of blue-green algae; a heterotrophic region, often containing diatoms, flagellates, bacteria, nematodes, and annelids; and an anaerobic region of black sediments containing free hydrogen sulfide, sulfate reducing bacteria, and reduced organic matter. Armstrong and

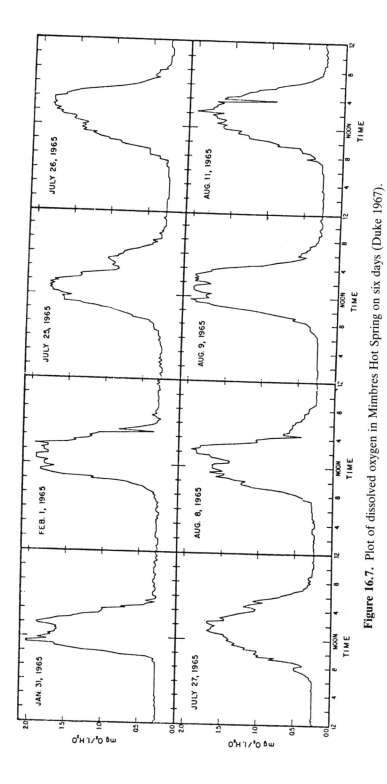

Figure 16.7. Plot of dissolved oxygen in Mimbres Hot Spring on six days (Duke 1967).

Odum (1964) described a redox potential difference of 0.2 to 0.5 V between platinum electrodes above and below the photosynthetic layer in an algal mat microcosm. This potential difference displayed a diurnal variation, with the higher values occurring during the daylight hours. They explained the data by postulating that photosynthesis in the upper layer of the mat was responsible for the high positive oxidation potential above the mat. During the night, as oxygen concentration decreases, the potential above the mat approached that of the highly reduced zones beneath (Figure 16.8).

Sollins (1970), who obtained his hypersaline mats from petroleum effluent holding ponds near Corpus Christi, Texas, placed $100\,cm^2$ fragments in plastic dishes under fluorescent lighting. About 6.5 klx of square wave input was provided. Using diurnal oxygen curve methods, he estimated gross primary production to range between 1.2 and $1.6\,g/m^2$ per day, with a range of redox potential difference of 0.3 V. Nighttime respiration was linearly correlated with oxygen concentration (Figure 16.9). The assumption was made that daytime respiration was similarly correlated, and the gross production was recalculated on that basis. The recalculated values averaged $2.7\,g\,O_2/m^2$ per day.

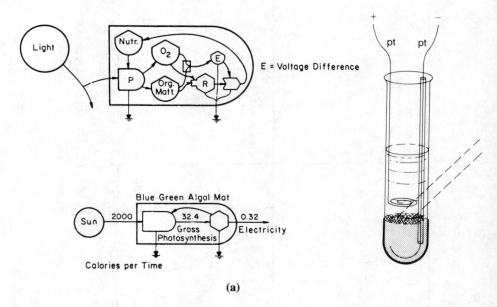

(a)

Figure 16.8. Photoelectric algal mat microcosm (Armstrong and Odum 1964). (a) Sketch with platinum wires used to draw electrical power, energy systems diagram, and aggregated diagram with energy flow values (Odum 1983a); (b) Diurnal record of metabolism and power.

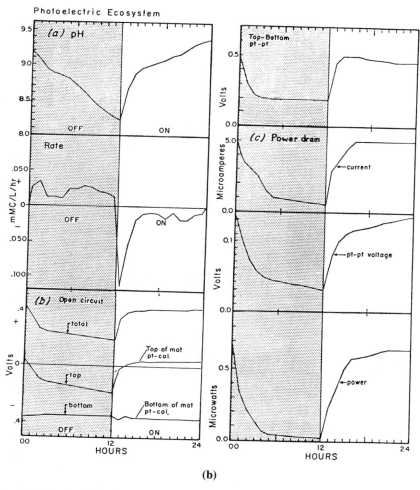

Figure 16.8. *Continued*

Great Salt Lake Degradation Microcosms

Paterek (1983) cultured organisms indigenous to hypersaline ecosystems: brine shrimp, halobacteria, and halophilic algae. He added these to autoclaved aerobic and anaerobic microcosms prepared with brine and sediments from the Great Salt Lake. Aerobic microcosms were housed in 125-ml cotton-stoppered flasks and anaerobic microcosms were kept in 50-ml serum bottles. Systems were incubated in the dark at 37°C for 40 days. Methane production and the concentration of the methanogenic precursor trimethylamine, were greatest in both types of microcosms with

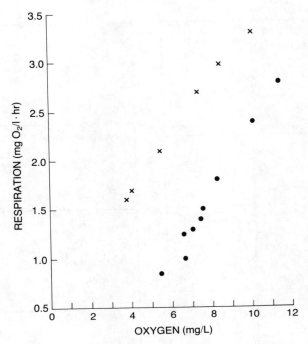

Figure 16.9. Night time respiration as a functioned of measured oxygen concentration in two hypersaline algal mat microcosms (Sollins 1970).

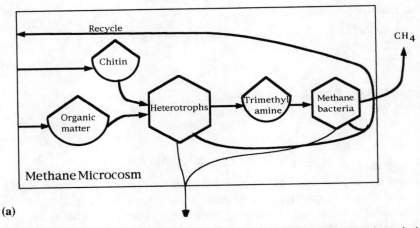

(a)

Figure 16.10. Methane production in briny microcosms with microbial degradation of indigenous halophilic macroorganisms (Paterek 1983). (a) Energy systems diagram; (b) micromoles of methanogenesis in anaerobic and aerobic microcosms from decomposition: solid line, brine shrimp; dotted line, algae; dashed line, bacteria.

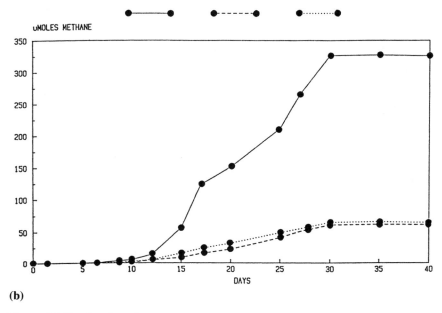

(b)

Figure 16.10. *Continued*

brine shrimp present (Figure 16.10). Both compounds were produced in microcosms enriched with halophilic algae and halobacteria, but at much decreased concentrations. These results demonstrated that the decomposition of normal components of the ecosystem did not continue to carbon dioxide and water, but stopped at methane.

Salt Pan Microcosms

Nixon (1970) fabricated defined microcosms consisting of 1 alga and 6 strains of bacteria. These microcosms were prepared from pure cultures of components isolated from the salt pans at La Parguera, Puerto Rico. The microcosms showed an unusual diurnal pattern of dissolved oxygen levels. The levels of oxygen fell for a few hours, beginning just after the lights came on. This phenomenon was attributed to photorespiration. Such light-accelerated respiration could be an adaptive mechanism for the elimination of mineral cycling blocks in the system, where large organic storages accumulated levels of dissolved organic carbon approaching 0.5 g/l and inorganic nutrients became limiting. Increasing the size of the inorganic pool through fertilization increased chlorophyll *a*, net photosynthesis, and protein content (Figure 16.11). Diurnal curves assumed a

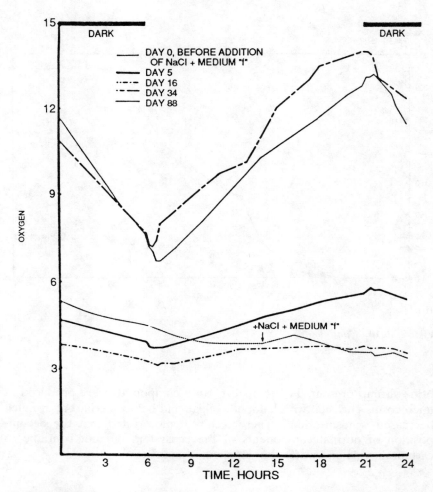

Figure 16.11. Diurnal oxygen curves of a salt pan microcosm before the salinity was increased from 15% to 18% with sodium chloride and the system was fertilized with nitrate, phosphate, vitamins, and trace elements (Nixon 1970).

normal pattern for a short time, but system metabolism oscillated widely and the unusual patterns reappeared.

Patterns of diurnal oxygen with photorespiration were simulated with the model in Figure 2.16. Brine microcosms are also described with Figure 3.5.

Part 3
Microcosms for Society

Part 3 includes those microcosms directly contributing to human society and economy. Some are small laboratory tests designed to help us understand larger utilitarian systems. Some are mesocosms, which do work for society and nature.

Many of the earth's old methods of developing food, salt, and other products have been supplied from complex, enclosed ecosystems. Partially enclosed ecosystems of various sizes are a common means of returning wastes to the environment. Such mesocosms contain domesticated ecosystems in symbiotic relationship with the environment and the human economy. Developing more symbiotic ecosystems and a better fit between humanity and the biosphere is the province of the evolving field of ecological engineering.

The role of humans within ecosystems is the increasingly important focus of those considering the symbiosis of society and its planetary life support system on Earth. With hopes for maintaining humans in space for longer times, the necessity for self-regulating life support systems is becoming apparent. Developing mesocosms that contain complex ecosystems and people is the only direct way to find a life support ecosystem for human beings and test its carrying capacity.

Operating microcosms and mesocosms are becoming important for public exhibition, education, and recreation as the public continues to learn that there is more to nature than just individual species. Adey and Loveland (1991) have published a book summarizing the knowledge concerning ecological mesocosms for public exhibition. Birmingham (1990) describes new Mesocosms and aquaria in Osaka, Japan called *Ring of Fire* designed by Peter Chermayeff and Cambridge Seven Associates, Cambridge, Mass.

Part 3 includes Chapter 17 on food production microcosms and Chapter 18 on waste processing microcosms. Chapter 19 reviews efforts to include human beings within enclosed systems on earth and in space. Finally, Chapter 20 examines possibilities, needs, and values of ecological microcosms for the future.

Chapter 17
Food Microcosms and Mesocosms

Systems of food production for the world's people are sometimes studied in microcosms, and some self-organizing, ecological mesocosms actually yield food products. Agriculture, fisheries, and aquaculture supply humanity its food. Some of these sources are intensively managed agroecosystems and others less-managed environmental systems. Agricultural and aquacultural microcosms have been used for controlled experiments to study larger food production systems under controlled conditions. The processes of food production and systems of waste decomposition are basic to human existence. Just as photosynthesis and respiratory consumption were coupled in a balanced aquarium (Chapter 2), the equivalents of these processes in the human economy are also coupled with the circular, causal symbiosis shown in Figure 17.1.

This chapter considers multispecies systems which generate food for human use, including practical and experimental crops, mushroom culture, aquaculture, etc. See also Chapter 19 for studies on crops in microcosms designed to help us learn how to support humans in space enclosures.

Ecological Engineering and Agriculture

The term "ecological engineering" is the study and management of ecosystems that are employed in partnership with human economy. Many useful, domesticated ecosystems have been known since ancient times for aquaculture, organic agriculture, and the manufacture of salt. Although some ecological engineering systems having wide utility are macroscopic, covering hundreds of acres, microcosm versions are often studied in laboratories. Depending on their size, domestic ecosystems isolated in chambers constitute microcosms or mesocosms. Many scientific studies

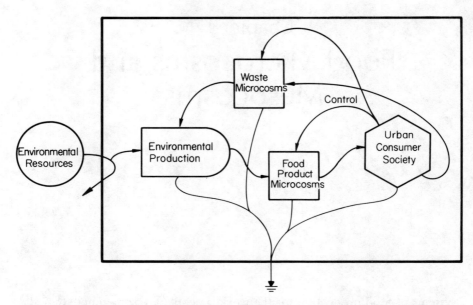

Figure 17.1. Role of food and waste microcosms interfacing with environmental systems, on the left, and the human economy, on the right.

are made of semi-isolated chambers containing ecosystems of possible or proven use. Some domestic ecosystems become unit processes in yet larger systems.

Research studies in ecological engineering not only consider the packaged ecosystems whose use has been long-established, but also involve the creative process of discovering new systems through the stimulation of new self-organization. Whereas molecular initiatives and biotechnology tend to increase energy requirements by substituting technology for natural self-management, ecological engineering decreases costs, reduces energy and technological inputs, and is better adapted to the coming world of declining resource availability. Systems of past or potential use can be analyzed for their self-sufficiency by calculating the Emergy ratios between purchased inputs and environmental inputs. Currently the U.S. ratio is 7:1 whereas systems that use more ecological contributions have lower values.

Turf and Crop Microcosms

Increasingly, experimental studies are putting agroecosystems into microcosm form. It is a small step from the standard growth chambers used for plant physiology, to a semi-enclosed crop ecosystem with its materials more self-contained and its gases recycled and monitored. Besides providing better controlled conditions for study, the crop microcosms allow

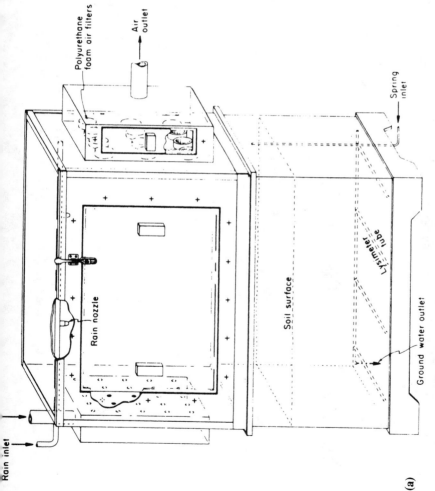

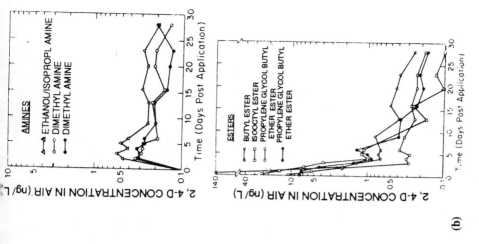

Figure 17.2. Terrestrial microcosm apparatus used to study the fate of chemicals introduced into synthesized ground ecosystems (Gile, Collins and Gillett 1982. Reprinted with permission from J. Agric. Food Chem., copyright 1982, American Chemical Society). (a) Apparatus; (b) record of 2,4-D in air above simulated wheat microcosm (Gile 1983).

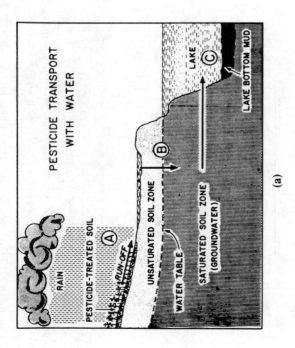
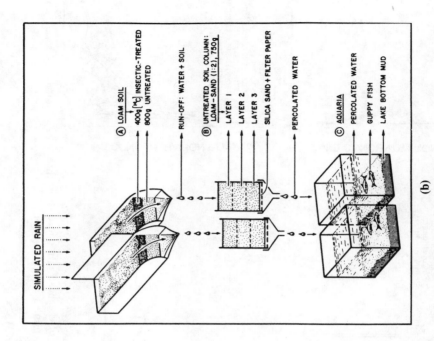

Figure 17.3. Microcosm apparatus for study of pesticide transport and fate (Lichtenstein and Liang 1987. Reprinted with permission from J. Agric. Food Chem. 35, copyright 1987, American Chemical Society).

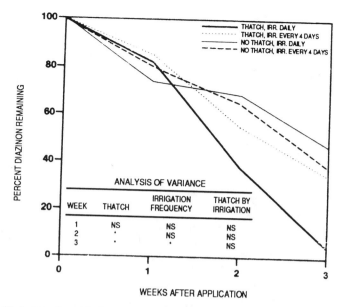

Figure 17.4. Decline of the pesticide Diazonon in grass microcosms (Branham and Wehner 1985, Reproduced from Agronomy Journal 77(1):101–104 by permission of American Society of Agronomy, Inc).

study of the behavior of volatile compounds, animal–plant dynamics in more natural circumstances, pesticides, radioactive substances, etc. See examples in Figures 17.2–17.5.

As suggested in Figure 6.1, terrestrial components tend to be larger and have longer development times than those in aquatic microcosms, and longer times are required for self-organization. Short-run crop microcosms tend to be a study of what has been introduced into the system. However, the question of time for self-organization does not arise with crop microcosms, since the microcosms have as much time as the crops in the field to develop their relationships with soil and other organisms. Examples include the study of productivity, the fate of chemicals, and toxicities.

In a simulated wheat field microcosm, added 2,4-D was lost into the air in 2–4 days (Figure 17.2) whereas amines generated by metabolic reactions lasted longer (Gile 1983). Malanchuk and Joyce (1983) applied 2,4-D to microcosms of agricultural sod taken from old hay fields and planted with oats. They measured nitrogen fixation and carbon dioxide evolution. A reduction of half in the amount of carbon dioxide was temporary (less than 7 days). There was little if any effect on nitrogen fixation.

Using the microcosm apparatus shown in Figure 17.3, Lichtenstein and Liang (1987) found that cover crops reduce runoff of labeled fonofos

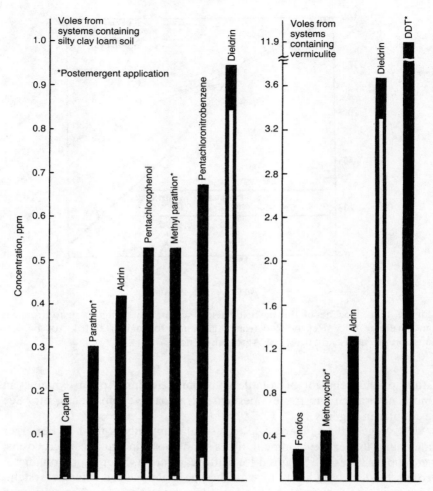

Figure 17.5. Fate of ^{14}C-tagged pesticides (white bars) and their residues (black bars) in the animals of crop microcosms after 5 days (Metcalf et al. 1979).

compared to fallow soil microcosms. Detoxified metabolites were found in plants. Simulated rain transported the toxic pesticides carbofuran and fonofos into aquatic microcosms just after application. However, these chemicals were barely measurable after 36 days. Lichtenstein et al. (1982) found treatments that influenced the rate of denaturation of fonofos and parathion by Fungl in soil–plant microcosms.

Branham and Wehner (1985) applied diazinon to microcosms including matted grass turf, and followed its decrease due to vaporization and metabolism (Figure 17.4). Metcalf et al. (1979) standardized and tested a terrestrial microcosm with Illinois soils and crop plants in 19-liter widemouth glass carboys at 26°C with 12 h simulated daylight. Microcosms

Table 17.1. Costs of Crop Microcosm Evaluation of Pesticides (1979 $ per individual model ecosystem): Metcalf, Cole, Wood, Mandel, and Milbrath, 1979.

Radiolabeled pesticide	$100.00
Costs of microcosms	8.03
Chemicals and supplies	135.41
Supplies to rear animals	12.18
Miscellaneous consumable supplies	25.70
Total	$282.66

To this add labor, and depreciation of growth chambers, equipment etc.

supported *Zea mays, Glycine max, Lumbricus terrestris, Limax maximus, Armadillidium vulgare, Estigmene pipiens quinquefasciatus, Physa spp.,* and *Gambusia affinis* in the leachate water. The fate and toxicities of ^{14}C-labeled pesticides were studied, including DDT, methoxychlor, aldrin, dieldrin, fonofos, phorate, parathion, methyl parathion, simazine, trifluralin, 12,4,5,T-isooctyl ester, hexachlorobenzene, pentachlorophenol, pentachloronitrobenzene, and captan. Large differences were observed in food-chain accumulation of compounds (Figure 17.5). The type of base soil material (vermiculite versus silty clay loam) correlated with some differences. Costs were given (Table 17.1) to help compare microcosm studies with more traditional field-plot studies and other toxicity assessment methods.

Algal Culture

Following two decades of physiological studies of algae, cultures were scaled up in size for food production. The larger cultures were not completely uni-algal, contained microbial consumers and some of the characteristics of microcosms and mesocosms. The balance of photosynthesis and respiration was a function of the rate of supply of new nutrients. Without nutrient influx, respiration increased to balance photosynthesis as part of successional changes in culture characteristics. With nutrient flux, pilot plants arrested their succession in net production stage. The general photosynthetic efficiency of food product was similar to that of agriculture (Burlew 1953). Considerable energy-using technology was required to collect and concentrate the algae for food.

Aquaculture

Fish ponds, fish hatcheries, and other aquaculture practices are food-yielding mesocosms which often possess a high degree of self-organization. Among the most successful examples are the shrimp ponds of Ecuador, a

364   17: Food Microcosms and Mesocosms

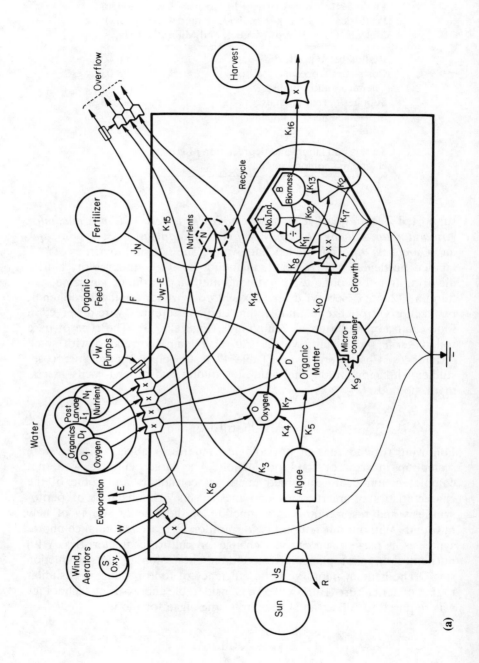

(a)

Aquaculture

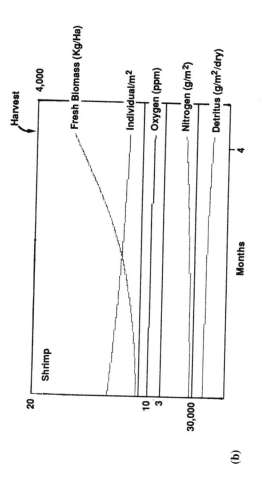

Unused available light, $R = J_S/(1 + k_0N)$;

Available nitrogen, $N = (K_1OD + K_{17}DI(B/I) + K_2B + J_N + J_W + N_1)/(1 + K_3R)$

Oxygen, $DO = K_4RN + K_6W(S - O/Z) + O_1J_W - K_7OD - K_8B - K_{15}(J_W - E)(O/Z)$

Detritus, $DD = K_5RN - K_9OD + J_WD_1 + F - K_{10}DI(B/I) - K_{14}(J_W - E)(O/Z)$

Individuals, $DI = J_WI_1 + J_1 - K_{11}I - XK_{16}I$

Biomass of shrimp, $DB = K_{12}DI(B/I) - K_{13}B - XK_{16}B$

Mortality, If $O/Z < 2$ then $X = 1$

Figure 17.6. Systems diagram of the simulation model SHRMPOND for short term shrimp pond mariculture as practiced in Ecuador. (a) Systems diagram with equations: (b) typical simulation with harvest based on individual shrimp size after four months.

very practical kind of mesocosm. A simulation model (Figure 17.6) may be helpful in elucidating the main principles involved in aquaculture, where the simplicity of early self-organization is used to channel productivity into a single-product yield.

Shrimp Mariculture Ponds of Ecuador

Within a period of 10 years, the entire coast of Ecuador, 350 miles long, has been covered with shrimp ponds, in a fevered movement not unlike a gold rush. Millions of dollars flowed into the region for construction and operation, and economic yields were large for those with the early ponds. Most of them were built within the mangroves or next to them. The losers were those whose livelihoods had been dependent directly and indirectly on the mangroves. Most of the ponds were 1–25 ha in area and about 1 m deep, but some were as small as 10 m across and some as large as 1,000 ha. Totaling 130,000 ha, the thousands of ponds are a special kind of microcosm, which, for the most part, are filled and seeded with post-larval shrimp, and drained in 3–6 months. The main species, *Peneus vanamei*, apparently is eurythermal and euryhaline, and thus able to develop in a wide variety of conditions. The wild populations are often associated with fresh water rivers carrying detritus, but are also found in mangroves. These ponds vary enormously with the different kinds of water pumped in, different kinds of bottom soils and organic matter content, and different kinds of fertilization and supplemental feeding. Post-larvae are either caught in the estuaries and on the beaches as they are swept in by the long-shore currents and tides, or they are raised from eggs in hatcheries, feeding during their early stages on cultured algae and brine shrimp nauplii.

The huge shrimp industry based on these ponds has a high level of knowledge of procedures that have usually worked, but very little has been written in the scientific literature. Published studies do not answer clearly such questions as what the dominant species, *Peneus vanamei*, eats in the ponds and in the natural waters. In some ponds there are plankton algal blooms producing detritus and microzoa; oxygen levels may be above saturation. In other ponds, turbid with organic detritus, the food chain appears to start with organic matter. These ponds and those with heavy additions of fishmeal or other feeds have oxygen below saturation. To avoid a low-oxygen kill, ponds with large, intensive additions of food are aerated. However, most operators pump in 10% of the pond water each day and let the waters overflow through screens. Thus, there is an open question as to how much of the food chain comes from efforts to culture or feed, and how much is derived from the pumped input. Through this turnover, the ponds aquire the safety feature of outside waters which used to be conditioned by the natural mangroves. Now, however, in many places most of the mangroves are gone, all the tidal prisms have been reduced by the pond construction, and tidal circulation is reduced. In areas where ponds are concentrated, each pond operator is pumping in the outflow from the neighboring ponds. It is clear that part of the success of these ponds depends on quickly developing a food source for the developing larvae, excluding as much as possible the larvae and juveniles of other consumer species, and harvesting before the remainder of the ecosystem can get started. Draining clears out competing species and predators. In other words, the ponds with several months of growth prior to harvest, are microcosms of early succession based on photosynthesis and/or organic nutrition to which an effort is made to couple a monoculture of consuming shrimp.

Computer Simulation of Shrimp Pond Mesocosms

Figure 17.6 depicts a systems diagram of the shrimp pond microcosm for simulation, calibrated with numbers supplied by Javier Duenas in Bahia, Ecuador. Note the four alternative sources for organic detritus: the organic matter on the bottom, left after the last draining of ponds (initial conditions for simulation), the water pumped in (10% per day), the production of the algae, and the added organic feed. In the model, the detritus pool includes the microzoa which can develop in a few days to turn the detritus into a more balanced food for the shrimp. The model is somewhat similar to the model for diurnal oxygen simulations (Figure 2.14), except that there are additional inputs, faster water turnover, and, especially, the seeding of a largely shrimp population.

Simulations of this model were run to generate the results of aquaculture operations for a few months. The program stops when the biomass per individual reaches the value set. Figure 17.6b depicts typical simulation results, with detritus and oxygen levels falling off at the end. The rise in nutrient accumulation in the system helps maintain oxygen production from the algae. For larger ponds with a moderate aeration coefficient, oxygen decreases some what as the population biomass becomes large, but the oxygen does not fall to the 2 ppm level which increases mortality. In other runs with heavier seeding of larvae, feeding, and a smaller aeration coefficient, the simulation did generate results with a low-oxygen mortality rate consistent with observations. The reasonable agreement between the model and typical pond conditions helps one understand the production of the very high protein yields. This system is a monoculture of edible consumer species based on a high-diversity microscopic detritus ecosystem. See Odum's (1962) general statement about the suitability of detritus ecosystems for human use.

Success in this aquaculture is based on having an edible "pioneer" species adapted to fast growth which occurs before a more complex food chain of larger competitors and carnivores can develop and share resources. The system may be vulnerable to any disruption of river and mangrove inputs which seed and stabilize the detritus food chain in the ponds. The same inputs support the wild parent stock of shrimp from which post-larvae are obtained for stocking the ponds.

Fish Hatchery Ponds

Fish hatchery ponds contain a partly self-organizing ecosystem managed for yield of a stock of young fish. The study by Culver (1988) on Australian ponds raising Golden perch (*Macquaria ambigua*) illustrates the way a rapid plankton succession with pulses of phytoplankton followed by zooplankton leads finally to growth of the stock for high yield (Figure 17.7). As with the shrimp ponds, the yield is based on rapid development of fast-turnover food-chain elements after filling, with the first pulse of

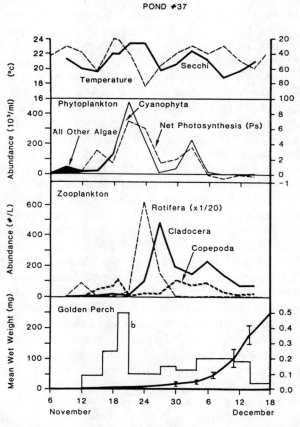

Figure 17.7. Pulse of productivity supporting fish growth in Australian fish hatchery ponds (Culver 1988).

transferred energy going into the stocked population, resulting in growth before other populations can develop enough to divert resources.

Appropriate Technology Mesocosms

The movement to develop lower-energy life-styles in a still-technological world, produced many new microcosm and mesocosm initiatives, sometimes adapting ancient ways of providing foods with an infusion of modern technology and its additional energy requirements. Some of these efforts were included in research programs.

Solar Heated Fish-Pool Aquaculture

As part of initiatives for energy conservation and appropriate technology, a group named "New Alchemy" studied aquaculture in small fish pools in a greenhouse shelter in Falmouth, Massachusetts, winter and summer. The tanks were arranged close to human dwellings so that the solar energy captured by the greenhouse tanks contributed to household heating, while also contributing protein food in the form of cultured *Tilapia*. Some experiments (Figure 17.8) included recirculation of high-nutrient fish pool waters to hydroponic vegetables and green plants for the human living environment. Fish were fed Purina developer trout chow (7.8% nitrogen) or rabbit food pellets, since the solar area was not large enough to support the fish without the outside organic feeding. The enriched waters supported large algal populations even in winter, but their metabolism per cell was small at that time. As the fish food was added, high levels of nitrogen accumulated in the waters in the form of ammonia, nitrite, and nitrate, with accompanying nitrogen bacteria. At times, the nitrification of ammonia to nitrite lowered the pH. The high nitrogen levels caused bacterial metabolism to prevail over algal growth. Various management alternatives were tried in order to decrease nitrogen effects which might limit fish growth. Figure 17.9 shows the growth of fish under various conditions.

Simulation models were developed for the system dynamics, including water quality, algae, and fish growth (Figure 17.10). Simulation results for diurnal variations (Figure 17.11b) and for the main features of growth were reasonably similar to the observed data (Figure 17.12b).

Using data supplied in the report by Zweig et al. (1981a, 1981b), the solar Emergy of the yield was compared with that put into the system by the sun and the purchased inputs (feed, electricity, amortized costs of tanks and space, labor). Solar Emergy is the solar energy equivalent required to make a product. Putting every input and yield on this common measure allows comparative evaluation. The Emergy of the yield divided by the Emergy put into the process by the operators (net Emergy yield ratio) was only 0.05, whereas most fisheries generate 20 times this or

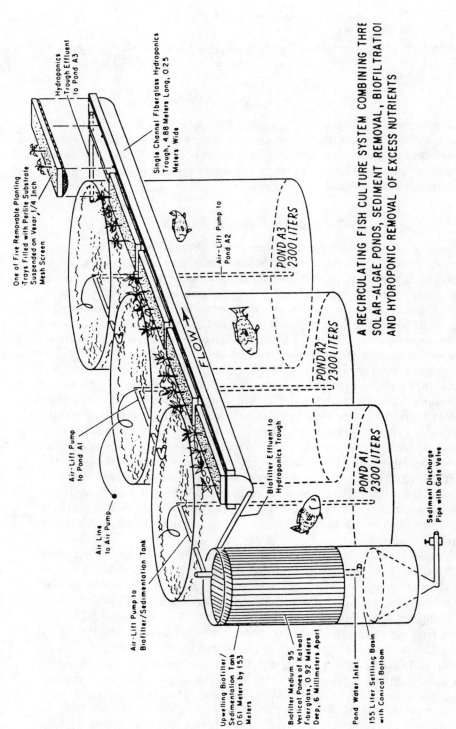

Figure 17.8. New Alchemy fish pool system (Zweig et al. 1982).

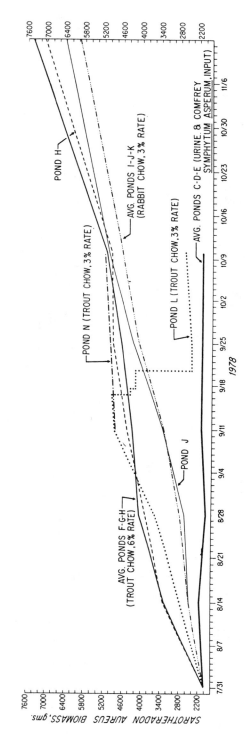

Figure 17.9. Growth of fish in pools in Figure 17.8 with various management strategies (Zweig et al. 1981).

17: Food Microcosms and Mesocosms

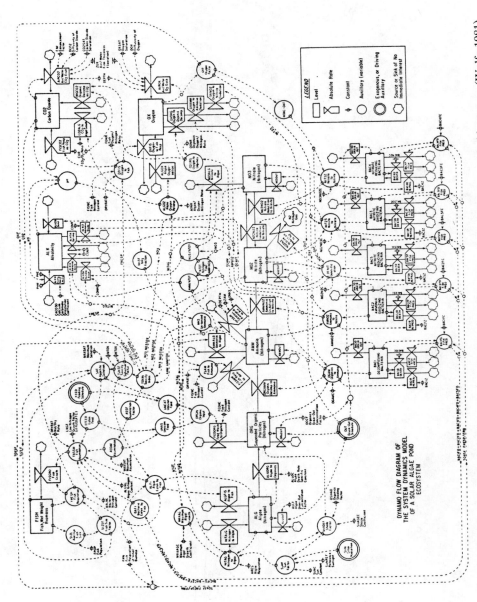

Figure 17.10. Simulation model of fish pool system written in Forrester Dynamo symbol language (Wolfe 1981).

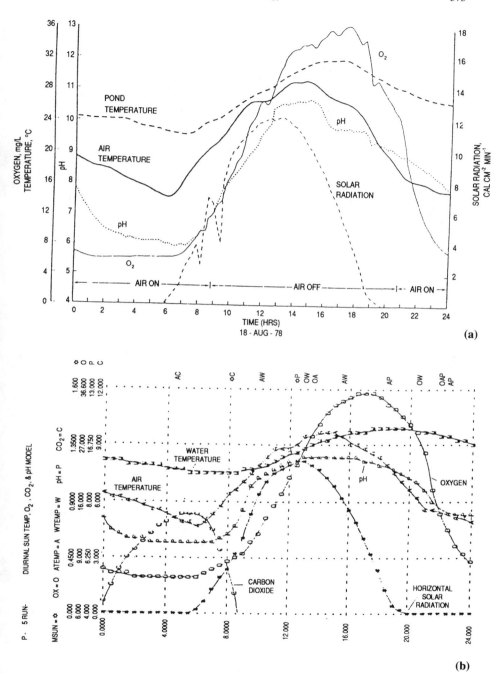

Figure 17.11. Diurnal variations in fish pools August 18, 1978. (a) Observed data; (b) simulation results (Zweig et al. 1981).

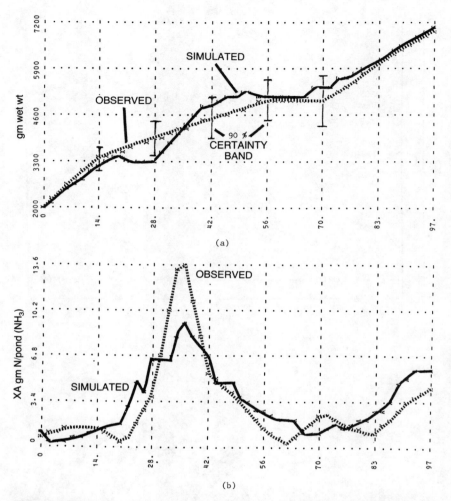

Figure 17.12. Comparison of data from fish pools with simulation of the model in Figure 17.10. (a) Fish biomass; (b) nitrogen (Wolfe 1981).

more. However, systems do not have to yield net Emergy to be useful. To be economical, a system must use free resources to match and justify the purchased inputs. To compete in the American economy, the ratio of purchased Emergy to free Emergy (Emergy investment ratio) should be about 7:1. The shrimp aquaculture ponds of Ecuador have a ratio of about 4:1 (Odum and Arding 1991). The intensively managed New Alchemy pools had an investment ratio of 20. This system used very little of the free sunlight compared to the large and costly inputs from society. The direct sunlight contributed only a fraction of one percent of the yield

Emergy. For basic food production, high investment ratio systems can not compete with those which use more sunlight, depend less on purchased inputs, and have a low cost per area. However, some intensive agriculture in the U.S., with investment ratios as high as 25, competes successfully in the high-priced, luxury products markets, such as winter vegetables. Evaluating a similar, small-area, solar greenhouse and basement fish aquaculture tank in an urban household (Mueller home site, Evansville, Indiana), Burnette (1978) also found a high investment ratio, although lower ratios would result if aesthetic and educational benefits, and the savings on waste disposal were included.

Chapter 18
Waste Microcosms and Mesocosms

Ecosystems for waste consumption are found everywhere in wild nature and in practical use by human societies. They include the waste treatment plants of urban civilization, which are really domesticated mesocosms. Waste consumption units close the material cycle from human economic society to the biosphere's life support system (Figure 17.1). Some units use the energy of the waste, whereas others use solar energy and/or physical stirring energy as well. Many alternative systems of waste disposal have been studied in microcosms. Food production microcosms and waste recycling microcosms are the main interfaces between the world's environment and the economic patterns of society (Figure 17.1). Because the material output of one is the input of the other, there is cooperative symbiosis between autotrophic producing units and heterotrophic consuming units.

Engineered systems for waste consumption often consist of unit processes in separate chambers connected with pipes and controlling valves. Another kind of system has all the processes working together in one chamber. Multi-species ecosystems require little control from outside and carry much of their own means for homeostasis. Managing self-organizing systems for useful purposes is the new field of ecological engineering (Mitsch and Jorgensen 1989). In this methodology, the first step in finding an efficient waste disposal system is to place new kinds of wastes in replicated microcosms seeded with many species. Almost any ecosystem may be a candidate to become an alternative ecotechnology, a new self-organizing unit process.

Receiving less attention than systems of production, humans have often allowed the consumption systems to self-organize on their own. Disdain for mess and smell has left this function to natural process. Like other self-organized ecosystems, decomposition systems often have higher diversity of species and processes than the unit processes of chemical industry.

Enclosed ecosystems that serve human society by processing "waste" include sewage disposal systems, composting, solid waste disposal, etc. Waste consumption systems generally have more respiration than photosynthesis and so they consume organic wastes, returning nutrient raw materials as an output. By means of heterotrophic units, wastes become useful byproducts and biogeochemical cycles are closed, all of which is necessary for whatever larger system prevails. Waste consumption units are necessary for fitting humans or other concentrations of life to their environment. These ecosystems are subsystems in the larger patterns of the biosphere.

Sewage Microcosms

Sewage plants are decomposition ecosystems accelerated in their activity by physical arrangements that supply oxygen. They operate within concrete boxes, and are controlled by valves on inputs and outputs. Perhaps prematurely, the world's urban civilizations standardized on relatively few kinds of sewage ecosystems for municipal and industrial use.

Bacterial Beds of Trickling Filters

With a bed of gravel as the substrate, percolation of organic-laden wastewaters generates a living filter, including a food chain web. See, for example, the organisms found in a trickling filter (bacterial bed) in Figure 18.1. The extensive literature concerning such filters was summarized by Hawkes (1963) and some of his conclusions follow (Figures 18.2–18.4).

Hawkes (1963) provided a network diagram of main components and processes (Figure 18.2a) for which we provide an energy language diagram in

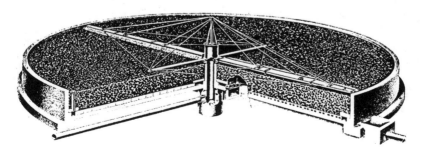

Figure 18.1. Typical Bacterial bed (trickling filter) system used for sewage waste treatment (Rich 1963).

378 18: Waste Microcosms and Mesocosms

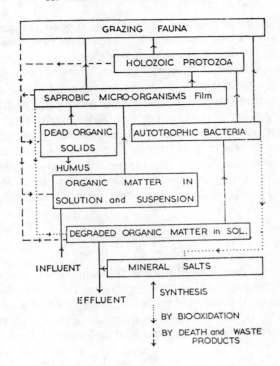

(a)

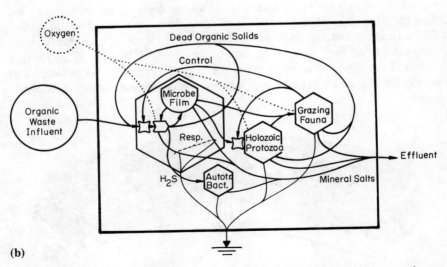

(b)

Figure 18.2. Systems diagram of bacterial bed ecosystem. (a) Diagram of processes provided by Hawkes (1963, Copyright 1963, Pergamon Press); (b) energy system diagram of the network in (a).

Figure 18.2b. Organic waste influents support a microbial film as the base of a food chain which includes holozoic protozoa and "grazing" animals that control the system. Organic solid byproducts are recycled for further consumption within the system. The effluent carries out the mineral nutrients remaining after consumption. This system uses oxygen, sulfates, and nitrates to oxidize the organic input fuel. The heavier loading causes the consumer system to develop great thickness in the filter.

The food chain started with zoogleal bacteria and associated nitrogen-fixing bacteria, and fungi (*Fusarium, Oospora, Sepedonium, Ascoidea, Subbaromyces, Sporotrichum, Penicillium*). Small microzoan consumers within the film included protozoa such as colonial *Carchesium*, nematodes, and rotifers. Larger animals which "graze the films" included snails, oligochaete worms (*Lumbricillus, Enchytraeus, Lumbricus, Eisenia, Dendrobaena*), springtails (*Achorutes, Tomocerus, Folsomia*), cercyon beetles, flies (*Psychoda, Hydrobaenus, Metriocnemus, Anisopus, Paracollinella, Scatella, Spaziphora*), and arachnids (*Lessertia, Porrhomma, Erigone, Platyseius*). In different sized beds, in different conditions, in different microhabitats, and in different seasons, different species were dominant. Clearly, much of the success of waste decomposition systems comes from the many species present which continually self-organize and fine-tune the network's ability to use influent.

As shown in Figure 18.3, an oscillation of the organic film alternated with oscillations of the consumer animal populations that grazed that film. The rise and fall is something of a prey–predator type oscillation. Anaerobic conditions may develop when organic loading is large. The trickling filters were operated with a rotating boom so that fluid was added at a frequency that depended on the speed of rotation. For the same organic loading, the faster rotation speed gave inputs that were frequent, smaller pulses compared to the infrequent, larger pulse provided by the slow rotation. By controlling the time scale of decomposition conditions, the hierarchical pattern was also controlled. As shown in Figure 18.4a, high frequency input resulted in pulses of smaller organisms, whereas the lower frequency input caused larger animal components to have a greater role. The type of spray nozzle also affected the food chain (Figure 18.4b).

Activated Sludge Ecosystem

Another type of system, activated sludge, is one of the most common sewage treatment systems and is based on a decomposer microbial system that organizes around sewage that is physically mixed with air. "Flocs" of organic material being consumed by bacteria are locally anaerobic, but continuously mixed with injected oxygen and recycled so that the process does not become limited. Whereas the trickling filter, with the aid of the higher animals, can consume most of the bacterial bodies and residues,

18: Waste Microcosms and Mesocosms

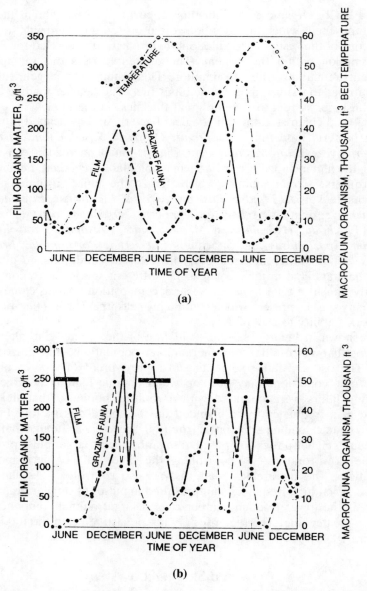

Figure 18.3. Seasonal fluctuations of film and fauna in bacterial beds treating wastewaters (Hawkes 1963, Copyright 1963, Pergamon Press). (a) Domestic sewage; (b) industrial sewage.

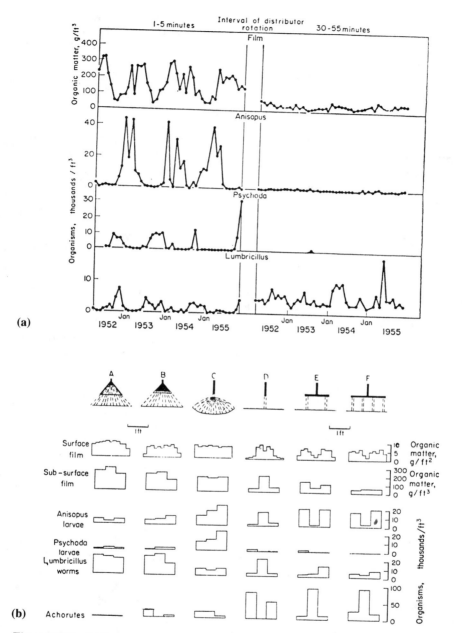

Figure 18.4. Differences in bacterial bed ecosystems produced by altered frequency of input of organic influent (Hawkes 1963, Copyright 1963, Pergamon Press). (a) Oscillations due to different frequencies of rotation of input spray boom; (b) ecosystem differences due to different local frequencies from different spray nozzles.

a sludge of mainly dead, bacterial bodies and undecomposed residues is removed for drying, burning, or other disposal. As a microbial ecosystem, with fast turnover and without the control of higher organisms, the activated sludge ecosystem is dependent on human control and vulnerable to control failures. Various kinds of toxic chemicals which reach such systems upset the microbial ecological food chain web and cause organic matter to accumulate without decomposition.

Hawkes (1963) reviews the ecology of the many species of bacteria, protozoa, and fungi interacting to accomplish fast decomposition of organic matter. Principal components of such systems are a dozen non-nitrifying bacteria, nitrifying bacteria which use ammonia and nitrite, species that form "zoogleal" floc, filamentous bacterial strands such as sulfur bacteria, nematodes, rotifers, and many protozoa. The essence of the process is the building of an autocatalytic floc on which polysaccaharides accumulate and entrap organic matter and more bacteria, facilitating organic removal and decomposition. As with other systems, maximum power is obtained by building a complex structure that facilitates and reinforces the process. Hence the name "activated" sludge.

The rapid succession that develops, as in many other ecosystems, results in an increase in size of the units of organic substances using up the very reactive organic molecules first with larger particles emerging later. Consumption by protozoans helps keep turbidity low. The activated sludge microbial ecosystem can be operated in continuous flow microcosms so as to relate organic concentrations to the chamber turnover time (dilution rate) and the influent loading (Figure 18.5; Herbert et al. 1956, quoted by Rich 1961). The kinetics of these clustered units of multi-species ecosystem organization are not unlike those of pure cultures in chemostat systems. As Figure 18.5 shows, if the flushing rate (D) gets too large, there is a system washout.

High Metabolism Microcosms

Academically interesting is the question: what is the highest level of metabolism possible in an ecosystem? The metabolism of raw sewage produces heat equivalent to that of a fire. Compost pile decomposition systems generate intense heat, sometimes catching fire. Maximum power theory predicts highest rates of energy processing per unit space when organized structure is allowed to develop. Sometimes this is the structure of fluid vortexes, and sometimes it is the structure of species diversity.

When there are high concentrations of organic matter, high rates of metabolism may develop. Characteristically, these systems in nature or in microcosms become anaerobic, not in the sense of being too short of oxygen to metabolize, but in the opposite sense of maximizing metabolism by developing a steep oxygen gradient for increasing rates of inward diffusing oxygen. By requiring less space, the sludge system is

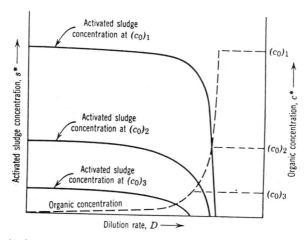

Figure 18.5. Activated sludge flocs as a function of influent organic loading and chamber turnover flushing time in continuous culture (Herbert et al. 1956).

adapted to the high intensity, waste generating cities. Examples of anaerobic sediment microcosms are given in Chapter 14.

Waste Mesocosms and Solar Energy

Many waste disposal systems take advantage of the homeostasis of ecosystems with photosynthetic production and organic matter consumption. Often, the autotrophic producers and heterotrophic consumers are intimately symbiotic within one ecosystem. For example, in a marine microcosm with eelgrass production and detritus decomposition, Harrison (1978) grew discs of the green alga *Ulva fenestrata* and marine bacteria in 250-ml flasks with nutrient-depleted sea water and detritus derived from *Zostera marina* leaves (Figure 18.6). With *Zostera* fragments present, dry weight increased 500% in 14 days, but without detritus, dry weight increased only 200%. With inorganic nitrate and phosphate and *Zostera* fragments, the weight of an *Ulva* disc increased by over 800%, compared to 400% in controls. In an additional experiment, 475 units of penicillin G were added to inhibit the bacteria (Figure 18.6). The antibiotic caused a reduction in algal growth in cultures containing detritus. This is an example of the enhancement of autotrophic growth by the metabolic activities of the heterotrophs. Because bacteria metabolize phosphorus and nitrogen in rapid, nearly closed cycles, Harrison concluded that some other organic or inorganic nutrient, such as magnesium, calcium, potassium, manganese, iron, or zinc, was more important.

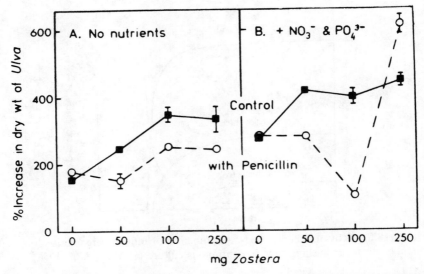

Figure 18.6. Percent increase in dry weight of *Ulva* discs incubated 14 days with *Zostera* detritus, showing the effects of penicillin G (Harrison 1978).

Sewage Ponds with Algal Photosynthesis

In some sunny, rural areas in the central United States, ponds that develop algal-microbial ecosystems are used for sewage treatment. Oswald et al. (1957) studied metabolism in sewage waste ponds with different retention times (Figure 18.7). With short retention time, respiratory consumption was greater than photosynthesis (Figure 18.7b). If held longer than two days, photosynthesis exceeded consumption (in ponds 30 cm deep). Deeper ponds (0.9 m) and ponds sampled in winter took longer. With longer retention, there was net production of organic matter. In the process, there was a transformation of organic matter from sewage to the organic matter of ecosystem biomass, which is a less active oxygen consumer in biochemical oxygen demand tests, hence the name "waste stabilization pond." Oswald et al. also studied the means and costs of extracting and drying algae, including flocculation and centrifuging, to use as food for animals.

Bartsch and Allum (1957) studied raw sewage ponds in the northern Plains area. In winter, organic matter decreased with sedimentation and decomposition, keeping oxygen levels low, especially under ice. In summer, photosynthesis increased organic matter, and only 8% of the oxygen demand was from the sewage. Light penetrated only 0.1–0.7 m, due to intense algal blooms. Oxygen concentrations near the top of the pond were super saturated to 30 ppm, but were zero on the bottom.

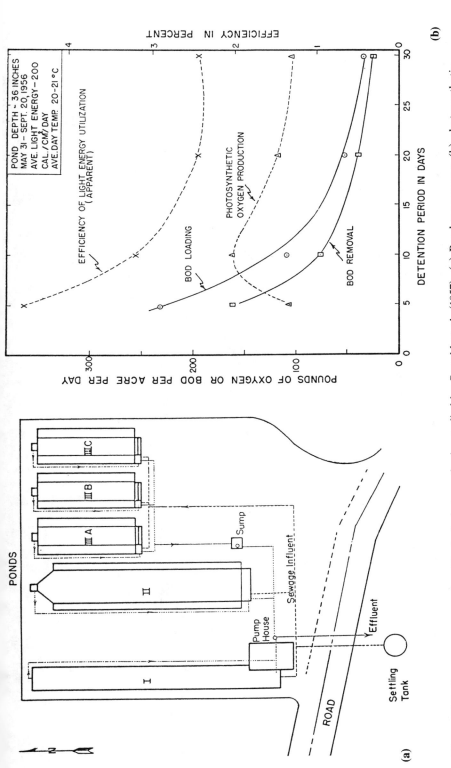

Figure 18.7. Sewage waste ponds with algal photosynthesis, studied by Oswald et al. (1957). (a) Pond mesocosms; (b) photosynthetic production rate, efficiency, respiration rates (BOD removal), and organic matter (BOD loading) as a function of water detention time.

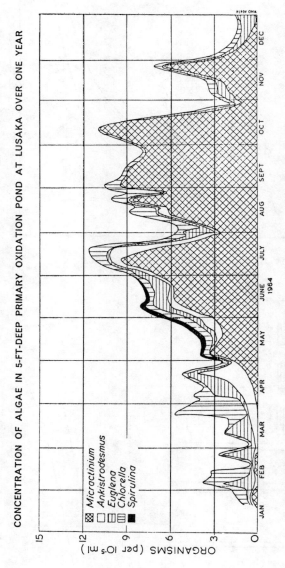

Figure 18.8. Algal populations in waste stabilization ponds in Zambia (Marais 1966).

Marais (1966) studied a 1.5 m deep waste stabilization pond in Zambia, which had a multispecies algal system (Figure 18.8). There was some daytime stratification, with oxygen ranging from 1 to 19 ppm. With long retention times through more than one pond, coliform bacteria, typhoid bacteria, and cysts and ova of *Bilharsia* were mostly removed. After several months' adaptation, the pond converted 200 mg/l organic influent to 25 mg/l effluent.

Taub (1964), considering microcosms for space, used fish-containing microcosms for disposal of dried human feces. *Tilapia* grew best when frozen brine shrimp were added as nutritive supplements.

Uhlmann and Cramer (1975) studied the metabolism of sewage microcosms, including still and flowing models. Diurnal changes in oxygen were used to evaluate photosynthesis, respiration, and their net contribution (Figure 2.8). Experiments included changing the length of daytime illumination and including effluent-generated oxygen regimes, like those observed in larger lakes and streams. The longer the turnover time of the aquatic system, the longer the time required to respond to changed input conditions. Oxygen patterns and metabolism were much less variable than populations of plankton. Activity per unit biomass was in inverse proportion to total biomass, a stabilizing mechanism.

The New Alchemy greenhouse tanks for aquaculture in Massachusetts were described in Chapter 17. Later work by J. Todd and Susan Petersen on similar tanks emphasized their role for waste processing. Septage, the sludge pumped out of septic tanks, was dispersed in water and processed through fiberglass tank mesocosms containing bacteria, snails, algae, aquatic plants, and woody plants in floating racks. The first stages were anaerobic, while the later stages were aerobic. Some tanks contained marsh communities and some had goldfish and shiners (Spencer 1990). Some effluents were processed through wetlands. Guterstam and Todd (1990) obtained higher levels of waste treatment by processing waters and suspended organic matter through a series of different mesocosms, using the diversity in time and series to maximize utilization. These serial systems are a kind of split trophic level and longitudinal succession.

Waste Treatment Mesocosms with Larger Aquatic Plants

In a long series of papers Wolverton and McDonald (McDonald 1981; McDonald and Wolverton 1980; Wolverton 1975, 1979, 1980, 1984, 1987a,b, 1989; Wolverton and McCaleb 1987; Wolverton and McDonald 1975, 1976, 1979a,b,c, 1980, 1981a,b, 1982a,b,c; Wolverton and McDonald-McCaleb 1986; Wolverton and McKown 1976; Wolverton et al. 1971, 1976, 1983, 1984a,b,c, 1989) described the processing of municipal wastes through mesocosms of reeds, hyacinths, and other macrophyte plants growing in rock filters and other arrangements. Better treatment was

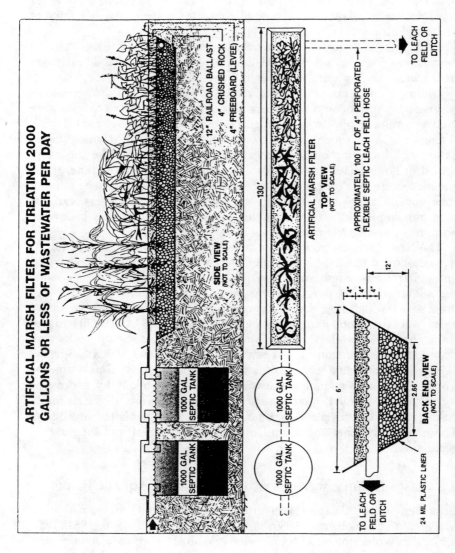

Figure 18.9. Artificial marsh mesocosm used for solar-aided waste treatment (Wolverton 1987a).

achieved with sunlight-driven photosynthesis and decomposition processes than with dark decomposition alone. Some of the microcosms were aerobic and some anaerobic (Figure 12.7, 18.9). These were effective in removing, decomposing, vaporizing, and binding toxic organic compounds, heavy metals, and nutrients into biomass and detritus. Some mesocosms were planned to fit at the end of septic tank outflows. For air purification, potted house plants were arranged with air blowing through soil. Sun, Liu, and Lin 1990 found unsterilized water hyacinth root microecosystem denaturing dye wastewater more than sterilized ones.

Microcosms Simulating Phosphate Mining Wastewaters in Estuaries

Carpenter (1971) investigated estuarine microcosms receiving phosphate-rich waters from the clay settling-basins of phosphate mines in North Carolina. Studies included observation of plastic pools with half of the water replaced weekly, and open ended U-shaped basins on the edge of Pamlico Sound, which received some exchange of estuarine water. Compared to controls, phytoplankton biomass and diversity was little affected, except that numbers of blue-green algae, *Anabaena*, *Oscillatoria*, and *Spirulina* (possibly nitrogen fixing) increased.

Waste Treatment on a Tidal Sandflat in a Microcosm

Kurihara (1983) studied the processing of sewage wastes in a tidal mudflat microcosm (Figure 18.10) to explore possibilities for coastal use. Activated sludge from the aeration tank of a sewage treatment plant was placed on the surface of sand in a tank, and thereafter on a more elaborate pilot test plant with an aeration tank. Polychaete worms, *Neanthes japonica* were studied as they consumed sewage sludge in continuous culture in the laboratory and then as part of the experimental microcosms. Chemical oxygen demand was reduced from about 25 mg/l in influent water to about 2 mg/l in treated waters in year-long operations. The sludge ingestion rate was $6 g/m^2$ per day. Rates were lower at winter temperatures, and populations were affected by changes in salinity.

Refinery Waste Systems

Busch (1963) studied the kinetics of biooxidation consumption of petrochemical wastes in an aerated microcosm. Over a considerable range of input loads and turnover times, the system consistently reduced waste concentrations to 8 mg/l. Copeland and Dorris (1964) studied metabolism in a string of pond mesocosms receiving oil refinery waste. The flow started (in the first ponds) with high rates of organic matter consumption ($20-40 g/m^2$ per day) and declined in the downstream ponds, there ac-

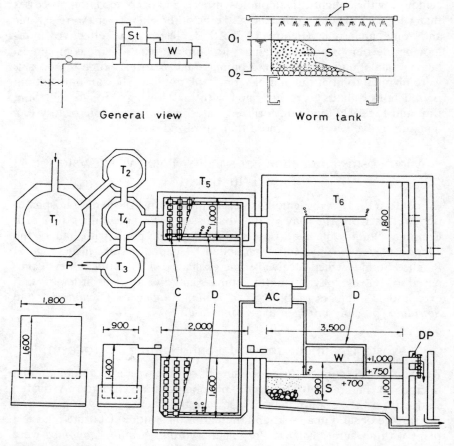

Figure 18.10. Marine polychaete-sand microcosm used to decompose sewage sludge. Above: St, seawater stock tank; W, worm tank; O_1, overflow; O_2, opening; S, substrates; P, perforated pipe; below: AC, air compressor; C, cubes of polyurethane foam; D, diffuser; DP, discharge pipe; P, primary sedimentation tank; S, sand bed; T_1, primary seawater tank; T_2, secondary sea water tank; T_3, sewage tank; T_4, mixing tank; T_5, aeration tank; T_6, sand tank; W, water head (Kurihara 1983).

companied by increased development of photosynthetic organisms in spring and summer. Organic production became as great as respiration. See Figure 18.11.

Solid Waste Microcosms

The world crisis with solid waste recycling is finally causing a change in concept of solids disposal. In many cities of the world, where garbage has been left in the open, a wide variety of birds and mammals have deve-

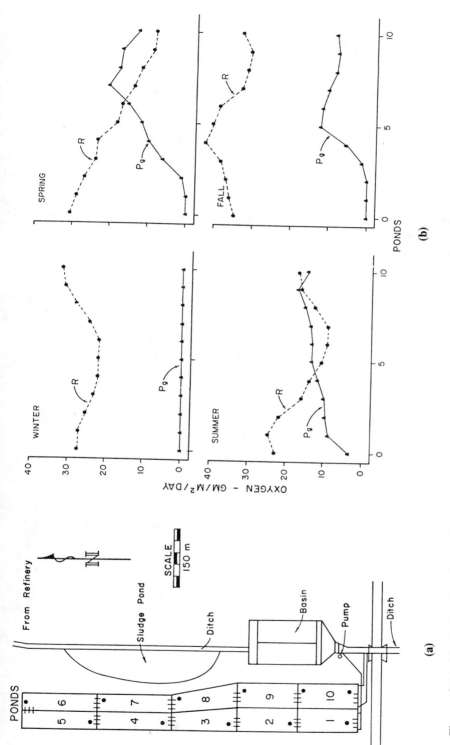

Figure 18.11. Metabolism in a series of pond mesocosms with refinery wastes (Copeland and Dorris 1964). (a) Sketch of pond Layout; (b) respiration, R, and gross photosynthetic rates, P.

loped as cheap and effective waste disposal processors, facilitating the growth of the microorganisms. Such areas are generally centers of wildlife concentration. However, the "ideal" of "neatness in our urban cultures" makes people disapprove of the uncontrolled complexity of this solution. National efforts at "cleanliness" have instead collected litter into dangerous dumps. These have become large, impromptu experiments with self-organizing decomposition, which are now being more carefully studied in experimental microcosms.

Landfill Microcosm

Coutts et al. (1987) used the apparatus in Figure 18.12 to model the anoxic processes in a landfill. Although structured like a compound chemostat, the flow rates were very small and large biomasses were able to develop on surfaces, as in the real landfills.

Fly-Ash Microcosms

Byproducts of fuel combustion include the acid vapors issuing from smoke stacks and the very alkaline solid residue ash, analogous to but different from the lava of a volcano. The solid materials scrubbed from smoke

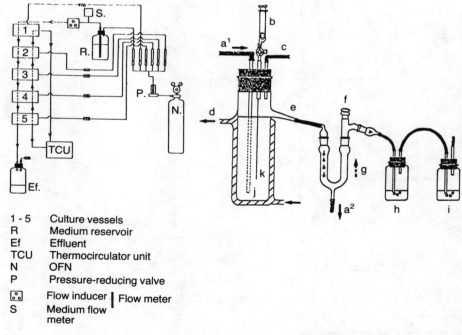

1 - 5	Culture vessels
R	Medium reservoir
Ef	Effluent
TCU	Thermocirculator unit
N	OFN
P	Pressure-reducing valve
[..]	Flow inducer │ Flow meter
S	Medium flow meter

Figure 18.12. Microcosm for processes in an anoxic landfill (Coutts et al. 1987).

stacks are acidic fly-ash. Ultimately, these byproducts must be reincorporated into natural earth cycles through ecosystems. The ash from power plants, formerly dispersed over the landscape, is now often filtered and scrubbed from smoke stacks. The heat-separated acid vapors released from the stack leave behind alkaline solids.

Malanchuk et al. (1980) combined fly-ash waste with soil to form a terrestrial microcosm. Trefoil was planted and inoculated with nitrogen-fixing bacteria. Deionized water was added each week, simulating rain. Leachates were analyzed chemically, examining the role of the micro-ecosystem as a filter. Higher loading rates of fly-ash caused higher rates of leaching of boron, magnesium, and potassium, but zinc was mostly retained and less affected by higher loading rates. (Loading was 224–672 tonne/ha.) Biomass yields were similar at the various loading rates. These microcosms were not given the opportunity to self-organize an input of high-diversity seeding, which might have provided indications of the most appropriate living components. The high levels of boron and heavy metals may be toxic to some species.

Van Voris et al. (1982, 1983) used transplanted terrestrial soil microcosms (Figure 11.4) to study the effects of recycling fly-ash into terrestrial processes. Tubes open at the top received rain, and the leachates coming out the bottom were analyzed. Biomass in the tubes was somewhat similar to that in similarly treated field plots (Figure 18.13a) Compared to controls, the microcosms with fly-ash released more nutrients, some of which may have been the extra nutrients of the ash (Figure 18.13b).

Tolle and Arthur (1983) compared the response of and uptake by alfalfa-, timothy hay-, and oat-vegetated microcosms with field plots, finding similarities in uptake enrichment ratios for 25 chemical elements. Boron was taken up by plant tissues in toxic concentrations. Levels of Mo, Se, and As were high enough to make the vegetation hazardous to animal consumers. Microcosm testing was less expensive than field plots.

Ground-Water Decomposition Microcosm

Some of the organic wastes of society are injected into sedimentary strata, there mixing with ground-water, supporting microorganisms, with consumption and depletion of oxygen. Parsons et al. (1984), Parsons and Lage (1985), Parsons et al. (1983), and Barrio-Lage et al. (1987a) studied the decomposition processes of organic substances in microcosms containing sediment, rocks, and water. The experimental protocol represented processes that occur in the ground-water strata.

Trichloroethene was added and microcosms held dark and undisturbed for 18 months. Microcosms with crushed calcareous rock and water depleted all the spike of Trichloroethene in 21 months, but less than 10% was recovered as dichloroethene. Ammonia increased as nitrates were reduced; sulfate and ferrous ions oscillated out of phase, but H_2S was not

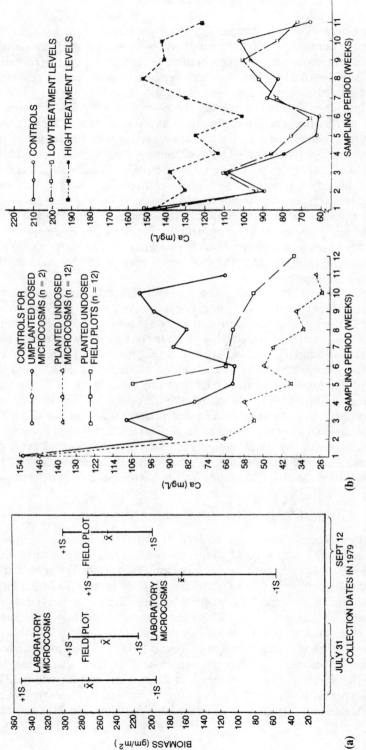

Figure 18.13. Effects of fly-ash on biomass and soil leaching in terrestrial microcosm tubes illustrated in Figure 11.4 (Van Voris et al. 1983). (a) Biomass of alfalfa–clover from 12 laboratory microcosms and 12 field plots; (b) calcium in leachates from microcosm tubes, with several fly-ash treatments.

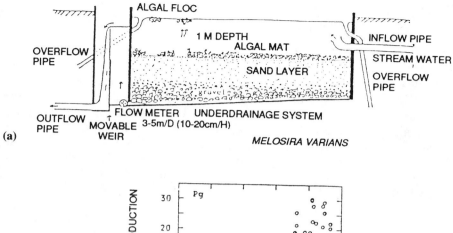

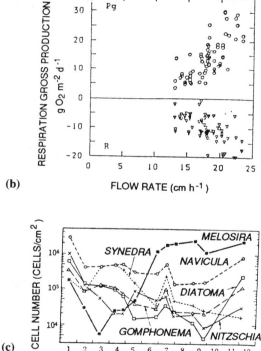

Figure 18.14. Sand filtration microcosm (Nakamoto, 1990). (a) Apparatus; (b) relationship between flow rate and metabolism, April 25–May 6; (c) seasonal changes in algae.

detected; trichloroethane was transformed to cis and trans dichloroethene. Chloroform formed from more chlorinated compounds. Reductive dehalogenation occurred in microcosms with marl, muck, and mud. Rates of depletion of trichloroethene were dependent on the amount of organic carbon present in the sediments. Transformations were much less in autoclaved experiments.

Water Treatment Microcosm

Nakamoto (1989) studied community metabolism of slow sand filter for water purification (Figure 18.14a), measuring the diurnal oxygen changes between inflow and outflow. Gross production rates were higher than the community respiration rates, both increasing with flushing rate (Figure 18.14b). Chlorophyll concentrations and metabolic rates were greatest at intermediate temperatures (10 to 20°C). Photosynthetic production was by algal mats, dominated by *Melosira* in the summer and autumn (Figure 18.14c). Algal mats required about 20 days to develop.

Chapter 19
Human Microcosms and Space

Designing mesocosms to include people has become a major goal of research in ecological engineering as a prelude to life in space and a means to understand the carrying capacity of the earth for humans. Closed systems for humans were a natural extension of the intellectual study of microcosms, models, and the biosphere. Previous chapters have described the creative genesis and resilience of self-organizing ecosystems in isolation when given materials, energy, and biodiversity. To adapt microcosms to people requires a symbiotic human microculture and technology, in partnership with a compatible mesoecosystem.

However, three decades of controversy remain unresolved about what sorts of human-containing systems can be sustained. No long-term human life support systems have yet been demonstrated. In some proposals, all components share a single chamber, whereas in others, components are isolated as unit processes and joined by pipes. At issue are the same questions about the future of humankind on earth. How much of the hierarchy of the biosphere is required? How much technology and biodiversity? How much energy and information?

By one view, the task of making a self-contained universe for people is to miniaturize the earth's biosphere, which has been called "space ship earth." As illustrated by Figure 19.1, the earth supports life with solar-driven processes of photosynthetic production and organic consumption which may balance each other over long periods. In the current economic frenzy, consumption exceeds production as previous storages of fuels, forests, and soil organic matter are used up. The earth system also includes a complex of other chemical cycles driven by the energy flows of geologic, oceanic, and meteorological systems. We take these for granted until it comes time to supply their functions in a microbiosphere.

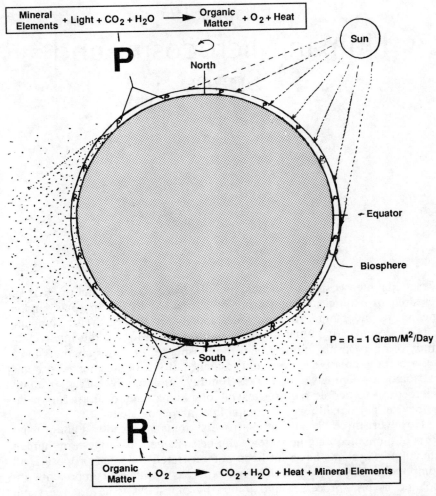

Figure 19.1. Production and consumption on spaceship earth. In the current age of consumption of fossil fuels, forests, and soils, total respiration exceeds photosynthesis with accumulation of carbon dioxide.

Most of the proposals for life support systems have assumed that efficiency and stability would come from keeping systems simple, but those studying microecosystems suggested the opposite, that self-organizing complexity is necessary. The prototypes and tests developed so far for human life support are considered in seven categories, A–G as follows.

Figure 19.2. Large terrarium, Climatron, of the Missouri Botanical Garden at St. Louis.

A. Climate-Controlled Chambers

Sometimes called climatrons, chambers are isolated enough to control climate, but energy, air, and water are freely exchanged with the outside, and people have free access. A famous example is the climatron of the Missouri Botanical Garden (Figure 19.2). A tropical rainforest is maintained, partly by the transplanting of species and partly by the self-organizing influences of birds, insects, and microbes which move freely, interacting with humans visiting for educational purposes. There are large electrical requirements for pumping in air from outside, providing artificial rain, and maintaining temperatures and humidities in the range for the moist tropics. The view of the ecosystem from above shows a kind of hierarchy of organization among the plants around the walkways with heavy human use. Whether the large amount of photosynthesis is equivalent to the total respiration has not been reported, but balance is not essential since air composition is maintained at a stable level by the flow from outside. The climatron was important to thoughts about a more self-contained system which would have closure and recycle air and water flows.

B. Stored Supplies from Earth

For much of the exploration in space so far, life support has been provided by supplies from earth. This system, represented in Figure 19.3a, includes transport of food, water, and equipment. Electricity is

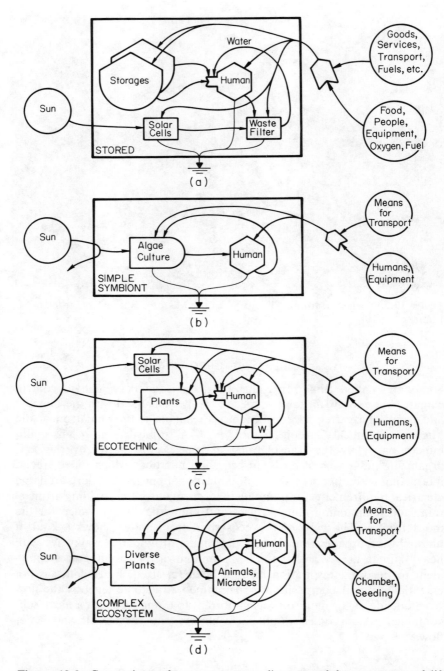

Figure 19.3. Comparison of energy systems diagrams of four concepts of life support for humans in space. (a) Transported storage; (b) simple symbiosis with a plant species; (c) ecotechnic system, using solar-based ecosystem and electricity from solar cells; (d) complex ecosystem containing humans without much technological input.

Table 19.1. Daily Requirements and Byproducts for a Human (167 lb). Myers, 1963b.

Item	Input need	Output
Main requirements		
Water	2200 g	2540 g
Oxygen 25 liters/hour	860 g	—
Organic matter	520 g food	60 g waste
Carbon dioxide	—	980 g
	3580 g	3580 g

Molar respiratory quotient $CO_2/O_2 = 0.825$
120 kilocalories/hour = 0.19 horsepower

Minor requirements (Rich, Ingram and Berger, 1961).
Amino acids: tryptophan, 0.50 g; phenylalanine, 2.2 g; lysine, 1.6 g; threonine, 1.0 g; methionine, 2.2 g; isoleucine, 1.4 g; valine, 1.6 g.
Vitamins: A, 5,000 I.U.; D,E,K; ascorbic acid, 71.6 mg; thiamine, 1.62 mg/riboflavin, 1.62 mg; niacin, 16.2 mg pyridoxine, 2 mg; pyridoxine, pantothenic acid, folic acid, biotin, inositol, B12, traces.
Mineral elements: calcium, 0.78 g; chlorine, 6.04 g; sodium, 3.90 g; phosphorus, 1.17 g; iron 11.7 mg; copper, 1.17 mg; iodine, 0.26 mg; potassium, magnesium, and sulfur;
Trace minerals: cobalt, manganese, molybdenum, zinc.

Superceded by the diet testing within Biosphere 2 (Nelson et al. 1993)

derived from sunlight, using solar cells, and wastes are stored. This system is almost all technological, the humans being the only living component. Some of the requirements for a human being are given in Table 19.1. As Figure 19.4 shows, the weights of materials that must be transported increase with time, making long-period support by simple transport from earth unfeasible. The space programs of the twentieth century were limited because of the excess weights and costs required by the "supplies from earth" life support system then in use.

C. Simple Symbiosis

Another life support system extensively tried in the early days of the space program, used a pure culture of a photosynthetic plant species symbiotic with humans. This system is diagrammed in its simplest form in Figure 19.3b. One plant species was linked with the human consumer to make a symbiotic pair. All other species were kept out. The idea was to keep the system so simple that all its mechanisms could be understood, the medium defined, and fewer resources and less space would be used.

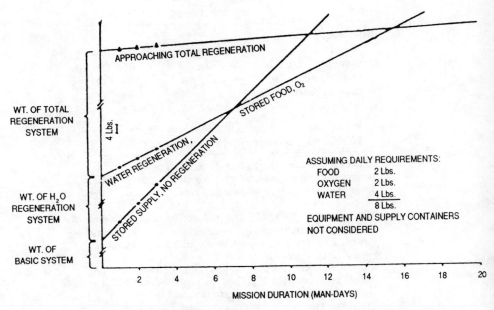

Figure 19.4. Weight requirements for time in space (Taub 1963; Myers 1963).

This monocultural approach was attractive to many scientists accustomed to isolating species in cultures or in agricultural plots.

From 1954 to 1968 several studies were made of closed systems with an algal culture coupled to mice, as a miniaturized test of a simple system for humans based on algal cultures. However, the ones tried by Myers (1954, 1940) coupled gaseous needs only (Figure 19.5). *Chlorella* was maintained in continuous culture in a rapid net growth stage (arrested succession). Oxygen requirements were 25 l/h per man and the algae provided 11–45 l/kg per h. Therefore:

$$\frac{25 \text{ l/h per man needed}}{11 \text{ l/kg per h supplied}} = 2.3 \text{ kg } \textit{Chlorella} \text{ culture needed per man}$$

Area needed was 21.6 m² (240 ft²). The bright light efficiency was 5%, and the light requirement was 2,400 kcal/h.

Pure cultures of algae (*Chlorella*) consumed carbon dioxide and generated oxygen while the mouse consumed oxygen and generated carbon dioxide. However, these amounts did not match because of the differing proportions of carbohydrate, protein, fat, etc. The fact that these proportions were different in the algae and the mouse meant that the ratios of oxygen used to carbon dioxide produced and vice versa were not com-

Simple Symbiosis

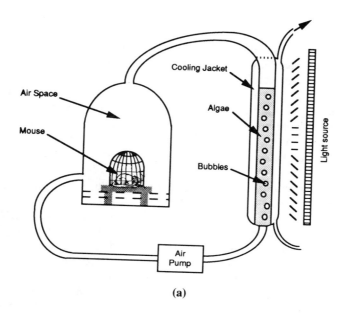

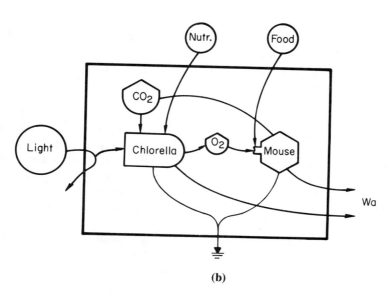

Figure 19.5. System of a mouse and algal culture, used by Myers (1960). (a) Sketch; (b) system diagram, goods and service requires omitted.

plementary. In other words, the respiratory quotients (CO_2/O_2) did not match.

Quoting Oleg Gazenko, Allen (1991) wrote the following about a Russian experience with the human–*Chlorella* system: "Evgenii Shepelev had been the first man to live in a closed system, only he and the *Chlorella* algae . . . Eventually Shepelev and his co-workers got a man to live in one for a month. It was heroic. The toxic gas percentages rose to borderline levels, carbon monoxide especially. The algal food, said Shepelev, was not tasty nor tolerable in quantities above one ounce per day, but the *Chlorella* did succeed in regenerating the required water and oxygen."

In the experimental system, called microterella, tested by Oswald and Golueke (1964) and shown in Figure 19.6, the algae generated oxygen for the mammal, and the mammal wastes, including urea, went into multi-species microbial decomposition area which released nutrients for the plants. Vapor from the mouse was condensed, conserving water. Algal cells were extracted from the culture medium, dried into a green powder, and returned as potential food for the mouse. However, some food supplements were required to balance necessary nutrition for the mouse and some accessory nutrients for the algae. Outside goods and services and electricity were required for these processes and for cooling. Oswald and Golueke wrote: "Perhaps the most surprising finding that was revealed in using the microterella was the self-regulating properties of the system. Despite increments or decrements in mouse, algal, or bacterial population, it remains in balance. Only drastic overloads cause a sustained imbalance. Thus, it may be said that a closed ecological system of the microterella type has homeostatic properties, a valuable buffer against sudden changes in the internal environment."

Duckweed, a somewhat edible small plant, was yet another kind of simple system tried. Duckweed normally floats with its small roots in the aquatic medium from which it draws its nutrients, but it gets its carbon dioxide from the gaseous phase through leaf pores (stomata). Duckweed links aquatic and terrestrial phases of a system. Plants were grown (Wilks 1962) on both sides of sheets of hydrophilic sponge material reinforced with stainless steel wire mesh. Air was circulated over the plants, and there were controls for vapor and cooling. A suction head moved over the surface, harvesting excess duckweed. Unfortunately, duckweed contains substantial amounts of oxalic acid, making it unsuitable for human consumption.

Kirensky et al. (1967) coupled humans to algal cultures (Figure 19.7) and reasonably stable oxygen and carbon dioxide concentrations were maintained over 30-day periods. Food was separately supplied. They reported that the problem of matching respiratory quotients of humans and algae was solved by changing the diet of the humans to match the needs of the algae.

Simple Symbiosis

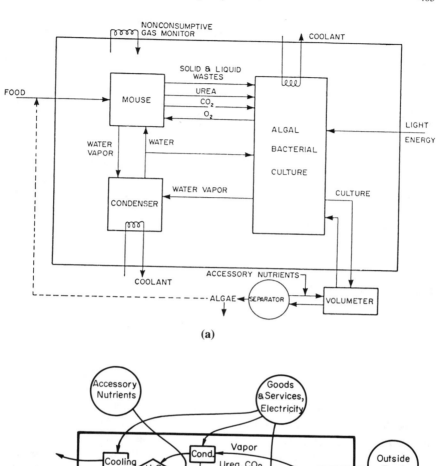

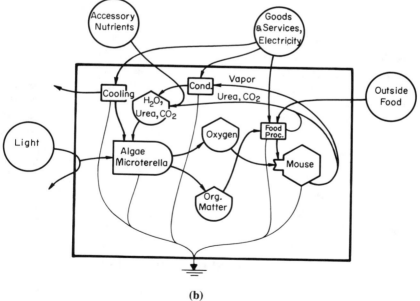

Figure 19.6. System of a mammal, algal culture, and bacteria, used by Oswald and Golueke (1964). (a) System plan; (b) energy systems diagram.

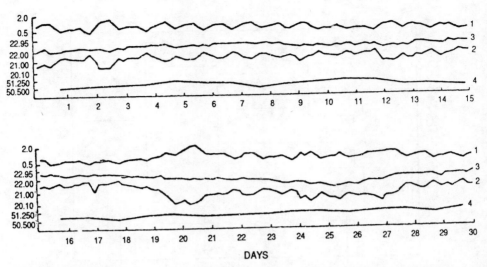

Figure 19.7. Graphs of carbon dioxide, oxygen, and human body weight maintained in a system coupling human respiration to *Chlorella* photosynthesis (Kirensky et al. 1967). 1, Carbon-dioxide in %; 2, absolute oxygen in percent; 3, oxygen concentration (scaled to oxygen in calculations) in gas phase; 4, human body weight in kg.

Allen (1992) writes about later experiments with BIOS chamber, led by its director Dr. Josef Gitelson, of the Institute of Biophysics in Krasnoyarsk, Siberia: "His series of BIOS-3 experiments from 1972 to 1984, in which three people had survived in good health for periods up to 6 months, was in an enclosed plant environment of vegetables plus *Chlorella*. Their wastes were handed out and protein and make-up water were sent in, but the air, virtually all the water, and well over fifty percent of the food was recycled."

Despite partial success in supplying some of the human requirements through plant symbiosis, there were problems with instabilities of pure cultures and low diversity, extensive management was needed, and supplements from outside were required. The simple system of Figure 19.3b was never realized.

D. Connected Unit Processes

Traditional in chemical and environmental engineering, are systems of unit processes connected with pipes, which process the output of one unit as the input of another. Major efforts were made to develop sustainable life support systems for space with connected process units, many con-

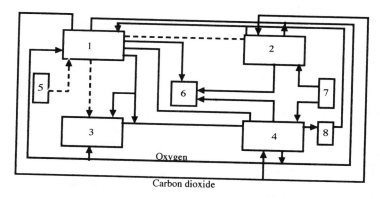

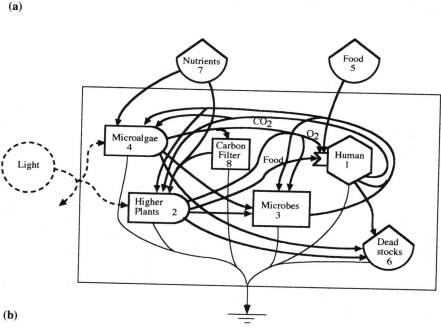

Figure 19.8. System of algae, food plants, and humans, operated for 3 months (Kirensky et al. 1970). (a) Published plan: 1, hermetic cabin with a man; 2, phytotron with higher plants; 3, microbe cultivator; 4, microalgae cultivator; 5, food stock; 6, deadlocks; 7, biogen element stock; 8, carbon filter; (b) energy diagram of system in (a). Goods and services not shown. Dotted lines were added.

taining living populations. For example, Figure 19.8 details a system which had a unit for crop plants, units for microbes and algae to aid waste recycling, a technological unit (carbon filter), and a unit for humans. Some systems were actually operated, but many more were quantitatively

408   19: Human Microcosms and Space

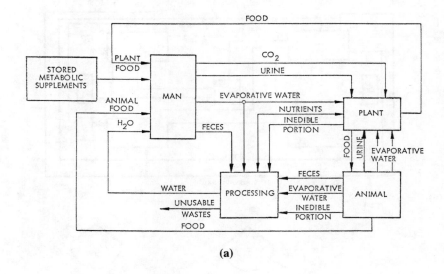

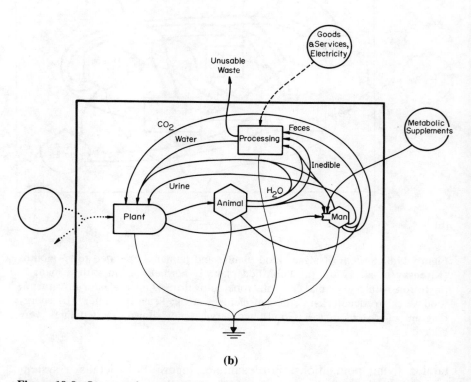

Figure 19.9. System of mass balance of plant, animal, and humans (Jagow 1967). Light added, services were not included. (a) Published diagram; (b) energy language diagram.

considered by means of computations in order to estimate feasibility regarding mass balances, energy, and space requirements (Figures 19.8–19.10).

Efforts in the United States and in the Soviet Union included food crops to provide food and oxygen to animals and to humans, and humans to supply nutrients (in the form of wastes), carbon dioxide, labor, and control. For example, Rich et al. (1959, 1961), using data from hydroponic cultures, estimated the amounts of mixed vegetable crops (peanuts, lima beans, soybeans) that could supply human food and gaseous needs in 196 m^3 of space. As with algae, much of the organic matter produced is not edible. For a closed system, the plant stalks must be consumed and recycled by other means (microbes, chemical oxidation, fire, animals, etc.).

With the system, translated to energy diagram form, in Figure 19.8, Kirensky et al. (1970), supplied light at a level one-third that of daylight, and found 1.5 kg *Chlorella* balancing a 70 kg human body. Appropriate levels of oxygen and carbon dioxide were maintained, with humans present, for three months. Part of the food supply was from terrestrial plants grown in another cell at the same time. The instabilities caused by ultraviolet light and a mutant alga were studied. Physiological performance of algae was modeled using limiting factor equations.

Another way investigators sought to balance nutrition was to use an animal intermediary in the food chain, thus making a slightly more complex ecosystem. Plants were to provide food and oxygen to animals and to humans; animals provided a balance of nutrition to humans; animals and humans returned carbon dioxide, wastes, and agricultural labor to plants. By 1967, concepts of plants, animals, and humans in an ecosystem were equivalent to those in Figure 19.8 and 19.9, for which several mass balances were evaluated.

The waste processing unit was a complex, multi-species microbial ecosystem, a subsystem within the larger mesocosm. Domesticated decomposition ecosystems have been important since ancient times (Chapter 18). Their utility and stability is part of the evidence favoring complex ecosystems for human support.

A very useful summary of the older report literature on life support systems was provided by the Ames Research Center (Jagow 1967). Jagow and Thomas (1966) has systems diagrams, rates of mass flow, weight, and power requirements.

In the nineteen eighties, the crop unit-process approach was again supported by NASA in the Controlled Ecological Life Support System (CELSS) program with very elaborate, energy intensive units for plant growth and processing. Chambers had lights to grow the plants, the electricity to come from solar cells. Folsom (1985) discussing the CELSS program with insight on the complexity required for life support wrote "The simpler the biological components of the system become, the more externalized and complete the control must be." Cullingford and Novara

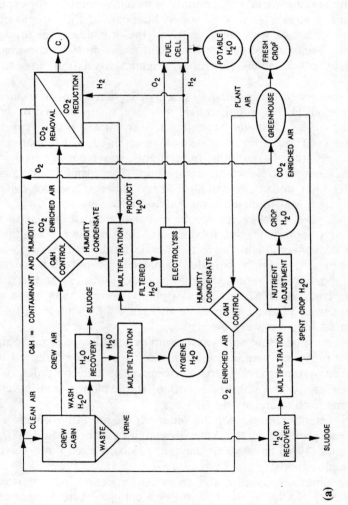

(a)

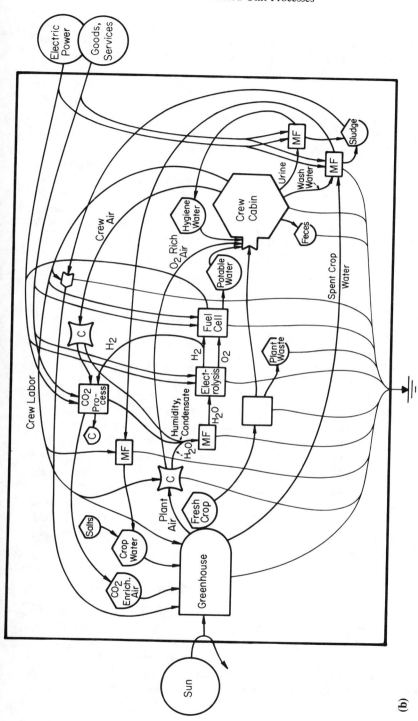

Figure 19.10. Life support systems designed in 1988 for future space missions. (a) Systems diagram from Cullingford and Novara (1988); (b) energy systems diagram of the system in (a), with energy sources and sinks added.

(b)

(1988) presented designs for the life support system in Figure 19.10a, which we diagramed in energy systems language (Figure 19.10b) for comparison with regular ecosystems. The newer work included simulation models for studying the performance and homeostasis of the designs, with results shown on video tapes. Schwartzkopf (1992) discusses concepts in the CELSS program for the moon, giving chemical analyses of lunar material, which contains calcium, potassium, and phosphorus for plant growth but little nitrogen and carbon. Proposals are for small area using high powered lights for plants. A plastic inflatable chamber of 2,356 cubic meters was also discussed. Pay-back on the dollar cost of setting up the system on the moon compared to the cost of supporting people on life-support supplies sent from earth was estimated at 2 years. While citing the poor match with human needs and the accumulation of ethylene, the NASA program was putting people in its high technology growth chamber for space supporting them with photosynthesis of wheat and soybean monoculture for two hours (Corey and Wheeler 1992).

E. Electro–Microbial System

Illustrated in Figure 19.11 was the interesting electro–biological system which was tried in the early space program. A microbial culture of *Hydrogenomonas* was coupled to solar cells and then to humans. Electric power from the solar cells caused electrolysis of water, with hydrogen gas and oxygen as products. The oxygen was available for human use, and the hydrogen was used as the energy source for the *Hydrogenomonas* cultures. These bacteria used the hydrogen gas, carbon dioxide, and nutrients to build bacterial bodies and carbohydrate. However, solar cells are no more efficient in the first step of photon reception than the green photovoltaic cells of plants. This ultimate attempt to support humans on sunlight was no different from growing plants in space, and the weight requirements and maintenance needs of electrolysis, maintaining pure cultures, and processing products was large. The lipid produced by these bacteria was indigestible by man, but a *Hydrogenomonas* system including animals to feed humans was considered by NASA contractors. See mass balances in Figure 19.11b.

F. Ecotechnic Mesocosms and Biosphere 2

Ecotechnological mesocosms are like the modern biosphere, a combination of complex living systems and electrically driven technology (Figure 19.3c). Whereas governments and space agencies made little real effort to test a closed system with a complex ecosystem including people, a new initiative emerged with private funding. Space Biospheres Ventures Corporation was founded by Margret Augustine and Edward Perry Bass, with staff from the Institute of Ecotechnics. At Oracle, Arizona, staff

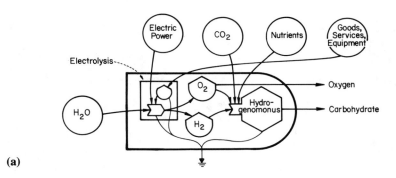

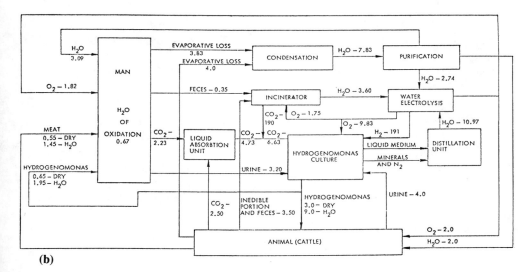

Figure 19.11. Electromicrobial system which converts electric power into organic matter and oxygen with the aid of *Hydrogenomonas* bacteria (Bongers and Medici 1968). (a) Systems diagram (Odum 1983a); (b) representative mass balance for system using *Hydrogenomonas* (Jagow 1967).

directed by John Allen and Mark Nelson developed and tested a mesocosm (Test Module) for a single person (Figure 19.12) and laid plans for the giant macrocosm called Biosphere 2·(Figure 19.13).

Test Module

During 1988–1989, while the large Biosphere 2 chambers were still under construction, a smaller chamber nearby ($27\,m^2$, $476\,m^3$) the size of a glass-walled house (Figure 19.12), was outfitted with monitoring equipment and supplied with a moderately high-diversity plant cover consisting of

Figure 19.12. Ecotechnic mesocosm, Test Module, Oracle, Ariz. (prototype for Biosphere II).

plants and small trees in various large pots with their soils. Human wastes went into an anaerobic decomposition chamber, which overflowed into an area populated with water hyacinths, before the water was used to irrigate the orchard trees and other edible plants, many of which were bearing fruit. Various staff members spent several days each in the chamber and Linda Leigh spent three weeks (Nelson 1989). The attention of the world's media was drawn to the experiment and to the concept of self-contained, high-diversity life support systems.

During experiments involving the Test Module, carbon dioxide levels went up at night and down in daytime, with a range of 150–700 ppm. With human presence, the carbon dioxide level was higher, but remained in a quasi-steady-state between 500 and 1,000 ppm. Occasional surges as high as 4,000 ppm resulted from human activity disturbing soils, cloudy days, etc. In the 1989 experiment, with arrangements for air filtration in the soil, no build-up of toxic gases was noted. The synthetic ecosystem appeared to be atmospherically self-regulating of its major gases, suggesting a microgaia process of homeostasis, which we described in Figure 2.10. The experiments with the Test Module were used to plan the big mesocosm. Details on the Test Module and data on its atmosphere are given in Appendix C.

Biosphere 2

The very remarkable, giant, chambered structure for Biosphere 2 (Figure 19.13) is located north of Tucson, Arizona, at an altitude of 4,000 ft among the natural desert vegetation of the mountains. Modular steel construction was fashioned somewhat after Buckminster Fuller's famous designs, and special glazing methods were used to seal the glass. The system covered 12,700 m² with 7,900 m² area available for plant photosynthesis. The structure had a volume of 180,000 m³ with 165,000 m³ air (Allen 1992). After some months of testing, 8 people entered the system on September 26, 1991 for a 2-year test of enclosed existence and system self-organization.

This ecotechnic system ran partly on green plants and partly on large flows of electricity. The idea was that electricity could be generated from

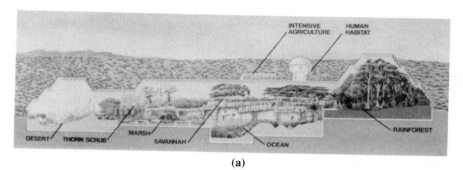

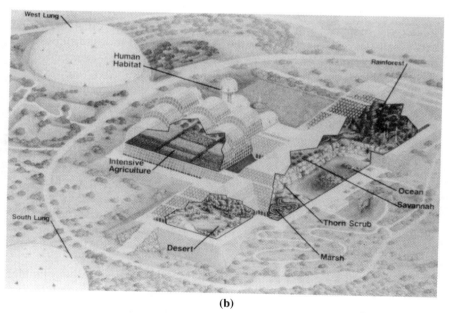

Figure 19.13. Large closed ecosystem completed in 1991 by Ecotechnics group (Space Biospheres Ventures, 1986). (a) Side view; (b) view from above showing biome areas; (c) Record of carbon dioxide and solar energy during the first months of the first test of Biospheres II (Allen 1992).

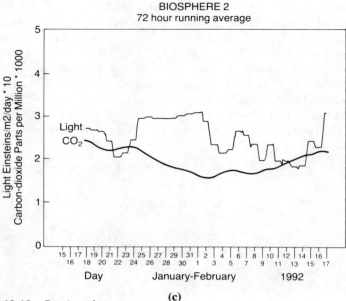

Figure 19.13. *Continued*

solar cells in space, although expectations for solar cell efficiencies may be unrealistic. The electricity in the Arizona operation was from a fossil fuel power plant, which was a much cheaper source than solar cells.

The living part of the system consisted of biome areas where the air conditioning systems set conditions suitable for species which had been transferred from earth biomes including rainforest, desert, savannah, ocean, and agriculture. Several years were spent gathering species from around the world, transferring them to temporary greenhouses with appropriate climate control, and getting ready to install these in the giant chamber. Many preliminary experiments were conducted involving the suitability of the living components so as to improve the likelihood of their survival. One area in the chamber was an agroecosystem with the crops to be managed by the human inhabitants for their own food and other support needs (Figure 19.13b). The area with food-producing crops was mostly separate from the larger areas of biomes with their high diversity of species.

The most striking part of the multimillion-dollar structure was its high technology systems for handling air, water, humidity, air conditioning, dealing with thermal expansion, maintaining slight positive pressure, etc. There was a large network of stainless steel tunnels for processing air and giving people access to pumps and other equipment underground. In addition, there was a complex network of sensors for monitoring, and controls which converged in a computer control center. The people inside

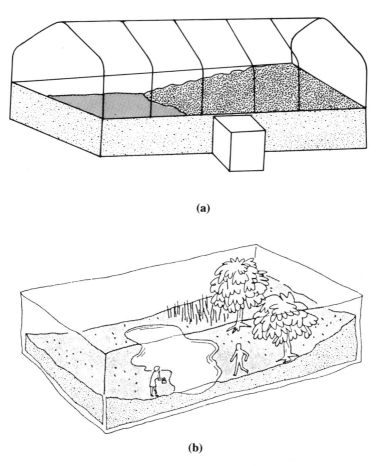

Figure 19.14. Concept of complex self-organizing human microcosm without technological inputs. (a) Experimental system in University of Texas proposal to NASA in 1960; (b) published proposal (Odum 1971).

and outside were connected electronically, with unlimited information exchange by television, telephones, etc. In other words, Biosphere 2 was somewhere between NASA's "all technology" life support (Figure 19.10) and the "all ecosystem" life support that we proposed and still hope to see (Figure 19.14). Because of the high technology electrical inputs, there was much more Emergy in support of the people in Biosphere II than was coming from the sun. This means that the occupants, through information exchange, could bring high Empower to bear on maintaining the ecosystem areas in semi-cultivation. The technology of the structure was as much a new invention as the ecosystems which emerged.

Ordinarily we might expect, based on microcosm experience, to see most of the "biomes" which were planted in Biosphere II to be displaced by self-organization generating a new biome, not like any on earth, but one adapted to this new special condition. Most of the species would become extinct, but a few, including some which were not even introduced on purpose, would become dominant. However, with the high input of electrical energy and information, a higher degree of management may be possible, as in agriculture and forestry.

Because of the massive technological input, Biosphere 2 attracted thousands of people uninterested in nature as well as those rallying about this symbol of a better earth. The systems being developed appeal to the high-tech culture of the United States and our fantasies about what can be done in space. In a way, this might be a shrewd way to move a nation in the right direction, even though it may not be the way to get a low-energy ecosystem self-organized for isolation.

At this writing (1993) the first Biosphere 2 test has been running over a year. The scenario was like that in small microcosms which were also started with large quantities of organic matter. As expected, self organization was observed with some species becoming more important, displacing others less well-adapted. Respiration exceeded photosynthesis causing quantities of carbon-dioxide and free nutrients to increase especially when cloudiness and winter conditions reduced available sunlight. Carbon-dioxide developed a quasi-steady state (Figure 19.13c), a balance between carbon-dioxide generation by soils, microbes, animals, and humans on the one hand and carbon-dioxide use by photosynthesis, removal, by reaction with carbonates in the ocean, limestone, and desert soils, and removal by a sodium hydroxide "scrubber" used for several months. Carbon dioxide ranged from 1,060 ppm in June to 2,466 ppm in December.

However, oxygen decreased from its initial composition of 21%, with photosynthesis not able to match the rate of oxygen utilization by respiration and other processes which may not be known yet. In January 1993 when oxygen was 15% oxygen was added to the biosphere from outside to protect the health of the biospherian econauts.

In developing a terrestrial mesocosm with people, three concepts may compete: (1) to minimize management and allow the self-organizing homeostasis to find its own gaseous levels; (2) to manage each biome section to match its earth prototype; (3) to manage for human sustainability. By trying to keep carbon-dioxide and ocean pH low and filter phytoplankton competitors to favor coral reef organisms, net photosynthetic potential was reduced. Oxygen production potential was also reduced when fast net producing weeds were removed to favor agriculture and the mature state biomes. That the system did not reach a state of net production in the first year suggests that for early self organization more than 0.16 ha per person (8 people in 1.28 ha) may be required. Light intensity inside is about half that outside. On earth there are 10 ha/person (5.14×10^{10} ha/5×10^9 people). Even so, the first period of Biosphere 2 was a tremendous success far exceeding the time of human life support previously demonstrated.

The world was fascinated. Because so much emphasis was placed on public relations at closure and relevance to space, news media treated it as a game to see how far things could go in complete isolation, as if they were on Mars. Commentators criticized the normal and correct ecological engineering procedure of making in-course corrections, such as adding calcium carbonate and auxiliary foods for human support while a workable agricultural system was found. It is likely to take years for self-organization to develop a fully adapted ecosystems making complete isolation possible. What is sure to result is a new appreciation for the marvelous homeostasis of our earth's life support system. Many scientists not trained in systems ecology were also critical, asking what research was taking place that could not be done by isolating units for separate study. Unfortunately, knowledge of the parts of an ecosystem provides little information about the operation of the whole. An ecosystem has to be run to determine which models are correct.

G. Complex Self-organized Mesoecosystem

The most promising, practical life support system for space, where concentrated energy sources are few and equipment maintenance difficult, may be an ecosystem that has self-organized to include human participation and only a minimum of technology (Odum 1963). We, and other microcosm scientists, have been advocating and submitting proposals to test this kind of mesocosm self-organization, containing humans, since 1955 (Table 19.2 and Figure 19.3d). However, a low-technology, complex, self-organized ecosystem large enough for humans has still not been tried. Whether a small, closed, diverse ecosystem is self-sustainable in space (without humans) is not even known. Microcosm tests in space are long overdue.

Work on microcosms in the 1950s and the development of the space program started the movement to use ecosystems for human life support. NASA, with emphasis on short-run needs, developed systems to transport requirements for a few days only. Except at some conferences, almost no support was found for developing ecosystems for space more complex than mixed microbial cultures. Although there is agreement that solar energy is necessary for long periods of support, there has been great disagreement about the area required per person and the role of plants, solar cells, biodiversity, and complexity in life support ecosystems. Space shuttle planners were discouraged away from the ecosystem concept because a substantial area of high-diversity photosynthetic ecosystem is required to support a human. Yet, that size may be essential for sustainable

Table 19.2. Milestones in Controversy over Human Life Support in Mesocosm.

1959, 1969, 1984 Research proposals for complex ecosystem module to National Aeronautics and Space Administration (Figure 19.14a).

1962 National symposium Aug. 28 sponsored by 6 national societies and published in Vol. 25 (6 & 7) of the American Biology Teacher (1963) contrasting the points of view on ecosystems or engineered life support. F. Taub organizer.

1966 At NASA conference on bioregenerative systems, Jenkins (1968) estimated that research and development spent by NASA, the Air Force, and other government agencies was $30 million since 1951. As the reports show, most funds went into single species biochemistry or organismic studies, not on ecological systems. He wrote:

> There have been about 15 years of fairly intensive research on algae, with occasional research studies on higher plants, and about 4 years of research on Hydrogenomonas bacteria and electrolysis. During this time, no efficient and reliable system capable of continued operation has been developed.

1977 With renewed interest NASA supported a small program: Controlled Ecological Life Support System (CELSS), including small project by C. Folsom and J. Hanson on sealed and defined microcosms (Chapter 20).

1982 Workshop on Closed System Ecology at California Tech (Jet Propulsion Lab) The workshop summary cited many examples of closed microcosm stability. Participants endorsed complex microcosms as important to basic science and to NASA. However, the question of large scale trials containing humans was not reported.

1986 Editorial on the need for realism in space (Van Allen 1986):

> In the arena of national space policy, I am deeply distressed by the dissipation and misdirection of our immense technical and human resources on enterprises that appeal to persons of a science fiction mind-set, but are otherwise ill-considered and fruitless.
>
> My own view is that our national predicament is the result of the clash between the mythology of manned space flight and the real achievements of space technology in practical applications to human welfare and the expansion of human knowledge.

1989 The second International workshop on closed ecological systems held at Krasnoyarsk and Shushenskoy, Siberia in September 1989 adopted a resolution that the term *biospherics* be used for the study of closed ecological systems (Nelson, Hawes, and Augustine 1990).

1991 Official start of Biospheres 2 human occupancy, Sept. 26.

support for a human by whatever means. Perhaps a photosynthetic surface with self-organizing components, which unfolds after reaching orbit, can support a human.

We would expand the procedure now standard for the smaller microcosms (without as much technology and cost as Biosphere 2) as follows: set conditions, mix in as many species as possible, have one or two people reside inside (rotating individuals in and out), and let nature (self-organization) take its course. The result should be a homeostatic, high-diversity, production and consumption ecosystem in which human participation has become a symbiotic part. This book has already reviewed the way small self-organized, complex microcosms become resilient, since unstable elements tend to be eliminated during the self-organizing period. See also Chapter 20. The complex ecosystem approach (Figure 19.2d) achieves stability by utilizing a complex, self-organizing ecosystem like that of the biosphere. As already discussed in Chapter 2, ecosystems self-organize in a process similar to learning. One adds a large variety of organisms, plus the mix of necessary physical and chemical ingredients, seals the system, exposes it to a sunlight energy source, and it starts its succession. Very quickly, those species prevail that can best couple with others in establishing the closed cycles of elements, in developing hierarchies, and in providing each other with reward stimuli. The criterion for survival in self-organizing systems is that a species must contribute something to another part of the system so that a feedback stimulus comes back to it. Whereas one species of plant does not match human needs metabolically, a mixed association is self-regulating to match human needs. Whatever is not used stimulates those plants that grow a little better with that extra resource.

These are the principles for complex self-organized life support (Odum 1963):

1. Small, complex microcosms have proved stable and self-regulating. Food chain diversity is necessary to these properties. Humans could not and should not try to operate at the base of the food chain (small area basis), because they require the many complex inputs for their nutrition and other needs.
2. Gross production is the appropriate measure of total performance in a light-driven ecosystem. Systems with maximum output (power) prevail through self-organization, achieving security from mutants and epidemics.
3. The efficiencies of gross production at maximum power loading are higher in complex microcosms than in simple cultures or photovoltaic cells operating with the same resources.
4. Proposed monoculture systems require digester units, which are really complex ecosystems, plus processing units not sustainable except with outside power and parts.

5. The quickest way to produce a working ecosystem including humans is to put them in a system with multiple seedings and let self-organization develop a high-performance system, since self-organization evolves towards maximum power performance. Such self-organization would include the tendencies of human psychologies and the human inclinations to manage.

Simulation Model for Closed Terrestrial Ecosystems with Humans

As already considered in Chapter 2, the understanding of how living models (microcosms) work is not complete without computer simulations. The simulations show whether the ideas about how the parts are working are consistent with each other and with observations. Terrestrial microcosm models were simulated in Chapter 2, representing in the short-run the gaseous homeostasis which rapidly develops when a terrarium is sealed.

To help understand what is involved with the main chemical cycles in closed ecosystems with humans, such as Biosphere 2, an overview model of a closed-to-matter terrestrial ecosystem, called BIOSPHER (Figure 19.15a), was run in simulation. The results of simulations of 2-week and 3-year periods showed the homeostasis of closed, coupled mineral cycles linked by biotic production and consumption. In the model, water and nutrients, including phosphorus, nitrogen, carbon, and oxygen, are in closed cycles and are the main variables. Calcium carbonate (in ocean waters, calcareous soils, and limestone) buffer the carbon dioxide. This model holds plants and consumers constant during any simulation run, by holding production and respiration coefficients constant. The first runs (Figure 19.15b,c) were provided with data for a terrestrial mesocosm 30 m high, supplied with assumed stocks of constituents. The simulation in Figure 19.15b started with an initial excess of organic matter. Respiration was much greater than photosynthesis, and carbon dioxide accumulated. The graph shows the daily fall and rise of carbon dioxide during alternating day and night. Because the total storage of oxygen is limited by the volume of air space, oxygen varies diurnally and seasonally much more than in the earth's atmosphere, which has a resevoir of oxygen with a much longer turnover time. If the model is calibrated with photosynthesis equal to respiration, a seasonally pulsing steady-state results (Figure 19.15c). As Figure 19.15c shows, adding a limestone ($CaCO_3$) reserve buffers the undulations of carbon dioxide, just as the ocean does on earth.

When Biosphere 2 was started, carbon dioxide levels rose rapidly and sodium hydroxide was added by means of a scrubber to convert some of it to calcium carbonate. Efforts were also made to raise the pH of the ocean module in order to encourage reef organisms accustomed to high pH. This management kept the carbon dioxide at a lower level than would have resulted otherwise. With less carbon dioxide, photosynthesis may have been carbon limited. Soon it was discovered that oxygen had declined from 21% to 16%.

In July 1992, data for Biosphere 2 were obtained, and the model in Figure 19.15a was recalibrated using a spread sheet template (EXCEL), resulting in the program BIOSPHER. Simulation runs were made with respiration higher than photosynthesis and with large initial storages of organic matter representing the soils and peats used to start Biosphere 2.

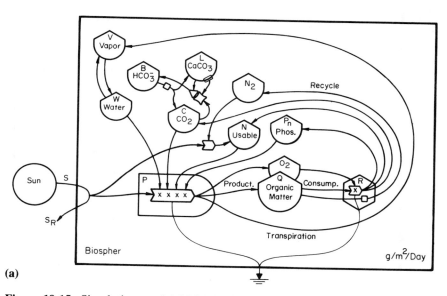

Figure 19.15. Simulation model BIOSPHER for a large, closed terrestrial mesocosm with cycles of carbon, oxygen, nitrogen, water, and phosphorus, buffered with limestone ($CaCO_3$). Equations:

$P_f = S_R WCNP_n$; $P = k_2 P_f$; $R = k_3 O_2 Q$; $dO/dt = dQ/dt = P - R$;

$S_R = S - k_0 P_f - k_1 S_R N_2$ and therefore $S_R = S/(1 + k_0 WCNP_h + k_1 N_2)$

$dV/dt = k_4 P_f + k_5 O_2 Q + m_9 W - k_6 V$; $dB/dt = m_4 C - m_2 B$

$dN/dt = m_1 S_R N_2 - k_9 P_f + k_8 O_2 Q$; $dL/dt = e_2 b - e_1 C$

$dC/dt = m_6 O_2 Q - m_7 P_f + e_4 B - e_3 C$

$dO_2/dt = k_2 P_f - k_3 Q O_2$; $W = TW - V - f_q Q$

f_q = water fraction within organic soils and biomass

(a) systems diagram; (b) simulation over a two week period; (c) simulation of the model in Figure 19.5a, calibrated with photosynthesis equal to respiration; (d) results of simulating the model in Figure 19.15a calibrated with data from Biosphere 2, with respiration in excess of photosynthesis. Left, without limestone; right, with limestone.

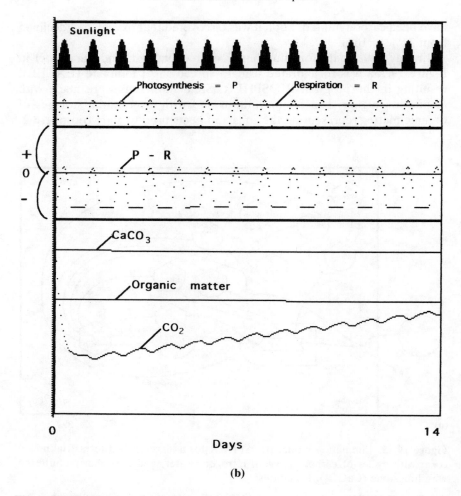

Figure 19.15. *Continued*

Although carbon dioxide was produced in excess, the effect of the carbon dioxide in the model was to increase photosynthesis, causing production to balance respiration. However, when the model was calibrated with respiration (15 g/m^2 per day) twice photosynthesis (8 g/m^2 per day), and with the calcium carbonate reserve present, the rate of photosynthesis was not enough to replace the oxygen, and the oxygen concentrations declined (Figure 19.15d, right graph). When the simulation was run without calcium carbonate, the higher levels of carbon dioxide were enough to stimulate photosynthesis, with production exceeding respiration and oxygen increasing (Figure 19.15d, left graph). The model runs calibrated for Biosphere 2 show how oxygen is linked to the carbonate

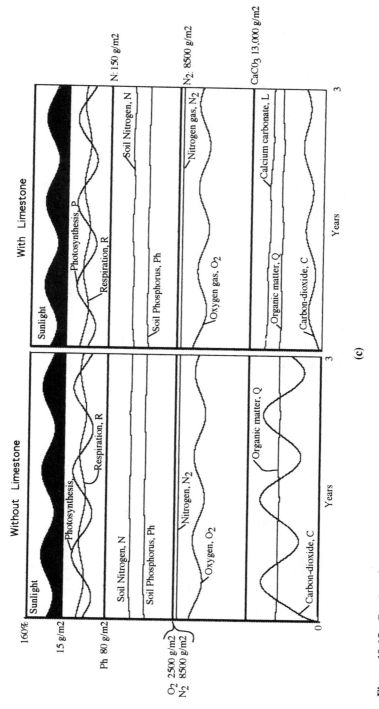

Figure 19.15. *Continued*

426   19: Human Microcosms and Space

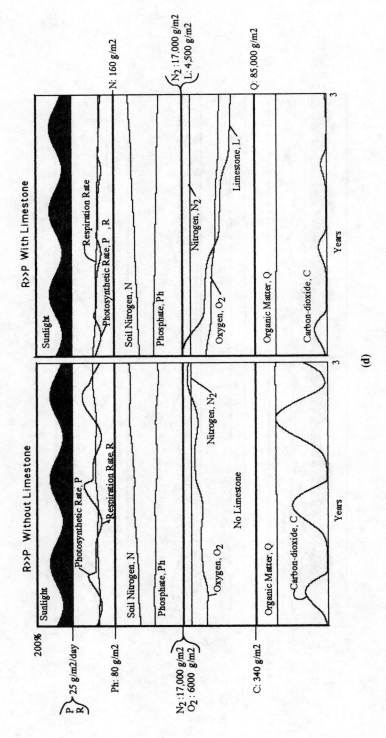

Figure 19.15. *Continued*

buffering, which tends to limit photosynthesis. The carbonate control of oxygen suggests new ways of thinking about the whole earth system. What proportion of carbonate maximizes whole-earth performance?

Microcosm results and their models reveal some of the self-regulating homeostasis in multiple connected chemical cycles, as their inputs and outputs flow between the organisms (Chapter 4). Where there are many complex, coupled cycles, anticipating results is not easy, but the insights emerging bear on planetary biogeochemistry. Walker (1991) simulated an overview model of the earth in which carbon, oxygen, and calcium cycles controlled carbon dioxide through biotic respiration and carbonate precipitation, but did not show the effect of carbon dioxide removal increasing oxygen.

The experience with Biosphere 2 suggests that letting self-organization find its own homeostasis may be better than human intervention, given the present stage of our knowledge. The experience shows the difficulty of trying to manage one biome module without regard for the gaseous network of the whole system. The way a complex network supports humans in a small system is becoming evident, whereas the intercoupling of the worldwide system of cycles is, as yet, little studied. The insights from Biosphere 2 already vindicate those who have worked so many years towards the ideal of a whole earth mesocosm. The concept global homeostasis is sometimes called the Gaia theory see Chapter 20 (Lovelock 1979).

Carrying Capacity

The critical question in considering alternative mesocosms for life support is the carrying capacity for a human. What area of sunlight and what other resources are required to sustain a human for a given level of existence (standard of living)? Resources of different kinds can be expressed on a common basis using the solar Emergy concept (solar energy equivalents required directly and indirectly) in order to estimate the carrying capacity. See Chapters 3 and 6 and recent papers (Odum 1986, 1988) for explanations of Emergy and transformity.

Since humans are high on the scale of required input supports, their requirements are much more than their gaseous life support. Several authors, examining only oxygen production and carbon dioxide uptake of rapidly growing plant cultures in early successional stages, claimed a human could be supported on $0.5\,m^2$ of sunlight or less (Kleiber 1961). This estimate was based on algae in a net growth stage with all the algal needs supplied by the laboratory at enormous unaccounted costs. Actual algal pilot plants showed more realistic yields, similar to agriculture (Burlew 1953). Support for humans, however, requires more than their net food and gaseous support; it requires the larger area necessary for the diversity that gives stability. Those considering the total needs of humans suggested 1.0 ha as minimal (Odum 1963). The resources necessary for humans depends on the level of support (standard of living).

Solar transformity is the solar Emergy required to generate one unit of another kind of energy. Coal is about 4×10^4 solar emjoules per joule (sej/J); electricity 1.6×10^5 sej/J, etc. For a human, the transformity ranges from 1×10^5 to 1×10^9 solar emjoules per Joule, depending on the level of the development of the economy, educational level, etc. In other words, the resources required for a human depend on the place they occupy in the energy hierarchy. Less Emergy is required for bare, primitive subsistence than for support of a human performing a high level of information and control service. Figure 19.16 graphs the area of solar energy required to support a person as a function of the person's solar transformity. Citizens in the U.S. receive more Emergy support on the average than Indian citizens, many of whom work without much education or technological aid. Multiplying the transformity by the human metabolic energy gives the amount of solar energy required per person. For example, using transformities from earth, one estimates that much more solar area is required per person for full support than is supplied from solar energy to Biosphere 2.

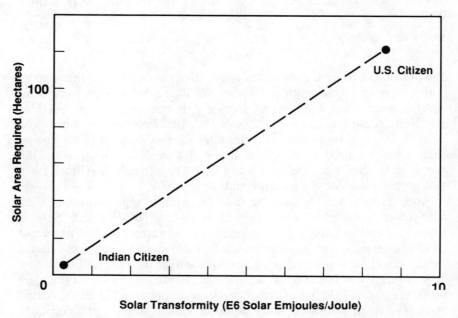

Figure 19.16. Solar area required as a function of the transformity of a human supported on solar energy in space with solar energy per day 4 times that on earth, 2.4×10^{10} solar joules/m^2 per year.

The following calculation evaluates the area in space required to support a human if there were no inputs other than the sun.

$$\frac{\text{(Solar transformity in sej/J)}(1.0 \times 10^7 \text{ J/day human metabolism})}{\text{(sej/ha per day solar insolation available in space)}} = \text{number of ha}$$

The available solar energy in space outside the earth is about four times greater than at the ground on earth because in space there is no atmospheric absorption or dispersion, and there is no night period without light.

Solar insolation:

$(4)(3870 \text{ kcal/m}^2 \text{ per day})(4186 \text{ J/kcal})(1 \times 10^4 \text{ m}^2/\text{ha})$
$= 6.48 \times 10^{11} \text{ sej/ha per day}$

For a high position in the energy hierarchy corresponding to an average human on earth now, with solar transformity of 5×10^6 sej/J, the solar area required is:

$$\frac{(5 \times 10^6 \text{ sej/J})(1.0 \times 10^7 \text{ J/day per person})}{(6.48 \times 10^{11} \text{ sej/ha per day})} = 77 \text{ ha}$$

This calculation shows the very large area required for the solar Emergy necessary to support a human in the style of humans in developed countries whose economies are largely based on the intense, high-Emergy, fossil-fuel use of this century.

Next, let's reverse the calculation and estimate the level of human existence which can be supported by one hectare of solar area in space. If the solar energy absorbed is divided by the human energy need of one person, the solar transformity would be:

$$\frac{6.48 \times 10^{11} \text{ sej/ha per day})}{(1 \text{ person/ha})(2,500 \text{ kcal/person per day})(4186 \text{ J/kcal})}$$

$= 6.2 \times 10^4$ sej/J of human metabolism

This transformity is equivalent to that of farm animals. In other words, life would be primitive and entirely geared to a low trophic level role if it were dependent entirely on a solar base with one person per hectare. Realism requires recognition of the larger surface area required to support humans on solar energy alone, such as a broad base on a moon or planet. For space vehicles, some kind of expansive surface would be required. Unfolding a large solar-receiving surface in space, without gravity, may not be out of the question.

In most life support plans now under consideration, a substantial energy subsidy from the earth is being planned in the form of electricity, external maintenance, the manufacture of structures, etc. Solar cells do not yet yield net Emergy, but are a way of transporting high Emergy into space to aid the manufacture of the high-transformity energy necessary to maintain humans above an animal level. In this way, the input solar emcalories (solar equivalents) per person can be higher. Biosphere 2 has large inputs of Emergy as electricity, as well as capital equipment and information.

Comparisons are made in Table 19.3 of the resources required to support a human being expressed in energy units of one kind (solar Emergy). The table includes the Emergy analysis of Sky Lab operations

Table 19.3. EMERGY Requirements for Human Life Support.

Note Item	Solar Emergy/person E15 sej/person/yr	Purchased Emergy Percent
1 Average world	3.6	49.00
2 U.S. Average	29.0	88.00
3 Complex, low tech ecosystem	333.0	99.81
4 Biosphere 2	4303.0	99.98
5 NASA Skylab	2,719,644.0	99.99

[1] On the average the world's renewable Emergy flux in rain, winds, and geologic processes supporting the land is free (9.44 E24 sej/yr), whereas the Emergy flux generated from non-renewable sources has to be purchased (9 E24 sej/yr).

[2] Odum, Odum, and Blissett (1987)

[3] Simple Enclosure, 1 person per hectare

$(1\,ha)(1\,E4\,m^2/ha)(\$500/m^2)(2\,E12\,sej/\$)/(30\,yr) = 0.33\,E18\,sej/person/yr$

Environmental inputs estimated on area of land basis:

$(9.44\,E24\,sej/earth/yr)(1\,E4\,m^2/ha)/(1.5\,E14\,m^2\,land) = 6.3\,E14\,sej/ha/yr$

[4] Installation evaluated as $35 E6 for 30 yr

$(35\,E6/30)(2.5\,E12\,sej/\$) = 2.92\,E18\,sej/yr$

Electric power, 3 E6 watts

$(3\,E6\,watt)(1\,J/sec/watt)(3.15\,E7\,sec/yr)(2\,E5\,sej/J) = 18.9\,E18\,sej/yr$

Operation: $(\$2\,E6/yr)(2\,E12\,sej/\$) = 4\,E18\,sej/yr$

Total annual Emergy $(2.92 + 18.9 + 4) = 25.82\,E18\,sej/yr$

per person: $(25.82\,E18\,sej/yr)/(6\,persons) = 4.3\,E18\,sej/person/yr$

Environmental inputs estimated on area of land basis:

$(9.44\,E24\,sej/earth/yr)(1\,E4\,m^2/ha)(1.28\,ha)/(1.5\,E14\,m^2\,land) = 8.0\,E14\,sej/yr$

[5] Energy analysis of space operations (Noyes, 1987)

$(44.5\,E9\,kcal\,fuel\,equiv./person/d)(365\,d/yr)(4186\,J/kcal)(4\,E4\,sej/J\,fuel)$
$= 2,719,644\,E15\,sej/person/yr$

Solar Emergy received negligible by comparison:

$(5.6\,kw\,solar\,electricity)(1,000\,watt/kw)(1\,J/sec/watt)(3.15\,E7\,sec/yr)$
$= 1.76\,E11\,sej/station$

per person: $(1.76\,E11\,sej)/(3\,astronauts) = 5.88\,E10\,sej/person/yr$

by Noyes (1977), showing the very large Emergy involved in costs of services for lift-off, equipment, ground support, etc. Much greater Emergy is required per person in space operations than on earth, with almost all of it furnished from the economy from earth, and very little from direct solar energy. Biosphere 2 also requires large amounts of purchased resources, compared to free environmental resources.

In the complex, low technology alternative, only the Emergy requirements of the chamber are included in the calculation (Table 19.3, item 3). Its Emergy requirement appears to be an order of magnitude less than the high technology Biosphere 2. Whether one person per hectare could be supported with this kind of system remains to be tested.

None of the proposed alternatives for humans in space show any possibility of being self-sustaining without regular support from earth, since the Emergy available from the sun is so much smaller than that required for a human in space. The conclusion would be different if rich, concentrated fuel resources were found on a planet. However, human-containing mesocosms can be important to the earth even though they require earth support. To compete economically on earth, a space operation, through its actions back on earth, must have as much positive effect in increasing Emergy production as alternative projects for the same money and resources.

Concluding Remarks

Earlier efforts to fit cultivated plants with animals without the mix of complex self-organizing life support were not successful. Various efforts showed that algae and mice, for example, were not complete partners. Putting people and farming in a microcosm without a good part of its contents devoted to self-organizing complexity will not achieve a stable system. This is an important principle for large microcosms containing people. A large variety of complex diversity needs to be added along with the people, so that the self-organizational process can proceed.

Self-organization on earth has produced a long series of energy transformations for the support of educated humans. Much more is needed than the food and oxygen required for body maintenance. For humans to be supported as on spaceship Earth, will require similar types of support. To support a human body in relatively small, closed systems may require that their functions be reduced to a much more primitive stage, commensurate with a narrow resource base. Sending human bodies without developed intelligence into space may accomplish less than sending the information of civilization into space by other means. The higher the transformity of information, the lower is its carrier's energy, and the greater may be the ease and flexibility of transmission. Perhaps it will eventually be possible to send high-transformity information into space, capable of regenerating and reorganizing on planetary surfaces already in space. This may not be people in space ships, but some essence of the genetic and cultural information of our evolution and civilization, capable of interacting with the other planet's resources, may generate our pattern elsewhere.

The study of closed systems including humans is an exciting approach to understanding humanity and nature on earth. It is good science and merits unified international support. That such systems can be self-sustaining in space on resources known so far may be only a dream. Thus, the controversy over the best application of society's efforts continues.

Chapter 20
Microcosm Perspectives

In our review of ecological microcosms we mixed models and measurements to help provide an understanding of the self-organization of systems. Many principles of ecology have been better illustrated in a microcosm than in the greater, global outdoors. While leaving out some phenomena, miniaturization has allowed organizational phenomena, normally requiring years, to show up in months. In this final chapter, the holism of microcosmology provides perspectives for our planetary processes at large. The principles of system design, energy transformation, mathematics, materials, complexity, and information, found in miniature, may be similar for the rest of our realms.

Increasingly recognized, the hierarchical structure of all of nature and humanity involves the scales of space, time, and energy. The phenomena in small microcosms can best be compared with phenomena of larger systems by increasing the scales' time and space. Position in the scale of energy hierarchy is given by the transformities (defined as the energy of one type required for each energy unit of another). For example, the phenomena in a pool may be compared to those in a lake ten times larger by multiplying the quantities and turnover time by ten. When a microcosm is constructed, the longer turnover time, and the larger parts of the hierarchy, are eliminated and the window of interest is trimmed to smaller dimensions. There is still a hierarchy to examine, but it is smaller and its components develop faster. Similarly, principles that apply to the smaller system can be compared with those for the larger, outdoor systems by slipping the window of interest upward, comparing small, fast compartments with the slower ones on the larger scale.

Perspectives on Principles of Self-organization

In Chapter 8, where we covered the History of the microcosm concept, we described the philosophic controversy within the science of ecology revolving about parts, wholes, and hierarchy. The resolution of differences and resynthesis of the field may hinge on the issue of how self-organization occurs. Microcosm research has been one of the main sources of information concerning this. Interpretation of the data depends on how natural selection is viewed.

Where resources are in excess and network controls have not developed, reproduction and growth are rapidly exponential, and competition is expressed by those species better adapted to such conditions, driving out others. Such competition accelerates the early incorporation of unused resources, and maximizes productivity and use (maximum power). Newly assembled microcosms have large rates of extinction, some of which is caused by competition for resources. Controversy concerns what happens next.

Some see the ecosystem continuing to be a little-organized, competitive struggle for existence. They believe the continual interplay of changing conditions keeps species from overgrowing each other. The organismal-oriented concept of natural selection is really "self-selection", where survival is believed to go to the maximum reproducer, thereby selecting its own kind over competitors. The system-oriented concept of natural selection is that units at one level are chosen by units at a higher level, usually through positive reinforcement of those contributing to the system beyond their own population functions. At a gathering at Biosphere 2 in 1990, E. P. Odum sought to explain this to people with an organismic orientation by using the phrase "network mutualism."

Many systems ecologists, including the authors of this book, see abundant evidence of organizing mechanisms on the larger scale. These mechanisms continually adapt their interspecies designs through reinforcement multiplier controls which tune the system to maximize power and commensurate efficiency (the maximum power principle). One of the purposes of this book was to assemble some microcosm examples and show the generality of that concept. Thus we mixed data on reinforcement with the diagrams and simulation models that show how reinforcement from one hierarchical level selects parts at another level maximizing the function of two or three levels.

Those taught to concentrate their research effort by studying only a part or a population, see the microcosm as they see their environment — as a mixture to be simplified. However, since self-organization can reinforce relationships that override such individual tendencies as runaway growth, population behavior in ecological microcosms differs from the behavior of those populations when isolated from main system processes. For example, prey–predator oscillations in a two-species study in isolation avoids the questions about system roles. Studies of the same oscillations within more complex microcosms show that all the species are affecting each other's performance. Often, when one population oscillates, they all do. No study of one or two species alone can predict much about how they will behave in the larger system.

Perspectives on Science and Microcosm Uses

Microcosms have been designed for many different conceptual purposes. Those set up for one purpose can be and have been valuable for others. To some, microcosms are a tool for studying large ecosystems. For others, they are new creations, worthy of study themselves. Experimental microcosms draw creative originality and new discovery out of their investigators.

Because science and engineering have emphasized taking apart and studying component mechanisms, many future advances are possible with the expansion of the neglected science of putting together. Microcosms are a means for synthesis. Materials, species, and energy inputs are combined, and the resulting self-organization is studied through measurements and models. The standard format for scientific experimentation in "take-apart", analytic science has been to isolate a process from all other variables in order to study its principles. For synthetic science (putting back together), experiments use the whole system, retaining everything except some one variable which is changed or omitted. In this way the effect of a part on the whole system is found. Similarly, the mathematical models for making such experiments understandable include the main processes of the whole system so that the effect of changing one part can be studied.

For understanding how parts work together, it is appropriate to ask how the whole ecosystem operates with a changed variable. For ecosystems, enough funds for replicated experiments on whole lakes, streams, wetlands, or forest tracts are rarely available. Even for the economically important questions, the best approach has been to develop microcosm experiments with the essence of the structure and operations of the larger system. Much of the recent microcosm work has been funded to answer questions of environmental impact, pollution, and resource depletion. As often in science, new discoveries and progress come about when investigators are forced to look at new situations and ask new questions. For many purposes, self-organizing systems are cheaper than human operations and engineering. Because of the miniaturization, self-organization, and homeostasis, microcosms and mesocosms may be more sustainable in a world of declining resources than systems where unit processes are separately managed.

Because of energy and cost realities, microcosmic methodology may be a cheaper biotechnology than gene engineering of single species. For needs on the scale of the human economy and environment, direct manipulations at the tiny microscale of genes may not be as cost effective as letting the larger units control the smaller units through the self-organizing network. Microcosmic engineering automatically uses biochemical microbial diversity from evolutionary variation and selection, plus the selective actions and controls of its species, as these species develop mutualistic reinforcement.

Perspectives on Scaling Results from Test Microcosms

Because they are smaller, faster, and lower in the energy hierarchy, the results of studies that use microcosms to test impacts on ecosystems differ from the results of full scale tests of larger, outdoor ecosystems. Because the oscillations due to larger–longer units and processes at the higher level are excluded, the longer–larger responses to treatments are not included (Chapter 6). Perez et al. (1991) verified these concepts with a study of three different sizes of microcosms, each treated with the pesticide kepone. Some oscillations and treatment responses on the larger scale were excluded in the smaller scale. They concluded that the smaller the microcosm, the greater the risk of underestimating the effects of the kepone treatment.

Our hierarchical theory (Figure 6.1), suggests that responses depend on the transformity of the impact, and each impact has a hierarchial level where its interaction is maximum. In general, events on the larger scale may be predicted by multiplying results found on the small scale by the size–time scale factor between the microcosm scale and the larger one of general concern. In this way perhaps the fast, small surges or oscillations in a small system can be used to infer the magnitude of those in the larger system. In other words, results of small microcosms should be scaled using hierarchical measures such as transformity, territory of influence, and replacement time. We need tests of this theory, which has the potential of great savings in the costs of environmental impact testing.

Energy transfer networks are accompanied by patterns of spatial hierarchy. Oscillations in time contribute to spatial patterns when pulses of consumption generate "gaps" as surfaces on which organic structures have developed are used up restarting microsuccession. Because containers often reduce the oscillations that might otherwise exist with larger consumer populations, microcosms held in constant conditions may lack larger scale oscillations. However, the small self contained oscillations observed in microcosms may provide better examples for understanding the principles governing oscillator systems. The worldwide interest in the "chaotic" oscillations that result at higher levels of available energy makes old and new data on microcosm oscillation important for restudy. Do oscillations increase the overall performance of production and consumption in the long run? Are there optimum frequencies of oscillation that maximize power?

Dwyer and Perez (1983) used MERLE tanks for a demonstration that processes were non-linear when cycles of various periods responded to a perturbation of turbulence applied with a regular 10 day sine wave. When the ecosystem was simplified by removing benthic community, the ups and downs of the fluctuations were simplified. The simplification was support for the principle that complexity in time is proportional to complexity of components and processes (Van Voris et al. 1980).

The comparative study of the many scales in the environment (Steele 1991; Holling 1992) is giving new impetus to general systems ideal: to find common plans for systems on different scales and for inter-scale relationships. Territory size, replacement time, oscillation frequency, and transformity appear to be quantitative measures for inter dimensional scaling.

Perspectives on Ecological Education

We know from experience how successful microcosms are as projects for students in the classroom at all levels from grammar school to graduate school. This is the best way for people to learn to visualize whole systems and thus their whole earth as well. After measuring parameters as they rise and fall with metabolism, succession, retrogression, or catastrophe, a person can no longer think that it is enough to study the small parts and mechanisms individually. Ecology courses at all levels can be organized around microcosms, with appropriate field trips to compare the small and the large. Some instructions for classroom work are given in Appendices A and B.

Living microcosms are attractive in homes and offices. Stimulated by Claire Folsome's spherical microcosms (see below), in 1986, commercial interests began selling small globes containing microecosystems with resistant species (Yensen 1988). A sealed commercial terrestrial microcosm called a "terasphere" included mosses and ferns. The ecosphere in Figure 20.1 contained blue-green algae and brine shrimp. Considering that these were shipped by mail, exposed to extreme temperatures, left in the dark, etc. it would not be unexpected that some had their animals go extinct. As we discuss in Chapter 16, the brine shrimp are normally part of a longitudinal or pulsed succession. Nixon (1970) found them hard to maintain in arrested steady-state microcosms.

In a television mediated world where buzz-words are temporarily important for the public and its interest in science, "Biospherics" has been suggested to keynote a new vista of holistic ecology for the public and the scientists alike (Allen and Nelson 1990). They visualize millions of people visiting the largest micro-macrocosm in Arizona to get excited and go home to do smaller micro cosmology as part of the emerging new social mission of the earth's people to finally become the stewards of the earth.

20: Microcosm Perspectives

Figure 20.1. Commercial microcosm.

Perspectives on Global Policy

Increasingly, public policy problems center on the complex environmental interactions of the human economy and the global environmental systems. Microcosms are already being used to study such problems, using the self-organizing complexity of the miniature, isolated microecosystem to gain insight on the larger system. Isolated waste systems in service of society were considered in Chapter 16. More global aspects of human recycling now under international policy consideration are amenable to microcosm study.

Among the global concerns is shading of available sunlight by atmospheric disruption or pollution. Burning wells after the Persian Gulf War of 1991 shaded Arabian ecosystems. Nuclear war may increase atmospheric turbidity with smoke and sediment (Turco et al. 1983; Ehrlich et al. 1983). If light to the biosphere is suddenly reduced, what will happen to the ecosystems? The following microcosm experiments provide some insight.

Copeland (1964) studied light reduction in a marine microcosm containing a turtle grass community in a growth chamber at constant temperature. After it had been acclimated to 1,500 ft-candles, the light was reduced six and one half fold to 230 ft-candles. At first, community

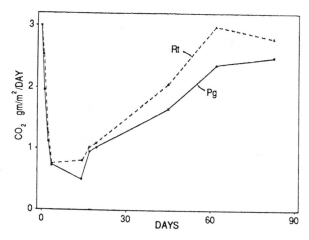

Figure 20.2. A plot of gross photosynthesis and total respiration of a marine microcosm subjected to light reduction (Copeland 1964).

metabolism decreased rapidly, but this was followed by a gradual increase in metabolism (both photosynthesis and respiration) until a level equal to the initial level was reached (Figure 20.2). The dominant producer changed from turtle grass to blue-green algae. Similar patterns were found in other microcosm studies (Chapter 7).

Tatrai (1982), investigated the effect of reduced light on sediment–water systems in aerated plastic boxes, six kept in total darkness and six held under a 12 h light–dark cycle. Chironomid larvae were added at densities from 0 to 10,000 per m^2. Every week, the overlying water was analyzed for phosphate phosphorus, total phosphorus, and total nitrogen in order to obtain daily rates of nutrient release. Nutrient concentrations were greater in the dark than in the light. The greater the number of larvae, the greater were the releases of nutrients. Algae grew in the lighted microcosms, but not in the dark ones.

Harte and Oldfather (1987), kept lake microcosms in complete darkness for 7 weeks and then gradually exposed them to light. Controls were maintained on a 12 h light–dark cycle. With the return of the light, the declining phytoplankton began to recover. There were shifts in population numbers among species of diatoms, rotifers, copepods, and cladocerans.

The results of these three experiments suggest that the result of reduced light on the biosphere would be a decrease in world-wide photosynthesis and a release of inorganic nutrients through heterotrophic metabolism. This would probably be accompanied by a die-off of many plant species, followed by the disappearance of many animals. Subsequently, the photosynthetic role would be taken over by other primary producers adapted to low light intensities. Over-all production and respiration could be expected to gradually rise to pre-light-reduction levels. This rise would be accelerated by the clearing of the atmosphere through the settling out of dust.

Perspectives on Sustainable Ecosystems

For humanity becoming more concerned with long-range sustainability of their own planetary system, the abstract questions become of greater academic and even public interest. How immortal can an ecosystem be? Are there principles to be found in the study of microcosms that will help us understand the sustainability of the human ecosystem of the earth?

Many people have kept sealed containers with living organisms incidentally in windows in their offices for many years. With intellectual roots from studies on the origin of life, and with emphasis on microbial processes, Claire Folsome and students at the University of Hawaii (Kearns 1983; Takano 1984; Folsom and Kearns 1981; Obenhuber 1985; Folsome and Hanson 1986) developed an organized research program on the self-organization and long-term sustainability of microecosystems by sealing living components in spherical glass chambers (Figure 20.3). Contents were systematically collected from different environments, freshwater and saltwater, in Hawaii and elsewhere, sealed and placed indoors receiving indirect daylight. Many of the containers had arrangements for extracting small samples for analysis without introducing additional microorganisms. Systems were highly reliable, apparently because of their self-regulating organization.

Folsome extracted small samples with a sterile syringe and to each added one of the substrates for 27 basic physiological functions. If the microbes were present for utilization of that substrate, it was recognized by the rapid utilization. Folsom found that the microbial functions for the 27 basic operations were maintained by the old microcosms year after year. Other monitoring was done with injected ^{14}C-labeled acetate, and the subsequent diurnal pattern used to verify operation of living biochemical processes. ATP, bacterial counts, and $_pO_2$ increased with time of closure requiring 5–6 months to reach a steady state. ATP concentrations exhibited oscillation, 0.2 to 1.2 ng/ml. Maguire (1980) reviewed the usual tendency for seeded animals to become extinct in closed microcosms (Chapter 3). However, Hanson was successful in maintaining the small crustacean, *Halocaridinia rubra*, in sealed a microcosm for five years with a mix of algae that had developed from an initial general seeding. Folsome and Hanson concluded, as we did from our studies, that different species predominated under slightly different conditions. However, all species would generate about the same metabolism at similar available energy levels. This is more evidence that self-organization fine-tunes for maximum power performance.

After Folsom's untimely death in 1989, Tabor MacCallum assembled his group of long-lasting microcosms in Tucson, Arizona (Figure 20.3).

Figure 20.3. A 1990 view of living microcosms sealed by C. Folsom 22 years before. (a) Single microcosm; (b) Set transferred from Hawaii to Biosphere 2 in Arizona (Photographs courtesy of Biospheres 2).

Some were as much as 24 years of age. Populations consisted primarily of microorganisms (algae, bacteria, and microzoa). All had developed pools of detritus; several dominated by blue-green algae. Some appeared little changed over long periods, but pulses and oscillations were also observed, even when external conditions were unchanged. Many pigment colors where represented: green, blue-green, yellow, pink, brown, and gray. Some of these developed during transport, when light exclusion may have increased the role of anaerobic bacteria.

Genetically Altered Organisms

Many questions arise about release of genetic engineered organisms into environments. Are the species competitive, replacing the wild varieties? Are the new engineered genes propagated into wild populations? It can be argued that an organism with genetic changes that increase some particular process, may draw the energies for the extra process from the general pool of available energy in that organism, thus making the species less competitive with wild populations when in the wild. Survival and propagation of genetic-engineered varieties have been studied in ecological microcosms (Armstrong 1987; Orvos et al. 1990; Bolton et al. 1990; Walter et al. 1991). Varieties of micro-organisms were propagated for a few days being displaced later.

Perspectives on Global Homeostasis

Starting with the examples and models of Chapter 2, we have emphasized the homeostasis which emerges in self-organizing systems in microcosms. Homeostasis of the earth has long been known to earth scientists finding similarities of pattern and process in the sedimentary record of a billion years of earth history. Data on world hydrology and biogeochemical cycles have generally shown continuing cycles, punctuated at times by oscillations and ice ages, and, over the long-term, affected by stages in living evolution. Lotka (1922a and b) described the evolution of maximum power as the fourth law of thermodynamics guiding the self-organization of the biosphere. Vernadsky (1929) wrote similarly: "The evolution of different forms of life throughout geological time increases the biogenic migration of elements in the biosphere."

The mathematical description of earth cycles, starting with Lotka (1925) and Kostitzin (1939), has helped us understand the different types of self-regulating system mechanisms. One of our early efforts was to apply Lotka's closed cycle regulatory concepts and data to strontium in fossils as evidence for the ocean and sedimentary cycle being in a homeostatic, stable, steady state (Odum 1950, 1951). Another global process studied in a geochemical microcosm was the formation of ocean water, with the

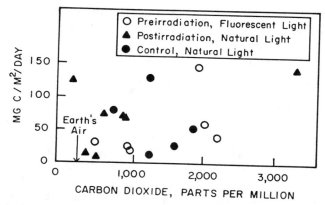

Figure 20.4. Average level of carbon dioxide maintained in rain forest microcosms after 6 months acclimation to forest floor conditions at El Verde, Puerto Rico in 1966 (Odum and Lugo 1970).

Goldschmidt reaction simulating the genesis of the early sea (Odum 1972):

basalt + volcanic acids → sediments (hydrogen silicates) + sea water

Early on we found some microcosmic life operating as a system and more stable to temperature change than the species operating separately (Figure 2.18). Even when exposed to 25,000 rads of ionizing gamma irradiation, our terrestrial rain forest microcosms in 1965 were found to be gaseously self-regulating, because production stimulates respiration and vice versa, as explained in Chapter 2. We found the carbon dioxide levels in these microcosms to be stable, rising and falling about a 24-hour mean, but each one was different, the exact curves determined by the composition of living components which had survived the initial seeding. Figure 20.4 shows the range from 300 to 3,000 ppm, the same range observed in Biosphere 2. We proposed then that evolution determines gaseous composition of the planet (Odum and Lugo 1970; Odum 1971).

In recent years another initiative by Lovelock (1979), considered the gaseous aspects of the earth's chemistry (nitrogen, methane, carbon dioxide, dimethyl sulfide, etc.), and also produced evidence for mechanisms of biospheric homeostasis. The name "Gaia" was applied, and it caught the world's attention. Roughly stated, the concept is that "life organizes the planetary chemistry for itself". This has encouraged public interest in the planetary ecosystem and is part of the new consciousness for global preservation.

Some opposition to ideas of living control comes from scientists who were trained with the assumption that physical processes controlled life, but not vice versa, perhaps because living energy was considered to be a smaller quantity. However, transformities measure the energy contributing to the ability to control, and the transformities of living organisms and their information are many orders of magnitude higher than those of the basic physical processes of the biosphere. Many examples of microcosms in this book have physical processes subordinate to control by living organisms. For example, the building of reefs controls the flows of water and wave energy.

Forty-four recent papers (Schneider and Boston 1991) consider ideas and evidences of Gaia. Many mechanisms of parts and processes of the earth are suggested as having homeostatic properties, mostly considered with semiquantitative inferences and some equations rather than simulations of the systems proposed. Many refer to a linear feedback concept rather than to the non-linear multiplicative autocatalysis of self organizational systems. Walker's chapter simulates a model with 24 equations which has photosynthesis, respiration, oxygen, carbon-dioxide exchanges with carbonate equilibria of the ocean, calcium, magnesium and phosphorus. It attempts for the earth over million year periods the same kind of overview of our simpler Biosphere model (Figure 19.15).

We find some Gaia expressions too anthropocentric, too organismal, and not going far enough toward understanding the systems dynamics (mathematical models). The geo-biosphere is a self-organizing system under principles of maximum power reinforcement, and like the microcosms, is evolving a self-regulated system of many hierarchical levels, a symbiosis between human economy, cultural information, the living cover, and the cycles of material. Living organisms, and especially the humans with social information capabilities, are constrained to contribute to the system that reinforces the symbiosis of humans and all of nature. Life and humans cannot operate the system for itself, except by doing what the system requires of them for the sustainable future of the whole. In Vernadsky's sense of Noosphere, the human role is to process information in service of the earth.

Appendix A
Directions for Classroom Microcosms

Much practical ecology can be taught on the high school, undergraduate, or graduate level with simple, inexpensive microcosms. Some practical suggestions follow for setting up several kinds of microcosms.

Pond Type Microcosm

Visit any reasonably healthy body of natural water to obtain material for a still-water microcosm. Dig up some bottom sediments, with the animals and plants that live in them, and place them in a plastic bucket. Organisms can be taken from the water column with a dip net or concentrated with a plankton net or seine. Finally, enough water should be collected to fill the planned container or containers. On the way home, avoid overcrowding of larger animals and avoid the use of metal pails.

Almost any container can be used to house a microcosm, but some precautions need to be taken. New plastic or wood can sometimes bleed volatile, poisonous organic compounds into the water. New concrete will leach basic compounds, raising the pH to unacceptable levels. Aged containers of these materials are frequently satisfactory. Glass aquaria sealed with silicone rubber are the most reliable containers.

Upon arriving at the location where the microcosm is to be established, the sediments and plants should first be gently transferred to the receptacle. Pond water should be added without suspending the sediments. Placing a glass or beaker on the bottom and slowly overflowing it is a good method. Lastly, the organisms should be added. A gentle stream of air bubbles will often cut down on mortality over the first few days. Water lost through evaporation should be replaced with distilled water or rain water.

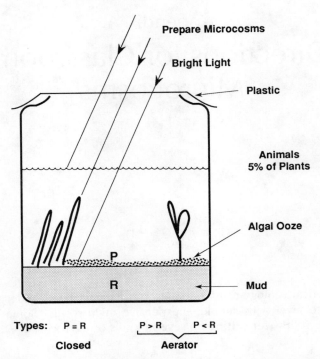

Figure A.1. Diagrammatic representation of preparation of microcosm for school use. The three relationships between photosynthesis (P) and respiration (R) are indicated along with the conditions under which they are achieved (after Odum 1960).

Avoid chlorinated tap water, as it often has toxic chlorine and metal substances and unfavorable basic conditions.

Three types of microcosms can be started with the same natural material.

1. One kind has more animals than can be supported by plants, as in the usual aquarium full of tropical fish. In this case, respiration exceeds photosynthesis (R > P), and oxygen comes in from the air continuously. Food for the excess animals must be supplied and carbon dioxide removed by aeration (see Chapter 8).
2. The second type is constructed with bright light, plants, and plant nutrients (the soluble "plant food" fertilizers readily available at garden

---

Figure A.2. Closed systems supported entirely by sunlight and their overall reactions. (a) Aquatic closed system; (b) terrestrial microcosm; (c) biosphere; (d) cycle of materials between production (P) and respiratory consumption processes (R); (e) energy diagram for the systems represented in a through c (after Odum 1971).

Pond Type Microcosm

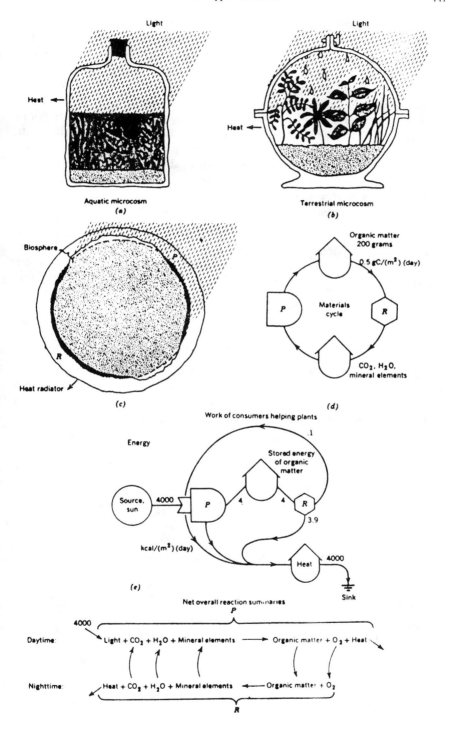

centers). Such a microcosm with have photosynthesis exceeding respiration ($P > R$) after a couple of weeks' development in bright light. If the microcosm is under aeration, excess oxygen will be carried off in the stream of bubbles and carbon dioxide brought to the plants. Rooted aquatic biomass will visibly increase or phytoplankton will bloom and the water will turn green.

3. A third type is a balanced system where photosynthesis equals respiration ($P = R$). This situation may result if the container is sealed. An open aquarium or other container can be sealed with plastic wrap (Figure A.1). Polyethylene film is permeable to oxygen and carbon dioxide, while Saranwrap is not. The microcosm may also be constructed in a carboy and sealed with a cork or cap. An absolute seal can be assured by pouring liquid paraffin around the cap and allowing it to harden.

A balanced system ($P = R$) represents the entire biosphere (Figure A.2) where ultimately, the production of the plant kingdom must balance the respiration of the animals (including ourselves) and the decomposers. If humans venture very far into space in the twenty-first century, life support systems will also have to balance (see Chapter 19).

Directions for Preparing a Stream-Type Microcosm

A simple stream microcosm is diagrammed in Figure A.3. Support a stream bed within an aquarium above the level of water. Use a simple bubble pump to lift water from the aquarium to the head of the artificial stream. The return is screened to prevent clogging. The supports and

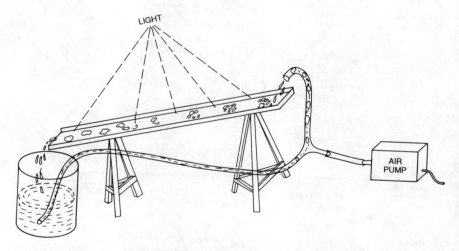

Figure A.3. Diagram of the construction of a simple stream microcosm.

stream bed should be constructed of some inert material such as brick, fiberglass, ceramic roofing tiles, etc. The stream bed should be filled with small rocks, gravel, leaves, and small sticks from a natural stream, to provide seeding and a place for the attachment of organisms. The aquarium water simulates lakes above and below the run of the stream. Glass and tygon tubing are suitable for use; metals are not. Excess organic matter at the start will tend to make respiration exceed photosynthesis ($R > P$).

Directions for Preparing a Terrestrial Microcosm

Terrestrial microcosms are good means for studying the principles of ecology. Making a tiny biosphere illustrates on a small scale, nature's way of sustaining the whole biosphere. A school-room with its windows lined with microcosms is a fascinating place to watch life develop, seeing something change each day. Usually, each microcosm turns out to have something interesting and different from all the others in the room. See the example in Figure A.4. It is fun to make small microcosms and study these while also having group discussions on the events taking place in the giant Biosphere 2 in Arizona. Watch these little worlds develop, record dates on which things were done and the dates on which interesting things happened. Here are some instructions for preparing a terrestrial microcosm.

1. Chose an open or closed type. Decide if you want an open-to-air microcosm or a sealed one. The open ones are easy to observed and manipulate. By having a free exchange of air, they are automatically prevented from getting too much carbon dioxide or running out of oxygen.

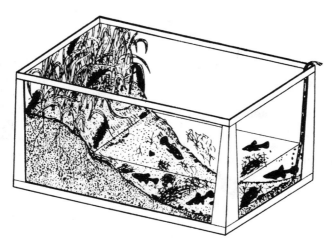

Figure A.4. Example of terrestrial microcosm (Metcalf, Sangha, and Kapor 1971. Reprinted with permission from Environ. Sci. Technol. 5(8), copyright 1971, American Chemical Society).

However, organisms that you do not want may invade, animals may crawl out, and the microecosystem is less a model of space ship earth because it is less self-regulating. A sealed-container microcosm is forced to develop some degree of balance between the photosynthetic processes using carbon dioxide and producing oxygen, and the nighttime processes which are the reverse, using oxygen and making carbon dioxide. If well-fertilized and lighted, the plants in a sealed microcosm will fill all the space with leaves. One microcosm in a spherical container looked like a cabbage after 6 months.

2. Select a container, which can be of almost any shape, made of glass or plastic. It must hold water (not leak). At least part of it needs to be transparent so you can see what goes on. Even if you use a glass bottle, it should have a wide mouth big enough for a hand to go in and out. The cover (if there is to be a cover) can be tight enough to stop most exchange while letting a little air come and go as the air expands due to heat or shrinks as it cools down. For example, a regular aquarium container can be covered with a glass plate or plastic film held in place by a rubber band. Or, the terrarium can be tightly sealed by using a bottle with a tight screw top or rubber stopper. Pressure will increase inside when it warms up and decrease when it cools down at night.

3. Arrange light conditions. Indirect daylight (1,000 ft-candles) is best, because direct sunlight (10,000 ft-candles) may make the ecosystem too hot and stress the animals. However, it makes an interesting microcosm for studying deserts and other hot places. Indoor light (100 ft-candles) has too little energy. Microcosms in low light take a long time to develop and don't generate as much living activity. Microcosms can be made without light to illustrate what goes on in dark places such as caves or down in the soil. Such microcosms run on decomposing organic matter. For example, a half-decomposed piece of log from the woods, with all its interesting beetles, fungi, and centipedes, can be put in a rectangular aquarium-type container. Decomposer microcosms can be put anywhere in the room.

4. Decide on an ecosystem type and add some soils from that environment to the container. Think of a type of ecosystem you would like your microcosm to resemble. For example, the dead log microcosm (previous paragraph) illustrates the interesting forest floor phenomena. People can exert their creativity at this point. From the natural environment collect some typical soil materials. Interesting microcosms have been started from cattle dung, desert soil, a block of lawn grass turf, a sand bed with beach plants and animals, a patch of farm soil, peat, gravel, vermiculite, garbage, shredded plastic (to make a microcosm simulating a dump), etc. Arrange the soil materials on a slant, high at one side and lower at the other. This is so one side can be drier (further up from some water you put into the system).

5. Add a little water and possibly some fertilizing nutrients (if the system is not already from a fertile place). Small packets of "plant food"

with nitrogen, phosphorus, potassium and trace nutrients are sold in any nursery store or department store. Don't add so much water that the soil is saturated (unless you want your microcosms to simulate a wetland). If you have an open system type, then you need to add a little water every week (artificial rain), because the microcosm water will evaporate and transpire out through the leaves. In a closed microcosm, the evaporated and transpired water will condense on the top and sides and drip down, simulating rain.

6. Seed with as many kinds of plant and animal life as you are able from the environment you are using as a reference area. When you add soil you are probably already adding many kinds of microorganisms, tiny animals, and seeds of plants. Add some more. Don't put in big individuals. Let the system grow its own, appropriately-sized organisms. The idea is to add many more kinds of life than you will eventually have. The ecosystem goes through a self-organizational process called succession. Microcosm ecology is different from planting a garden where you decide what and where. The microcosm idea is to let nature do the designing. As time goes on, it gets more and more complex and interesting. You can continue to add seeds and small organisms so that there is plenty of genetic information, just as in the outdoors where winds, waters, and birds circulate the seeds and eggs of small animals.

7. **STUDY AND RECORD EVENTS**. One can just enjoy the events in the microbiosphere or one can study them, counting things, recording the dates on which organisms appear, estimating growth rates, etc. If you have to terminate the study, inventory everything in the container as you take it apart. However, there is great scientific interest in what happens in closed ecosystems after a long period of time. Put the microcosm in a window and leave it, even if you do not have time to watch it very often.

Appendix B
Directions for Monitoring Microcosms

Almost any measurement technique used in ecology can be adapted to use in microcosms. Some simpler procedures for schoolrooms involve measuring oxygen and carbon dioxide, nutrients, and species diversity. For more methods see *Standard Methods for Examination of Waters* (ASPHS, 1992).

Dissolved Oxygen Measurements

Measure dissolved oxygen as shown in Figure B.1. The reagents are prepared as follows:

1. Manganous sulfate: Place 480 g of $MnSO_4 \cdot 4H_2O$ in 1 liter of distilled water. This solution may be stored indefinitely.

2. Potassium hydroxide–potassium iodide solution: To 1 liter of distilled water add 700 g of potassium hydroxide. Use caution! The solution becomes very hot and is very caustic. Human skin is soluble in strong hydroxide solution. Then, add 150 g of potassium iodide to the potassium hydroxide solution. (Or use 500 g of sodium hydroxide and 135 g of sodium iodide.) This solution may also be kept indefinitely, provided it is tightly capped with a rubber stopper so it does not absorb carbon dioxide from the air. Ground glass stoppers are not acceptable as they will "freeze" in place.

3. Sulfuric acid: Use concentrated sulfuric acid (not fuming) as it comes from the bottle. Avoid getting the acid on skin, cloths, etc. as it will react with almost anything except glass, paraffin, or teflon.

4. Starch solution: Reagent grade soluble starch is preferable, however, almost any starch source can be used (corn starch, flour, etc.). An excess

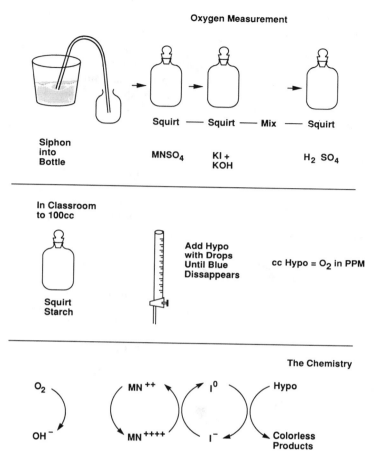

Figure B.1. Diagram of the Winkler oxygen technique. First two rows illustrate the procedure. Bottom row gives the oxidation–reduction reactions (after Odum 1960).

of starch should be boiled in distilled water and then filtered. Refrigerate so the solution can be kept for several weeks.

5. Sodium thiosulfate (hypo) solution: This solution is made by dissolving 3.102 g of $Na_2S_2O_3 \cdot 5H_2O$ in distilled water. The weighing out of this salt is most critical for accuracy in oxygen measurements. Do not use the anhydrous form. This solution will keep for several weeks in the refrigerator.

The procedure for measuring dissolved oxygen is as follows:

1. Siphon sample water into a sample bottle of 100–150 ml capacity. Let it overflow several times. Avoid bubbles. Excess water may be returned to the microcosm before reagents are added.

2. Add 10 drops (about 0.5 ml) of manganous sulfate solution and 10 drops of the potassium hydroxide–potassium iodide solution. A flocculent precipitate will appear. Stopper the bottle and mix with a rotating wrist motion. Keep the bottle sealed and avoid bubbles. At this point, the solution in the sample bottle can absorb oxygen from the air, increasing the possibility of error. Wait one min for fresh water and 15 min for salt water.

3. Remove the stopper and add 10 drops (about 0.5 ml) of sulfuric acid. Replace stopper and mix as above. The solution will clear and assume a light yellow to dark brown color as iodine is formed in proportion to the amount of oxygen present. At this point the samples may be stored for a day or two, preferably in a refrigerator.

4. Measure out 100 ml from the sample bottle with a graduated cylinder or pipette into a white cup, add several drops of starch solution, stirring to obtain a blue or blue-black color.

5. Titrate with the sodium thiosulfate solution by adding it drop wise from a burette or graduated pipette. Stir or swirl after each addition. When the titrant in the beaker clears, losing its blue color, the end point is reached. The number of milliliters of sodium thiosulfate added equals the amount of dissolved oxygen present in the original microcosm water.

Oxygen Saturation

The amount of oxygen dissolved in water in contact with air depends upon temperature, atmospheric pressure, and salinity. Clean water left in a flat, shallow container overnight in contact with the air will come to saturation. The value will probably be around 8 ppm, depending on the temperature. See Table B.1.

Microcosm Oxygen Metabolism

The total metabolism of a microcosm (P and R) can be measured using the oxygen technique. To measure increases and decreases, oxygen measurements should be made every three hours over the course of a day and a night. If water is sufficient, make duplicate or triplicate analyses. Plotting a graph of oxygen against time will often produce a curve similar to the one in Figure B.2. The rate of daytime rise indicates the net photosynthesis rate while the rate of nighttime fall represents the nighttime respiration.

Table B.1. Saturation values for oxygen in ppm.

Temperature (degrees Fahrenheit)	32°	50°	75°	90°
Fresh water	14.2	10.8	8.2	7.2
Salt water (35 ppt)	11.3	8.7	6.7	4.1

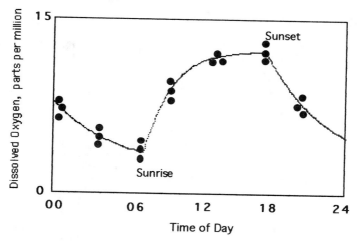

Figure B.2. Idealized plot of replicate measurements of dissolved oxygen concentration in a microcosm.

Carbon Dioxide Metabolism Determined from pH Changes

Carbon dioxide changes may be determined from changes in pH (Beyers et al. 1963), since the pH of any given water is a function of the amount of carbon dioxide dissolved in that water. A curve must be constructed by which pH changes can be translated into carbon dioxide changes. Different waters have different curves, each obtained from a titration of the given water with distilled water saturated with carbon dioxide. A known amount of gas in a known amount of distilled water is added to the experimental water and the change in pH is measured.

The apparatus and procedure for this titration are shown in Figure B.3a. Carbon dioxide flows from the carbon dioxide source (A) through an air stone into the flask (B) which contains approximately 1 liter of distilled water. The carbon dioxide source is an acid–calcium carbonate generator (Figure B.3b). After twenty minutes, the distilled water titrant becomes saturated with carbon dioxide under local ambient atmospheric pressure. The carbon dioxide continues to flow through the flask during the rest of the titration. Atmospheric pressure may be measured with a mercury or aneroid barometer, or approximated from the graph in Figure B.4 which gives the standard barometric pressure for various elevations above sea level. Also measure the temperature of the water in flask B. Place 1,500 ml of the microcosm water in beaker C (Figure B.3a), and raise its pH by bubbling for several hours with carbon dioxide-free air to

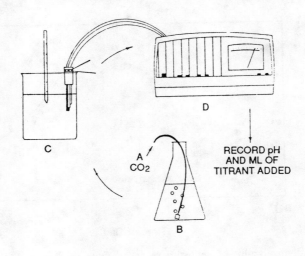

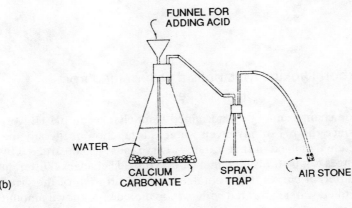

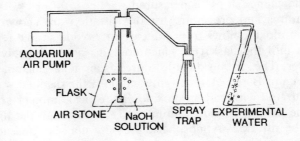

Figure B.3. Diagrams of procedure and apparatus for titration with carbon dioxide saturated water. (a) Titration; (b) carbon dioxide generator; (c) carbon dioxide free air generator.

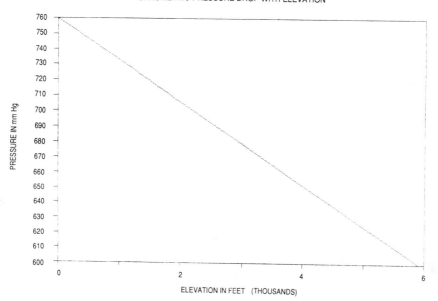

Figure B.4. Relationship of barometric pressure to elevation.

remove the atmospheric carbon dioxide dissolved in the water. Hard water may require 6 hours. Carbon dioxide-free air can be produced by passing air through a saturated solution of any soluble hydroxide (Figure B.3c).

The initial pH of the water in beaker C is measured with a pH meter and recorded. Withdraw 1 ml of titrant from flask B and allowed it to flow into beaker C. The tip of the pipette containing the titrant should be placed just below the surface of the water in the beaker. The contents of the beaker are then stirred gently with a stirring rod or magnetic stirrer. Take care not to break the surface of the water or to form bubbles. The pH is again measured and recorded. The procedure is repeated until the range of pH in the beaker covers that expected to occur in the microcosm.

Figure B.5 shows plots of saturation values of carbon dioxide in the titrant (millimoles per milliliter) at various temperatures and pressures. These values are named M factors. Use the M factor to plot a graph of pH versus the number of millimoles of carbon dioxide added to the microcosm water. A millimole of carbon dioxide is equal to 0.044 g of carbon dioxide. Next, plot a graph of the pH of the microcosm water and the amount of carbon dioxide dissolved in the water. This curve (Figure B.6) can be obtained through the use of the F factors given in Table B.2. The F factors are a series of constants, which when multiplied by the M

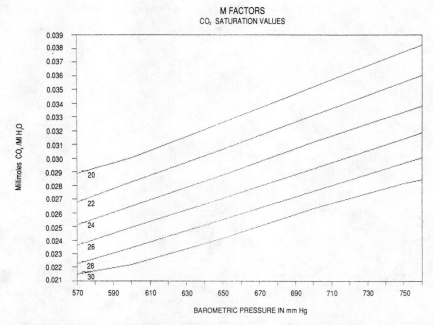

Figure B.5. Plots of M factors against barometric pressure at various temperatures (degrees Celsius).

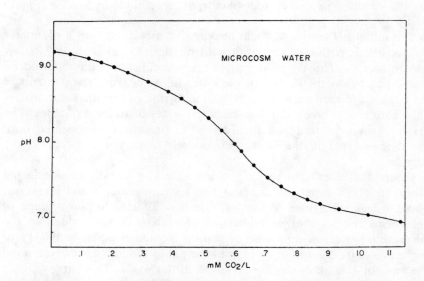

Figure B.6. Relationship between pH and dissolved carbon dioxide in water from a microcosm.

Table B.2. F factors. This series of constants is multiplied by the appropriate M factor and the product plotted against pH to form the titration curve. These values hold true only if 1,500 ml of the microcosm water are used and the titrant is added in milliliter increments (after Beyers 1964).

Titrant (ml)	F factor	Titrant (ml)	F factor
0	0.000	26	17.038
1	0.666	27	17.682
2	0.332	28	18.325
3	1.996	29	18.967
4	2.660	30	19.608
5	3.322	31	20.248
6	3.984	32	20.888
7	4.645	33	21.526
8	5.305	34	22.164
9	5.964	35	22.801
10	6.623	36	23.438
11	7.280	37	24.073
12	7.937	38	24.707
13	8.592	39	25.341
14	9.247	40	25.974
15	9.901	41	26.606
16	10.554	42	27.237
17	11.206	43	27.868
18	11.858	44	28.497
19	12.508	45	29.126
20	13.158	46	29.754
21	13.807	47	30.381
22	14.455	48	31.008
23	15.102	49	31.633
24	15.748	50	32.258
25	16.393	51	32.882

factor chosen from Figure B.5 for a given titration, will yield a series of values of carbon dioxide concentrations corresponding to each pH reading obtained in the titration. The products of the M factor and the F factor can be plotted against the pH readings to obtain the desired curve. For a more detailed description of F and M factors, see Beyers et al. (1963). The slope of the titration curve is directly influenced by the buffering capacity or hardness of the water. Soft water has a steeper curve than hard water.

460   Appendix B: Directions for Monitoring Microcosms

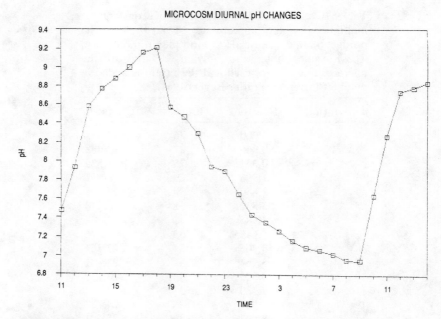

Figure B.7. Plot of pH readings against time in a microcosm.

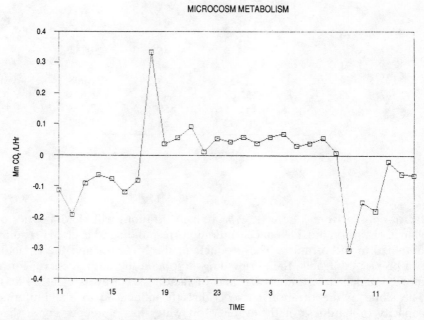

Figure B.8. Carbon dioxide diurnal rate of change curve for a microcosm.

Table B.3. Calculations. Calculation of a carbon dioxide diurnal rate of change curve for a microcosm (after Beyers 1964).

Time	Midtime	pH	Value from graph	Difference	Length of time	Metabolism
10:30 AM	11:00 AM	7.47	0.715	−0.113	60	−0.113
11:30 AM	12:00 PM	7.93	0.602	−0.192	60	−0.090
12:30 PM	01:00 PM	8.58	0.410	−0.090	60	−0.064
01:30 PM	02:00 PM	8.77	0.320	−0.064	60	−0.076
02:30 PM	03:00 PM	8.88	0.256	−0.076	60	−0.120
03:30 PM	04:00 PM	9.00	0.180	−0.120	60	−0.080
04:30 PM	04:52 PM	9.16	0.060	−0.060	45	0.333
05:15 PM	05:52 PM	9.21	0.000	0.416	75	0.036
06:30 PM	07:00 PM	8.57	0.416	0.036	60	0.056
07:30 PM	08:00 PM	8.47	0.452	0.056	60	
08:30 PM	09:00 PM	8.29	0.508	0.092	60	0.092
09:30 PM	10:00 PM	7.94	0.600	0.012	60	0.012
10:30 PM	11:00 PM	7.89	0.612	0.055	60	0.055
11:30 PM	12:00 PM	7.65	0.667	0.043	60	0.043
12:30 AM	01:00 AM	7.43	0.710	0.060	60	0.060
01:30 AM	02:00 AM	7.35	0.770	0.040	60	0.040
02:30 AM	03:00 AM	7.26	0.810	0.060	60	0.060
03:30 AM	04:00 AM	7.16	0.870	0.070	60	0.070
04:30 AM	05:00 AM	7.09	0.940	0.030	60	0.030
05:30 AM	06:00 AM	7.06	0.970	0.040	60	0.040
06:30 AM	07:00 AM	7.02	1.010	0.056	60	0.056
07:30 AM	07:50 AM	6.96	1.066	0.006	40	0.009
08:10 AM	08:50 AM	6.95	1.072	−0.408	80	−0.306
09:30 AM	10:00 AM	7.64	0.664	−0.150	60	−0.150
10:30 AM	11:00 AM	8.27	0.514	−0.179	60	−0.179
11:30 AM	12:00 PM	8.74	0.335	−0.020	60	−0.020
12:30 PM	01:00 PM	8.78	0.315	−0.059	60	−0.059
01:30 PM	02:00 PM	8.84	0.256	−0.062	60	−0.062
02:30 PM		8.98	0.194			

Metabolism of a Microcosm

Figure B.7 shows an example of a diurnal record of pH in a microcosm. The graph of pH versus CO_2 in Figure B.6 was made with the titration procedure. Hourly changes in carbon dioxide were calculated from pH changes using the data from Figure B.6 (Figure B.7). The raw pH data from the microcosm as well as the calculations to construct the diurnal rate of change curve are shown in Table B.3. It lists the midtime of each time period, the initial and final pH values of the period, the number of millimoles of carbon dioxide necessary to produce the particular pH change in 1 liter of microcosm water, the duration of the time period, and the change in carbon dioxide in millimoles per liter per hour.

For comparison with other microcosms and aquatic ecosystems of differing depths and volumes, this type of data may be converted to grams of carbon dioxide per square meter per day, if respiration and photosynthesis are expressed in millimoles and the depth in centimeters, by multiplying by the depth times 0.44. In this case, the microcosm had a depth of 17.0 cm. The net photosynthesis is calculated at $-9.52\,g/m^2$ per day and the nighttime respiration is 8.02 g. The shape of the diurnal curve shown in Figure B.8 is typical of microcosms under artificial light. There is an increase in CO_2 (high rate of respiration) immediately after the lights are extinguished due to the utilization of the stored products of photosynthesis accumulated during the day. There is a large decrease in CO_2 (high net photosynthetic rate) immediately after the lights come on due to the large amount of stored plant nutrients produced during the night by the process of respiration.

Appendix C
Experiments on the Closed Ecological System in the Biosphere 2 Test Module

Abigail Alling, Mark Nelson, Linda Leigh,
Taber MacCallum, Norberto Alvarez-Romo
and John Allen, Space Biospheres Ventures (SBV),
P.O. Box 689, Oracle, AZ, and Robert Frye,
University of Arizona, Tucson, AZ

Introduction

To prepare for the operation of Biosphere 2, a closed ecological facility, the "Test Module" was constructed in 1986 (Figures 19.12 and C.1). It was designed both as a test of the physical structure and engineering components for Biosphere 2 and as a test bed for developing ecological systems. After a series of tests and experiments on engineering and biological components had been conducted, closures were made in 1988–1989 with people included. This paper reports measurements during Test Module closures, including those when the test module was providing human life support.

These experiments represented the first opportunity to examine long-term changes in the composition of the atmosphere of closed ecological systems which included higher plants and soil supporting people. Previous short-term closures of small systems were reported by Odum and Lugo (1970), but prior to this, no work on long-term closures had been attempted. Long-term closures of small aquatic systems have been examined by Folsome and colleagues (Folsome and Hanson 1986; Kearns and Folsome 1981, 1982; Kearns 1983; Obenhuber 1985; Obenhuber and Folsom 1988).

Description of the Test Module Systems

The test module had a variable volume of about $480\,m^3$ ($17,000\,ft^3$). The "biochamber" had a footprint of about $7\,m$, lined with steel. Its sides and roof were of steel spaceframe and double laminated glass, which admitted an average of 65% photosynthetic active solar radiation into the structure.

Appendix C: Experiments on the Biosphere 2 System

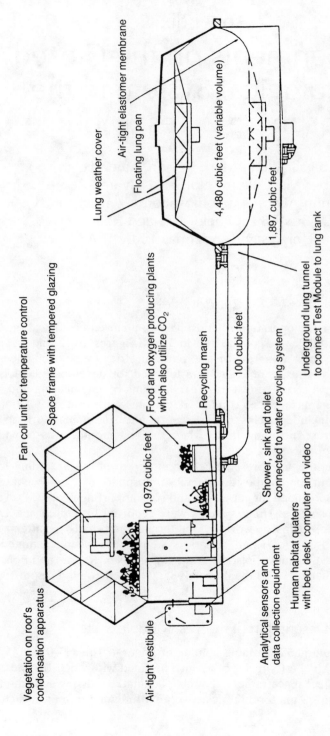

Figure C.1. Section view of the biospheric research and development test module with lung at Space Biospheres Ventures, Oracle, Arizona.

An underground air duct connected the biochamber to the building housing the variable volume chamber called a "lung" (Figure C.1) (Nelson et al. 1991a,b). The lung structure was developed by Space Biospheres Ventures (SBV) to minimize the effects of changes in internal temperature and external atmospheric pressure. With increased temperature in the Test Module or decreased barometric pressure in the outside environment, the variable chamber expanded; when temperature decreased or outside pressure increased, the chamber contracted. The lung reduced the possibilities that the Test Module might implode or explode, thus permitting a less reinforced and more sunlight-admitting structure to be utilized.

The weight of the pan in the lung structure permitted a positive displacement from inside the closed system to the outside of about 0.2 atmosphere. It also enabled leak rates to be determined by measuring the different in the observed level compared to that predicted for the temperature and pressure. A new system developed by SBV provided the seal for the glass–steel spaceframe structure. Underneath, a steel liner provided the ground seal in both biochamber and lung. An energy–utility room provided the energy for the mechanical systems of the Test Module: running heat exchangers, pumps, fans, lights, etc. (Dempster 1988, 1989). In experiments in 1990–1991 the Test Module had a leak rate of about 24 percent per year (Nelson 1991b).

Monitoring System

Sensors in the Test Module sent information to a hierarchical, five-level structure in a command center, located in the SBV Mission Control building. The five functional levels were (1) point sensing and activation, (2) local data acquisition and control, (3) system supervisory monitoring and control, (4) global monitoring and historical archive, and (5) telecommunications.

Systems included the G2 Gensym real time expert system and RTAP, a real time application platform. The G2 software controlled and monitored gases continuously and checked the analytic sensor calibrations. The G2 program has a provision for modeling carbon dioxide with real time interaction models and observed data. RTAP was the software used for data acquisition and control.

Developing adequately sensitive and dependable methods of monitoring gases with continuous sensors was difficult. In the first experiments there was significant sensor drift and noise; frequent calibrations had to be done to determine the actual concentrations present. However, these recalibrations showed that trace gases were at least one order of magnitude below the concentration levels for potential health concern. The third iteration of an analytical system achieved a far more reliable performance of continuous monitoring of the 11 trace gases we identified as being of primary concern: CH_4, total nonmethane hydrocarbons, NOx, O_3, NO, CO_2, O_2, H_2S, SO_2 NO_2, and NH_3. During human closures, air samples were passed through a gas chromatograph and ion chromatograph system once a day. Fifty three organic gases were identified in trace amounts within the enclosure during September 10–30, 1988 (Alling et al. 1990). Testing of the potable, recycled, and irrigation water quality was also done on a periodic basis.

Water Systems

The water recycling system consisted of three subsystems: potable water, waste water recycling from the habitat, and plant irrigation water.

The waste recycling system provided complete recycling of all human wastes. With this system, no wastes were removed from the Test Module; the sewage water was purified by the action of microbes and aquatic plants, and then used to irrigate the plants in the Test Module. The system was designed to clean 20–60 liters of effluent per day, and during all the human closures the $2.6\,m^2$ system effectively and without malodor cleaned the waste products, using both anaerobic and aerobic processes.

Potable water was distilled from the atmosphere by two dehumidifiers and sterilized with ultraviolet sterilizer systems. Potable water supplied all drinking water as well as a 0.9 l/min shower.

Irrigation water included all runoff-water from life systems and some potable water. Water was held in a reservoir and pumped to the plants through computer-controlled solenoid valves to various irrigation zones.

Life Systems

Many of the challenges faced in designing Biosphere 2 and the Biosphere 2 Test Module are analogous to problems faced in spacecraft or in tightly sealed, energy-efficient buildings. Materials are screened before selection to minimize potential outgassing problems, and systems must be developed to deal with those gases that are released.

The challenge in arranging life support was to provide adequate uptake of carbon dioxide to compensate the approximate 900 g (37 g/h) carbon dioxide exhaled by a person each day, to provide water purification through evapotranspiration and subsequent condensation, and to provide a complete range of foods for human nutritional needs.

Atmospheric CO_2 concentration was also dependent on soil metabolism — on the initial soil mass to plant mass ratio reported elsewhere for a wide range of sizes of closed ecological systems (Frye 1989; Frye and Leigh 1990).

The plant species chosen had a high growth rate, high photosynthetic rates, and were selected at a young growth phase to maximize the amount of carbon dioxide which could be utilized by each plant. These included a savannah canopy with C4 grasses, agricultural plants such as sweet potatoes, with a high photosynthetic rate, wilderness plants including sugar cane, a "ginger belt" of plants from the rainforest, including bananas and the fast growing plants of the order zingerberacae, and a marsh recycling system with water hyacinths as the dominant species, which have a photosynthetic growth rate of $31.35-42.40 \, g/m^2$ per day. A bio-accessions list, computer linked, inventoried all the plant species introduced. Measurements were made of soil and foliage biomass.

It was also important to maintain air clean of toxic substances. Benchtop research on soil bed reactors was conducted for SBV at the Environmental Research Laboratory of the University of Arizona, a consultant on the agricultural and engineering systems of Biosphere 2. These researchers studied the dynamics which followed the introduction of known quantities of trace gases to space Biosphere Reactors (SBRs) and the control systems. In addition, studies were conducted to determine optimal soil media, flow rates, and long-term effects on soil fertility and crop production in SBRs (Hodges and Frye 1990). Prior to the human closure experiments, a series of experiments in the Biosphere 2 Test Module were performed to examine the uptake of introduced gases such as methane and ethylene by SBRs and the effects of air-pumping on soil respiration levels.

On December 31, 1986, following engineering tests, closure experiments were begun to observe light levels, temperatures, responses of ecosystems, community structure, higher plants, soils and their interaction with the atmosphere, grass pollination, and the behavior of bees. Included was a five month closure test of the sustainability of plants. As a result of these tests, a life system of vegetation and soils was developed for the Test Module.

Human Closure Experiments

In September 1988, SBV included one human (John Allen, SBV's Director of Research and Development) in the closed ecological system for 72 h. In March 1989, Abigail Alling, Director of Marine Systems, SBV, lived in the Test Module during a five day closure experiment, and in November 1989, Linda Leigh, Terrestrial Biome Design Coordinator of SBV, lived in the Test Module for 21 days. By June of 1990, SBV had over 60 person-days of experiments in the facility (Figure C.2). During the experiments, the system was 100% closed with respect to water, food, and air. All waste materials were recycled in the Test Module using a marsh aquatic plant–microbial recycling system developed with consultation from Bill Wolverton of NASA Stennis Center in Mississippi.

Results

Measurements of physical and chemical properties within the test module during the periods of human occupancy are given in Figures C.3–C.18. Data are given for the 21-day human closure of November 2–23, 1989.

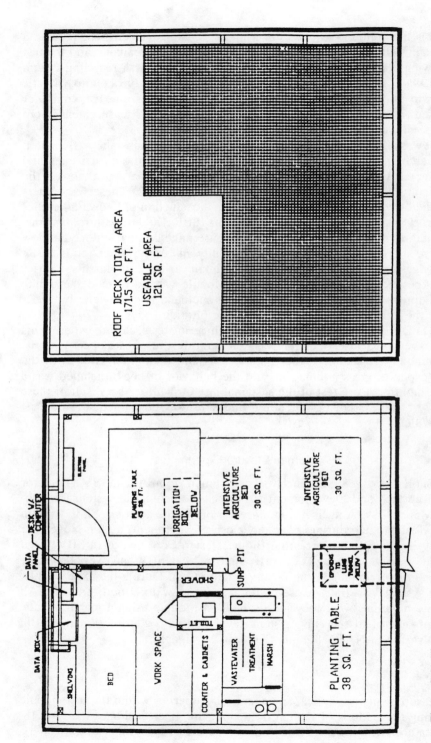

Figure C.2. Floor plan of the Biosphere 2 test module (excluding lung) during the three "human and ecosystem" experiments (Sept. 1988, March 1988, and Nov. 1989).

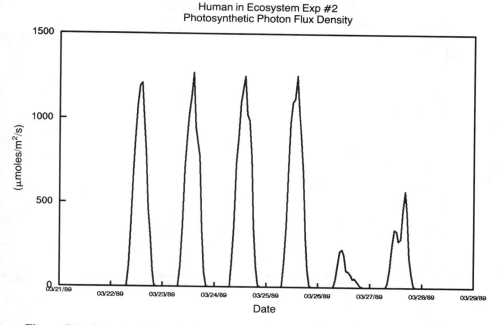

Figure C.3. Solar insolation in the test module March 22–28, 1988.

The Test Module was sensitive to changes in solar energy and ambient air temperature. Figure C.3 shows incident light energy input into the Test Module during the second human closure experiment. Notice the differences in temperature and humidity on cloudy days, March 26–27, 1989 compared to sunny days preceding (Figures C.4 and C.5). Figure C.6 shows nitrogen oxides (NOx) concentrations which ranged from 0.15 to about 3 ppm. Cautionary 8 h levels are considered to begin above 30 ppm. Figure C.7, ozone levels, shows the highest concentrations were 0.021 ppm. Cautionary levels begin at 0.1 and danger levels at 0.3 ppm. Figure C.8 is a graph of sulfur dioxide. Levels stayed below 0.005 ppm — well below the alert levels of 2–5 ppm. Figure C.9 depicts methane levels. The slight rise to about 150 ppm (still far below those of concern) during the human closure was similar to other experimental results obtained in trials conducted at SBV and at the Environmental Research Laboratory, University of Arizona. More time was required for the methane-metabolizing microbes to build up their populations in order to bring down atmospheric concentrations. It was an example of classic self-regulation with a negative feedback loop.

Figures C.10 and C.11 record levels of nitrates and phosphates in the aquatic waste processing system during the third closure experiment. They illustrate the effect of wastes being held in the anaerobic holding

470 Appendix C: Experiments on the Biosphere 2 System

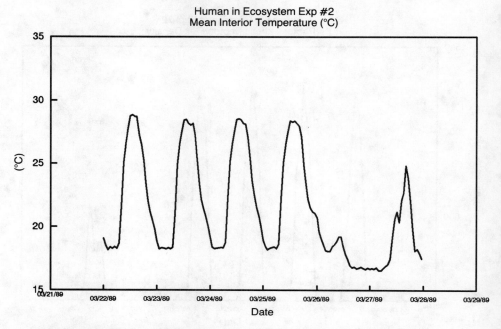

Figure C.4. Temperature in the test module March 22–28, 1988.

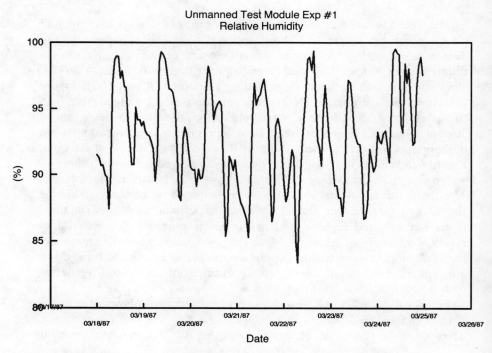

Figure C.5. Relative humidity in the test module March 22–28, 1988.

Results 471

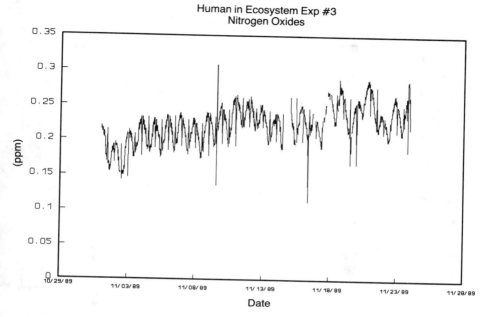

Figure C.6. Nitrogenous gases (NO_x) in the test module Oct. 29–Nov. 29, 1989.

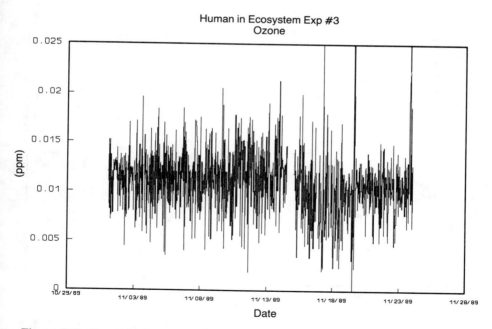

Figure C.7. Ozone in the test module Nov. Oct. 29–Nov. 29, 1989.

472　Appendix C: Experiments on the Biosphere 2 System

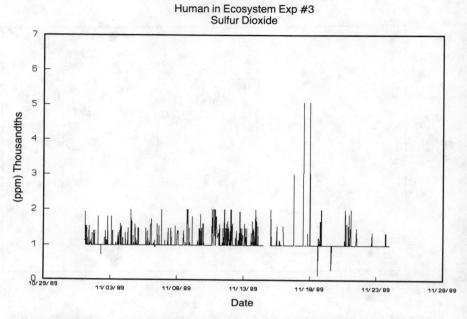

Figure C.8. Sulfur dioxide (SO_2) in the test module Oct. 29–Nov. 29, 1989.

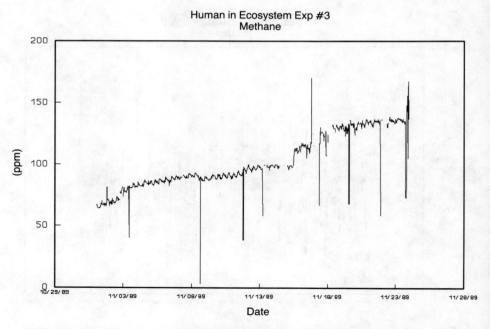

Figure C.9. Methane in the test module Nov. 1–Nov. 24, 1989.

Results 473

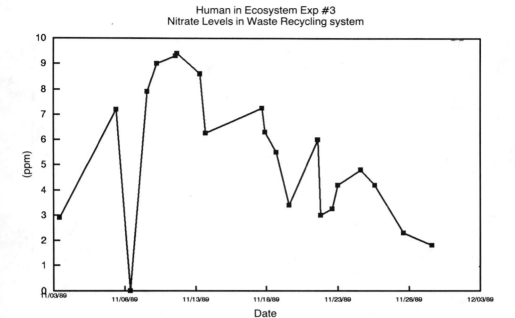

Figure C.10. Nitrate levels in waste recycling system in the test module Nov. 2–30, 1989.

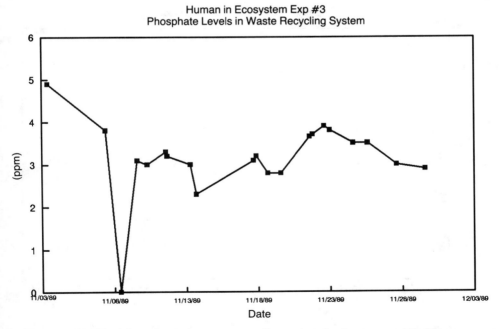

Figure C.11. Phosphate levels in waste recycling system in the test module Nov. 2–30, 1989.

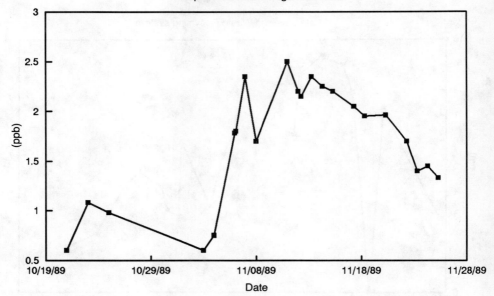

Figure C.12. Phosphate levels in irrigation water Oct. 18–Nov. 28, 1989.

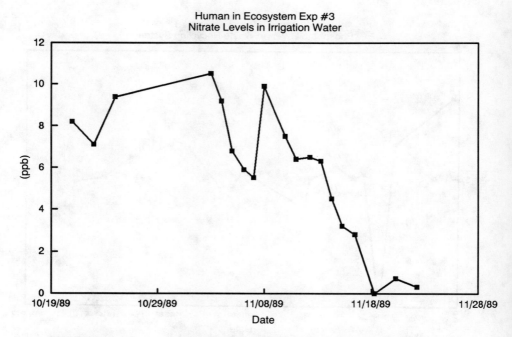

Figure C.13. Nitrate levels in irrigation water Oct. 18–Nov. 20, 1989.

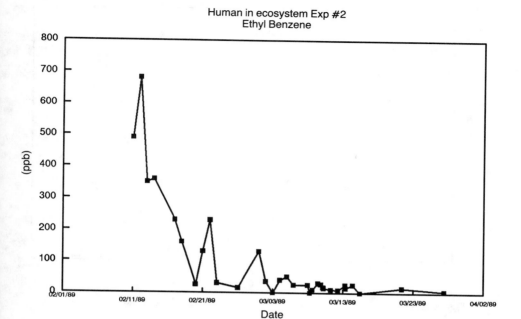

Figure C.14. Ethyl benzene in the test module Feb. 6–March 28, 1989.

tank where anaerobes start the process of purifying the water, which is followed by batch additions to the anaerobic tank where the aquatic plants and their symbiotic microbes continue the process, bringing nutrient levels down so that the water can then be routed to the irrigation water system, while producing an abundant increase in plant biomass. Figures C.12 and C.13 show the levels of nitrate and phosphate in the irrigation water supply. Concentrations rose after entry of the human into the system, and periodically with batch additions from the waste recycling system. Rises were followed by uptake by plants, decreasing concentrations again. Figures C.14–C.16 detail the initial appearance of toluene, ethyl benzene, and tetrahydrofuran, but levels were reduced to near zero after operation of the biological purification systems. These graphs are in parts per billion, while warning levels for these gases are from 100–200 ppm.

Carbon Dioxide

Figure C.17 shows hourly average CO_2 levels in the Test Module during the five experiments conducted without human occupation. While Experiments 1, 3, and 5 show evidence that the atmospheric CO_2 had reached an equilibrium, the CO_2 levels during Experiment 4 did not seem to

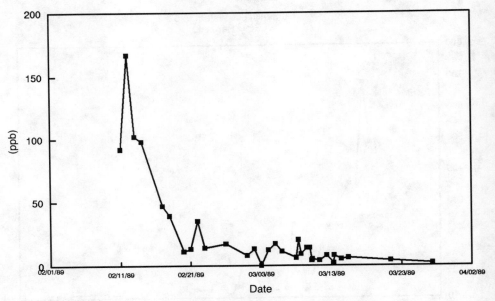

Figure C.15. Toluene in the test module Feb. 6–March 28, 1989.

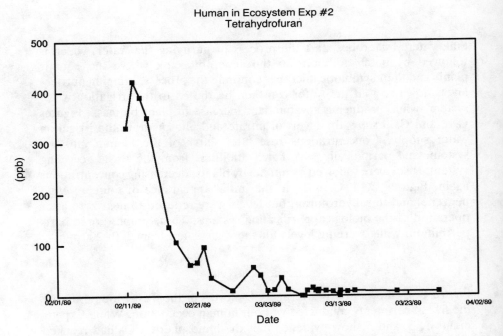

Figure C.16. Tetrahydrofuran in the test module Feb. 5–March 28, 1989.

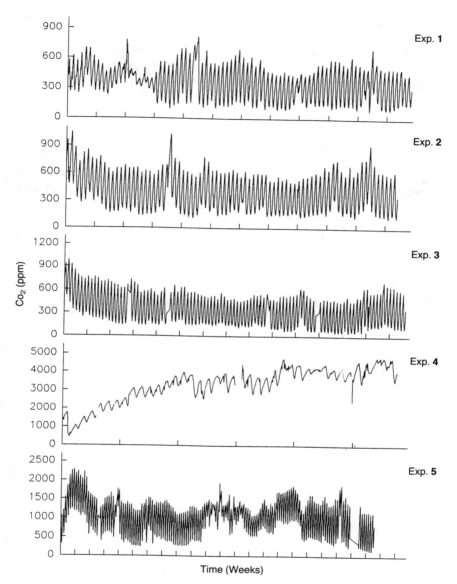

Figure C.17. Carbon dioxide levels in the test module during five experiments of 5–22 weeks each.

stabilize. While in Figure C.17 (Exp. 2) the curve appears to flatten out, this is an artifact, due to the upper limits of the (PRIVA) meter used to sense CO_2. The actual level of CO_2 achieved prior to termination was in excess of 9,000 ppm. The different behavior of CO_2 levels during

Experiment 4 can be explained by the change in the amount of soil inside the Test Module. Soil volumes during Experiments 1 and 3 were less than one-third the volume used in Experiment 4. In Experiment 4, the amount of soil was tripled because a soil bed reactor was installed to allow experimentation with the soil's ability to remove atmospheric contaminants. As can be seen from Figure C.17 (Exp. 4), this change in soil mass resulted in a dramatic change in the concentration of CO_2. For Experiment 5, the size of the soil bed reactor was halved. The result was a lower and more stable CO_2 level.

The behavior of CO_2 levels in the atmosphere of the Test Module was regular and predictable, largely determined by photosynthetic photon flux density (PPFD) and the temperature. Regular diurnal curves of CO_2 concentration were evident in all experiments. Maximum levels of CO_2 occurred in the hours just prior to sunrise, while minimum levels occurred just prior to sunset. These curves were the result of the alternating regimes of net system photosynthesis during the day and net respiratory consumption at night.

Change in carbon dioxide (net photosynthesis) was plotted as a function of light intensity (Figure C.18). The curve has the shape found in textbooks relating photosynthesis to light intensity and in papers reporting

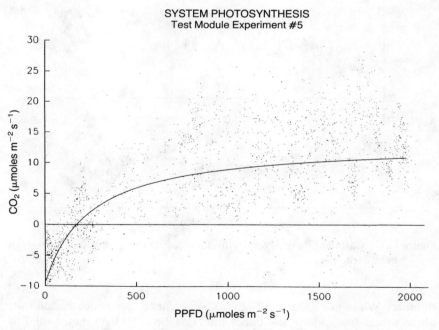

Figure C.18. Net photosynthesis rate in moles of carbon dioxide used as a function of light intensity. PPFD, photosynthetic photon flux density.

results of microcosm studies. Figure C.18 shows considerable scatter and variability, indicating other factors besides light were at work. Variation in rates of net photosynthesis may be due to variations in photosynthesis and/or variations in the concurrent respiration of soils, animals, and human participants.

Discussion

Previous work on closed ecological systems in the U.S. and the Soviet Union had demonstrated that algae and higher plants were able to regenerate most of the water and oxygen required by humans, while removing respired carbon dioxide (Shepelev 1972; Terskov et al. 1979). Our test module may have been the first system with so many of the needs for human life support supplied from the biotic components. Waste materials were treated within this materially closed ecological system, and main parts of the atmosphere were recycled. Food was supplied from fruits and vegetables grown inside (Leigh et al. 1987; Glenn et al. 1990). Higher plants and soils were used to regenerate atmospheric gases.

Technogenic Outgassing

For purification, air was pumped through a soil bed reactor, facilitating microbial metabolism of potentially dangerous trace gases from technogenic, biogenic, and anthropogenic off-gassing. The closure experiment showed that recycling systems were able to deal with the trace gases initially released. Trace organic gases and potential toxic gases were kept within acceptable concentrations for human and plant life (as defined by OSHA and the American Conference of Governmental Industrial Hygienists) during all the human closure experiments. Indeed, they were kept to levels far below that of concern (Alling et al. 1990).

The studies on the Biosphere 2 Test Module indicate its capability for either short or long term closure tests to investigate possible problems. For example, trace gas purification by the bioregenerative system was tested by inputting a fixed concentration of a gas and observing its uptake over time. For another example, concentration of one gas was changed to observe what effects this had on other gases present. The Test Module was used during 1992 to study problems that arose in Biosphere 2's operation starting September 1991, such as excess soil respiration and the ability of plants to respond to high CO_2 levels. In the future, the Test Module may be used for experiments related to near-term space application, such as space station life support systems, technologies for extended planetary missions, and initial lunar base requirements. These may include studies of "hybrid systems" which integrate both physical–chemical and bioregenerative subsystems.

Bibliography

References cited are marked with an asterisk*

*Abbott W (1966) Microcosm studies on estuarine waters. I. The replicability of microcosms. J Water Pollut Control Fed (February):258–270

*Abbott W (1967) Microcosm studies on estuarine waters. II. The effects of single doses of nitrate and phosphate. J Water Pollut Control Feb (January):113–122

Abbott W (1969) High levels of nitrate and phosphate in carboy microcosm studies. J Water Pollut Control Fed (October):1748–1751

Adams VD, Werner MD, Parker JD, Porcella DB (1985) Use of a three-phase microcosm for analysis of contaminant stress on aquatic ecosystems. In: Boyle TP (ed) Validation and predictability of laboratory methods for assessing the fate and effects of contaminants in aquatic ecosystems. (ASTM STP 865) American Society for Testing Materials, Philadelphia, pp 31–42

*Adey WH (1983) The microcosm: a new tool for reef research. Coral Reefs 1:193–201

*Adey WH, Loveland K (1991) Dynamic aquaria, the construction and operation of microcosms, mesocosms, and aquaria. Academic Press, San Diego, California

Adler D, Amdurer M, Santschi PH (1980) Metal tracers in two marine microcosms: sensitivity to scale and configuration. In: Giesy JP Jr (Ed) Microcosms in ecological research. (DOE Symposium Series 52) US Department of Energy, Washington DC, pp 348–368

*Ahmadjian V (1967) The lichen symbiosis. Blaisdell, Waltham, Massachusetts 152 pp

*Allee WC, Emerson AE, Park O, Park T, Schmidt KP (1949) principles of animal ecology. Saunders, Philadelphia

*Allen J (1991) Biosphere 2: the human experiment. Penguin Books, New York

*Allen J (1992) Biosphere 2: description, purpose, and conceptual design. Space Biospheres Ventures, and Synergetic Press, Oracle, Arizona
*Allen J, Nelson M (1986) Space biospheres. Synergetic Press, Tucson Ariz.
*Alling AL, Leigh L, MacCallum T, Alvarez-Romo N (1990) In: Nelson M, Soffen GA (eds) Biosphere 2 Test Module experimentation program in biological life support technologies. (NASA CP-3094) US National Aeronautics and Space Administration, and Synergetic Press, Oracle, Arizona
Allred PM, Giesy JP (1988) Use of in-situ microcosms to study mass loss and chemical composition of leaf litter being processed in a blackwater stream. Arch Hydrobiol 114(2):231–250
*Alongi DM (1985a) Effect of physical disturbance on population dynamics and trophic interactions among microbes and meiofauna. J Mar Res 43:351–164
*Alongi DM (1985b) Microbes meiofauna and bacterial productivity on tubes constructed by the polychaete *Capitella capitata* Mar Ecol Progr Ser 23:207–208
*Alongi DM (1987) The influence of mangrove-derived tannins on intertidal meiobenthos in tropical estuaries. Oecologia (Berlin) 71:537–540
*Alongi DM, Hanson RB (1985) Effect of detritus supply on trophic relationships within experimental benthic food webs. II. Microbial responses, fate, and composition of decomposing detritus. J Exp Mar Biol Ecol 88:167–182
*Alongi DM, Tenore KR (1985) Effect of detritus supply on trophic relationships within experimental benthic food webs. I. Meiofauna–polychaete (*Capitella capitata* (type I) *fabricius*) interactions. J Exp Mar Biol Ecol 88:153–166
Alongi DM, Tietjen JH (1980) Population growth and trophic interactions among free-living marine nematodes. In: Tenore KR, Coull BC (eds) Marine benthic dynamics university of south carolina press, Columbia South Carolina, pp 151–163
*Ambuhl H (1959) Die Bedeutung der Stroming als ökologischer Faktor. Schweiz Z Hydrol 21:133–264
*Amdurer M, Adler DM, Santschi PH (1982) Radiotracers in studies of trace metal behavior in mesocosms: advantages and limitations. In: Grice D, Reeve MB (eds) Marine mesocosms, biological and chemical research in experimental ecosystems. Springer, Berlin Heidelberg New York, pp 81–95
Andersen FO, Kristensen E (1988) The influence of macrofauna on estuarine benthic community metabolism, a microcosm study. Mar Biol (Berlin) 99(4): 591–603
Anderson JM, Ineson PA (1982) Soil microcosm system and its application to measurements of respiration and nutrient leaching. Soil Biol Biochem 14:415–416
Anderson JM, Huish SA, Ineson P, Leonard MA, Splatt PR (1985) Ecological interactions in soil. (Special Publication Number 4) The British Ecological Soc, pp 377–392
Anderson JM, Leonnard MA, Ineson P, Huish S (1985) Faunal biomass: a key component of a general model of nitrogen mineralization. Soil Biol Biochem 17(5):535–737
*Anderson RV, Gould WD, Woods LE, Camberdella C, Ingham RE, Coleman DC (1983) Organic and inorganic nitrogenous losses by microbivorous nematodes in soil. Oikos 40:75–85
*Anita NJ, McAllister CD, Parsons TR, Stephens K, Strickland JDH (1963) Further measurements of primary production using a large volume plastic sphere. Limnol Oceanogr 8:166–183

Ansell AD, Raymont JEG, Lander KF, Crowley E, Shackley P (1963) Studies on the mass culture of *Phaeodactylum*. II. The growth of *Phaeodactylum* and other species in outdoor tanks. Limnol Oceanogr 8:184–206

Ansell AD, Raymont JEG, Lander KF (1963) Studies on the mass culture of *Phaeodactylum*. III. Small-scale experiments. Limnol Oceanogr 8:207–213

Antsyshkina LM, Kirilenko N, Mamontov VY, Mel'nikov G, Ryabov B (1965) Experimental confinement of fish in hermetically sealed aquariums with and without *Chlorella*. Probl Space Biol 4:613–620

*APHS (American Public Health Association) (1992) Standard methods for examination of waters, Washington DC

Apostol S (1973) A bioassay of toxicity using protozoa in the study of aquatic environment pollution and its prevention. Environ Res 6:365–372

Armitage BJ (1980) Effects of temperature on periphyton biomass and community composition in the Brown Ferry experimental channels. In: Giesy JP Jr (ed) Microcosms in ecological research. (DOE Symposium Series 52) US Department of Energy, Washington DC, pp 668–683

*Armstrong J et al (1987) Microcosm method to assess survival of recombinant bacteria associated with plants and herbivorous insects Curr Microbiol 15(4):229–232

*Armstrong NE, Gloyna EF (1968) Radioactivity transport in water — numerical solutions of radio-nuclide transport equations and role of plants in Sr-85 transport. (Technical Report #14 to the US AEC) Center for Research in Water Resources, Environmental Health Engineering Research Laboratory, University of Texas, Austin

*Armstrong NE, Odum HT (1964) Photoelectric ecosystem. Science 143:256–258

*Atz JW (1949) The balanced aquarium myth. Aquarist 14:159–160, 179–182

Ausmus BS, Voris PV, Jackson DR (1980) Terrestrial microcosms: what questions do they address? In: Giesy JP Jr (ed) Microcosms in ecological research. (DOE Symposium Series 52) US Department of Energy, Washington DC, pp 937–953

*Ausmus BS, Eddlemon GK, Draggan SJ, Giddings JM, Jackson DR, Luxmore RJ, O'Neill EG, O'Neill RV, Rosstod DM, Van Voris PV (1980) Microcosms as potential screening tools for evaluating transport and effects of toxic substances. (EPA-600/3-80-042 Publ 1506) US Environmental Protection Agency, Washington DC

*Baker LA, Perry TE, Brezonik PL (1985) Neutralization of acid precipitation in softwater lakes. In: Lake and reservoir management: practical applications. (Proc Fourth Annual Conference of the North American Lake Management Society), pp 356–350

*Bakke T (1986) Experimental long term oil polution in a boreal rocky shore enironment. Proceedings 9th AMOP Technical Seminar, Canada, ISBN 0-662-14812-6

*Bakke T (1990) Benthic mesocosms: II. Basic research in hard-bottom benthic mesocosms. In: Lalli CM (ed) Enclosed experimental marine ecosystems: a review and recommendations. Coastal and Estuarine Studies, 37, Springer, Berlin Heidelberg New York, pp. 122–135

*Balch N, Boyd CM, Mullin M (1978) Large scale tower tank systems. Rapp P-V Reun Cons Int Explor Mer 173:13–21

Bale MJ, Fry JC, Day MJ (1987) Plasmid transfer between strains of *Pseudomonas aeruginosa* on membrane filters attached to river stones. J Gen Microbiol 133(11):3099–3108

Barrio-Lage G, Parsons FZ, Lorenzo PA (1988) Inhibition and stimulation of trichloroethylene biodegradation in microaerophilic microcosms. Environ Toxicol Chem 7(11):889–896

*Barrio-Lage G, Parsons FZ, Nassar RS, Lorenzo PA (1986) Sequential dehalogenation of chlorinated ethenes. Environ Sci Technol 20(1):96–99

Barrio-Lage G, Parsons FZ, Nassar RS, Lorenzo PA (1987b) Biotransformation of trichloroethene in a variety of subsurface. Mater Environ Toxicol Chem 6(8):571–578

*Barrio-Lage G, Parsons FZ, Nassar RS (1987a) Kinetics of the depletion of trichloroethene. Environ Sci Technol 21(4):366–370

*Bartsch AF (1961) Algae as a source of oxygen in waste treatment. J Water Pollut Control Fed 33(3):239–249

*Bartsch AF, Allum MO (1957) Biological factors in treatment of raw sewage in artificial ponds. Limnol Oceanogr 2(2):77–84

Bauer JE, Kerr RP, Bautista MF, Decker CJ, Capone DG (1988) Stimulation of microbial activities and polycyclic aromatic hydrocarbon degradation in marine sediments inhabited by *Capitella capitata*. Mar Environ Res 25(1):63–84

Beall ML, Nash RG, Kearney PC (1976) Agroecosystem — a laboratory model ecosystem to simulate agricultural field conditions for monitoring pesticides. In: Ott W (ed) Environmental modeling and simulation. (EPA-600/9-76-016) US Environmental Protection Agency, Washington DC, pp 790–793

*Beard P (1930) Adventures in dish gardening. AT DeLaMare Co Inc, New York

*Berger WJH, Parker FL (1970) Diversity of planktonic foraminifera in deep sea sediments. Science 168:1345–1347

Bej AK, MH Perlin RM (1988) Atlas model suicide vector for containment of genetically engineered microorganisms. Appl Environ Microbiol 54(10):2472–2477

*Bell SS, Coull BC (1978) Field evidence that shrimp predation regulates meiofauna. Oecologia (Berlin) 35:141–148

*Bellamy DJ, Rieley J (1967) Some ecological statistics of a "miniature bog." Oikos 18:33–40

Bengtsson G, Enfield CG, Lindqvist R (1987) Macromolecules facilitate the transport of trace organics. Sci Total Environ 67(2–3):159–164

Bengtsson G, Berden M, Rundgren S (1988) Influence of soil animals and metals on decomposition processes, a microcosm experiment. J Environ Qual 17(1):113–119

Bentjen SA, Fredrickson JK, Van Voris P, Li SW (1988) Use of intact soil-core microcosms for evaluating the fate and potential ecological effects from the release of *Azospirillum lipoferum* TN5 mutants (abstract). Abstr Ann Meet Am Soc Microbiol 88:284

Bentjen SA, Fredrickson JK, Van Voris R, Li SW (1989) Intact soil-core microcosms for evaluating the fate and ecological impact of the release of genetically engineered microorganisms. Appl Environ Microbiol 55(1):198–202

*Berger WJF, Parker FL (1970) Diversity of planktonic formainifera in deep sediments. Science 168:1345–1347

*Berounsky VM, Nixon SW (1985) Eutrophication and the rate of net nitrification in a coastal marine ecosystem. Estuar Coastal Shelf Sci 20:773–781

*Beyers RJ (1962a) The metabolism of twelve laboratory microecosystems. (PhD Dissertation) University of Texas, Austin
*Beyers RJ (1962b) Relationship between temperature and the metabolism of experimental ecosystems. Science 136:980–982
*Beyers RJ (1963a) A characteristic diurnal metabolic pattern in balanced microcosms. Publ Inst Mar Sci Texas 9:19–27
*Beyers RJ (1963b) The metabolism of twelve aquatic laboratory microecosystems. Ecol Monogr 33(4):255–306
Beyers RJ (1963) Balanced aquatic microcosms — their implications for space travel. Am Biol Teacher 25(6):422–428
*Beyers RJ (1964) The microcosm approach to ecosystem biology. Am Biol Teacher 26(7):491–498
*Beyers RJ (1965) The pattern of photosynthesis and respiration in laboratory microecosystems. Mem Ist Ital Idrobiol (Suppl 18):61–74
Beyers RJ (1966) Metabolic similarities between symbiotic coelenterates and aquatic ecosystems. Arch Hydrobiol 62(3):273–284
*Beyers RJ, Gillespie B (1964) Measuring the carbon dioxide metabolism of aquatic organisms. Am Biol Teacher 26:499–510
*Beyers RJ, Larimer JL, Odum HT, Parker RB, Armstrong NE (1963) Directions for the determination of changes in carbon dioxide concentrations from changes in pH. Publ Inst Mar Sci Texas 9:454–489
Bianchi TS, Rice DL (1988) Feeding ecology of *Leitoscoloplos fragilis*. II. Effects of worm density on benthic diatom production. Mar Biol (Berlin) 99(1):123–132
Bianchi TS, Dawson R, Sawangwong P (1988) The effects of macrobenthic deposit-feeding on the degradation of chloropigments in sandy sediments. J Exp Mar Biol Ecol 122(3):243–256
Biesinger KE, Stokes GN (1986) Effects of synthetic polyelectrolytes on selected aquatic organisms. J Water Pollul Control Fed (March):207–213
*Billings WD, Luken JO, Mortensen DA, Peterson KM (1982) Arctic tundra: source or sink for atmospheric carbon dioxide in a changing environment. Oecologia (Berlin) 53:7–11
*Billings WD, Luken JO, Mortensen DA, Peterson KM (1983) Increasing atmospheric carbon dioxide: possible effects on arctic tundra. Oecologia (Berlin) 58:286–289
*Billings WD, Peterson KM, Luken JO, Mortensen DA (1984) Interaction of increasing atmospheric carbon dioxide and soil nitrogen on the carbon balance of tundra microcosms. Oecologia (Berlin) 65:26–29
Binet FP, Trehen S, Deleporte (1988) Experimental approach by microcosm of an interactive soil–earthworms system. (Colloquium on the function of the soil and interactive biological systems, Part 2) Rev Ecol Biol Sol 24(4):703–714
*Birmingham J (1990) The wizard of wet. Continental Profiles, (July) 22: 49–52
*Boling RH Jr, Peterson RC, Cummins KW (1974) Ecosystem modeling for small woodland streams. In: Patten B (ed) Systems analysis and simulation in ecology, vol 3. Academic Press, New York
*Bolton H Jr, Fredrickson JK, Workman DJ, Bentjen SA, Li SW (1990) Field calibration of microcosms to assess the fate and ecological effects of a genetically altered *Rhizobacterium* (abstract). Abstr Ann Meet Am Soc Microbiol 90:314
Bolton H Jr, Fredrickson JK, Thomas JM, Li SW, Workman DJ, Bentjen SA, Smith JL (1991) Field calibration of soil-core microcosms: ecosystem structural and functional comparisons. Micob Ecol 21(2):175–189

Bond H, Lighthart B, Shimabuku R, Russel L (1976) Some effects of cadmium on coniferous forest soil and litter microcosms. Soil Sci 121:278–287

*Bongers L, Medici JC (1968) Chemosynthetic metabolism of *Hydrogenomonas*. In: Bioregenerative systems, NASA SP-165, National Aeronautics and Space Administration, Washington DC, pp 9–19

*Bovbjerg RV, Glynn PW (1960) A class exercise on a marine microcosm. Ecology 41(1):229–232

*Bowden RD (1991) Inputs, outputs, and accumulation of nitrogen in an early successional moss ecosystem. Ecol Monogr 61:207–233

Bowling JW, Giesy JP, Kania HJ, Knight RL (1980) Large-scale microcosms for assessing fates and effects of trace contaminants. In: Giesy JP Jr (ed) Microcosms in ecological research. (DOE Symposium Series 52) US Department of Energy, Washington DC, pp 224–247

*Boyd CM (1981) Microcosms and experimental planktonic food chains. In: Longhurst AR (ed) Analysis of marine ecosystems. Academic Press, New York, pp 627–649

Boyle TP (1985) (ed) Validation and predictability of laboratory methods for assessing the fate and effects of contaminants in aquatic ecosystems. (ASTM STP 865) American Society for Testing Materials, Philadelphia

*Branham BE, Wehner DJ (1985) The fate of diazinon applied to thatched turf. Agron J 77:101–104

Branham BE, Wehner DJ, Torello WA, Turgeon AJ (1985) A microecosystem for fertilizer and pesticide fate research. Agron J 77:176–180

*Brawley SH, Adey WH (1981) The effect of micrograzers on algal community structure in a coral reef microcosm. Mar Biol (Berlin) 61:167–177

*Breder CM Jr (1931) On the organic equilibria in aquaria. Copeia 2:66

*Breder CM Jr (1957) Miniature circulating systems for small laboratory aquaria. Zoologica 42:1–10

Breen AB, Reynolds L, Burtis B, Bellew G, Sayler E (1988) Introduction and effects of the PSS50 degradative genotype in four chlorobiphenyl lake water microcosms (abstract). Abstr Ann Meet Am Soc Microbiol 88:309

Bretthauer R (1980) Laboratory aquatic microcosms. In: Giesy JP Jr (ed) Microcosms in ecological research. (DOE Symposium Series 52 USA) US Department of Energy, Washington DC, pp 416–445

*Brock TD, Brock ML (1966) Temperature optima for algal development in Yellowstone and Iceland hot springs. Nature 209:733–734

Brockman U (1990) Pelagic mesocoosms. II. Process studies. In: Lalli CM (ed) Enclosed experimental marine ecosystems: a review and recommendations. Coastal and estuarine studies, 37, Springer, Berlin Heidelberg New York pp. 81–108

Brockway DL, Smith PD, Stancil FE (1984) Fate and effects of atrazine in small aquatic microcosms. Bull Environ Contam Toxicol 32(3):345–353

*Bruce HE, Hood DW (1959) Diurnal inorganic phosphate variations in Texas bays. Publ Inst Mar Sci 6:133–145

Burlew JS (1953) Algal culture from laboratory to pilot plant. Carnegie Institution of Washington Publication 600, Washington DC

*Burnette MS (1978) Energy analysis of intermediate technology agricultural systems. (MS Thesis) Environmental Engineering Sciences, University of Florida, Gainesville, Florida

*Burney CM, Johnson KM, Sieburth JM (1981) Diel flux of dissolved carbohydrate in a salt marsh and a simulated estuarine ecosystem. Mar Biol 63:175–187

*Burns L (1970) An analog simulation of a rain forest with high–low pass filters and a programmatic spring pulse. In: Odum HT, Pigeon RF (eds) A tropical rainforest. TID-24270 (PRNC-138) US Atomic Energy Commission, Oak Ridge, Tennessee, Appendix, p I-284

*Busch AW (1963) Process kinetics as design criteria for bio-oxidation of petrochemical wastes. (ASME paper No 62-pet-7) Trans American Society of Mechanical Engineers, Petroleum Division, 6 pp

*Butler JL (1964) Interaction of effects by environmental factors on primary productivity in ponds and microecosystems. Phd dissertation 99 pp

*Cairns J Jr (1969) Rate of species diversity restoration following stress in freshwater protozoan communities. Univ Kansas Sci Bull 48(6):209–224

Cairns J Jr (ed) (1971) The structure and function of fresh-water microbial communities. University Press of Virginia, Charlottesville, Virginia

*Cairns J Jr (ed) (1982) Artificial substrates. Ann Arbor Science, Ann Arbor, Michigan

Cairns J Jr (1984) Multispecies toxicity testing. Environ Toxicol Chem 3:1–3

Cairns J Jr (ed) (1986) Community toxicity testing. (ASTM STP 920) American Society for Testing Materials, Philadelphia

Cairns J Jr (1986) The myth of the most sensitive species. Bioscience 36(10):670–672

*Cairns J Jr, Yongue WH Jr (1973) The effect of an influx of new species on the diversity of protozoan communities. Rev Biologia 8(1–4):187–206

Cairns J Jr, Dickson KL, Slocomb JP, Almeida S, EU KI, Liu YC, Smith HF (1974) Microcosm pollution monitoring. Trace Subst Environ Health 8:223–228

Cairns J Jr, Pratt JR, Niederlehner BR, McCormick VA (1986) Simple cost-effective multispecies toxicity test using organisms with a cosmopolitan distribution. Environ Monit Assess 6:207–220

*Canfield JH, Lechtman MD (1964) Study of hydrogen-fixing microorganisms for closed regenerating biosystems. (Technical Documentary Report No AMRL-TDR-64-35) Aerospace Medical Research Laboratories, Wright Patterson Air Force Base, Ohio Megna Corporation, Anaheim California

Carlough LA, Meyer JL (1989) Protozoans in two southeastern blackwater rivers and their importance to trophic transfer. Limnol Oceanogr 34(1):163–177

*Caron DA, Sieburth JM (1981) Response of peritrichous ciliates in fouling communities to seawater-accommodated hydrocarbons. Trans Am Microsc Soc 100(2):183–203

*Carpenter EJ (1971) Effects of phosphorus mining wastes on the growth of phytoplankton in the Pamlico River estuary. Chesapeake Sci 12(2):85–94

Chandler GT, Fleeger JW (1987) Facultative and inhibitory interactions among estuarine meiobenthic harpacticoid copepods. Ecology 68(6):1906–1919

*Chaney WR, Kelly JM, Strickland RC (1978) Influence of cadmium and zinc on carbon dioxide evolution from litter and soil from a black oak forest. J Environ Qual 7(1):115–119

Chao WL, Ding RJ, Chen RS (1987) Survival of pathogenic bacteria in environmental microcosms. Chin J Microbiol Immunol (Taipei) 20(4):339–348

*Childres JJ, Fisher CR, Favuzzi JA, Sanders NK, Alayse AM (1991) Sulfide-driven autotrophic balance in the bacterial symbiont-containing hydrothermal vent tubeworm *Riftia pachyptila* Jones Biol Bull (Woods Hole) 180(1):135–153

Clarholm M (1985) Possible roles for roots, bacteria, protozoa, and fungi in supplying nitrogen to plants. In: Fitter AH, Atkinson D, Read DJ, Usher MB (eds) Ecological interactions in soil. Special Publication of The British Ecological Soc. Blackwell Scientific Publ Boston Massachusetts, pp 335–365

*Clark JR, Cherry DS, Cairns J Jr (1982) Food quality of aufwuchs from artificial streams receiving low levels of perturbations. Water Resourc Bull 18(5):761–767

Coats JR (1980) A stream microcosm for environmental assessment of pesticides. In: Giesy JP Jr (ed) Microcosms in ecological research. (DOE Symposium Series 52) US Department of Energy, Washington DC, pp 715–723

Cole LK, Metcalf RL (1980) Environmental destinies of insecticides, herbicides, and fungicides in the plants, animals, soil, air, and water of homologous microcosms. In: Giesy JP Jr (ed) Microcosms in ecological research. (DOE Symposium Series 52) US Department of Energy, Washington DC, pp 971–1007

Cole LK, Metcalf RL, Sanborn JR (1976) Environmental fate of insecticide in a terrestrial model ecosystem. Int J Environ Stud 10:7–14

Cole LK, Sanborn JR, Metcalf RL (1976) Inhibition of corn growth by aldrin and the insecticide's fate in the soil, air, crop, and wildlife of a terrestrial model ecosystem. Environ Entomol 5:583–589

Coler RA, Gunner HB (1970) Laboratory measure of an ecosystem response to a sustained stress. Appl Microbiol 19(6):1009–1012

Collos Y, Linley EAS, Frikha MG, Ravail B (1988) Phytoplankton death and nitrification at low temperatures. Estuar Coastal Shelf Sci 27(3):341–347

*Confer JL (1972) Interrelations among plankton-attached algae and the phosphorus cycle in artificial open systems. Ecol Monogr 42:1–23

*Conger GP (1922) Theories of macrocosms and microcosms in the history of philosophy. Columbia University Press, New York

*Connor MS, Teal JM, Valiela I (1982) The effect of feeding by mud snails *Ilyanassa obsoleta* (Say) on the structure and metabolism of a laboratory benthic algal community. J Exp Mar Biol Ecol 65:29–45

Conquest LL (1983) Assessing the statistical effectiveness of ecological experiments. Int J Environ Stud 20:209–221

*Cooke GD (1967) The pattern of autotrophic succession in laboratory microcosms. Bioscience 17(10):717–721

*Cooke GD (1971) Aquatic laboratory microsystems and communities. In: Cairns J Jr (ed) The structure and function of fresh-water microbial communities. University Press of Virginia, Charlottesville, Virginia, pp 47–85

Cooke JG, White RE (1987) The effect of nitrate in stream water on the relationship between denitrification and nitrification in a stream–sediment microcosm. Freshwater Biol 18(2):213–226

Cooke JG, White RE (1988) Nitrate enhancement of nitrification depth in sediment–water microcosms. Environ Geol Water Sci 11(1):85–94

Cooke GD, Beyers RJ, Odum EP (1968) The case for the multispecies ecological system with special reference to succession and stability. In: Bioregenerative systems. NASA SP-165 (Library of Congress Number 68-60345) US National Aeronautics and Space Administration, Washington DC, pp 129–139

*Cooper DC (1973) Enhancement of net primary productivity by herbivore grazing in aquatic laboratory microcosms. Limnol Oceanogr 18(1):31–37

Cooper DC, Copeland BJ (1973) Responses of continuous-series estuarine microecosystems to point-source input variations. Ecol Monogr 43:213–236

Coover MP, Sims RC (1987) The effect of temperature on polycyclic aromatic hydrocarbon persistence in an unacclimated agricultural soil. Hazard Waste Hazard Mater 4(1):69–82

Coover MP, Sims RC (1987) The rate of benzo-a-pyrene apparent loss in a natural and manure amended clay loam soil. Hazard Waste Hazard Mater 4(2):151–158

*Copeland BJ (1964) Evidence for regulation of community metabolism in a marine ecosystem. Ecology 46(4):563–564

*Copeland BJ (1967) Biological and physical basis of indicator organisms and communities. III. Biological and physiological basis of indicator communities. In: Olson TA, Burgess FJ (eds) Pollution and marine ecology. Wiley Interscience, New York, pp 285–288

*Copeland BJ, Dorris TC (1964) Community metabolism in ecosystems receiving oil refinery effluents. Limnol Oceanogr 9(3):431–447

*Copeland BJ, Gloyna EF (1965) Radioactivity transport in water — structure and metabolism of a lotic community. Technical Report to the AEC-EHE-02-6501 N66-33913 (NASA Scientific & Technical Information Facility Austin, Texas TID-22873) pt 1:i–51

*Copeland BJ, Odum HT, Cooper DC (1972) Water quality for preservation of estuarine ecology. In: Gloyna EF, Butcher WS (eds) Water resource symposium No 5. Center for Research in Water Resources, University of Texas, Austin, Texas, pp 107–126

Correll DL, Wu TL (1982) Atrazine toxicity to submersed vascular plants in simulated estuarine microcosms. Aquat Bot 14(2):151–158

Corey KA, Wheeler RM (1992) Gas exchange in NASA's biomass production chamber. Bioscience 42(7):503–509

Couteaux M (1984) Utalisation des microcosmes pur l'analyse des fonctions ecologiques des protozoaires de l'humus. Acta Oecologica Oecol Gen 5:71–76

Couteaux M (1985) Relation entre la densitt apparente d'un humus et l'aptitude a la croissance de ses cilies. Pedobiologia 28:289–303

*Couteaux M, Devaux J (1983) Effect d'un enrichissement en champignons sur la dynamique d'un peuplement camoebien d'un humus. Rev Ecol Biol Sol 20:519–545

*Coutts DAP, Senior E, Balba MTM (1987) Multi-stage chemostat investigation of interspecies interactions in a hexanoate-catabolizing microbial association isolated from anoxic landfill. J Appl Bacteriol 62:251–260

Coveney MF, Wetzel RG (1989) Bacteria metabolism of algal extracellular carbon. Hydrobiologia 173(2):141–150

*Cowles HC (1899) The ecological relations of the vegetation of the sand dunes of Lake Michigan. Bot Gaz 27:95–117, 167–202, 281–308, 361–391

*Craft D (1983a) Microsystem sediment–water simulation: a practical technique for predicting reservoir water quality. (Publ No REC-ERC-83-12) US Department of the Interior, Bureau of Reclamation, Denver, Colorado

*Craft D (1983b) A microsystem sediment–water simulation study for the proposed Jordanelle Reservoir, Heber City, Utah. (Publ No REC-ERC-83-131) US Department of the Interior, Bureau of Reclamation, Denver, Colorado

Cragg BA, Fry JC (1984) The use of microcosms to simulate field experiments to determine the effects of herbicides on aquatic bacteria. J Gen Microbiol 130:2309–2316

Crittenden JL, Mulholland RJ, Hill J, Martinex F (1983) Sampling and data analysis properties of Prony's method applied to model identification. Int J Syst Sci 14(5):571–584

*Cronan CS (1980) Controls on leaching from coniferous forest floor microcosms. Plant Soil 56:301–322

*Cronan CS (1985) Comparative effects of precipitation acidity on three forest soils: carbon cycling responses. Plant Soil 88:101–112

*Cullingford HS, Novara M (1988) Conceptual design of a piloted Mars sprint life support system. In: Engineering Soc Advancing Mobility Land Sea Air Space, 18th Intersociety Conference on Environmental Systems, San Francisco, pp 1–10

*Culver DA (1988) Plankton ecology in fish hatchery ponds in Narrandera, NSW Australia. Verh Int Ver Limnol 23:1085–1089

*Cumming FP (1969) An experimental design for the analysis of community structure. (MS Thesis) Department of Botany, University of North Carolina, Chapel Hill, North Carolina

*Cumming FP, Beyers RJ (1970) Further laboratory studies of forest floor microcosms. In: Odum HT, Pigion R (eds) A tropical rain forest. TID-24270 (PRNC-138) US Atomic Energy Commission, Oak Ridge, Tennessee, pp Chapter I-3 Appendix B:I-54–I-56

Curl EA, Rodriguez-Kabana R (1971) Microbial interactions. In: Wilkinson RE (ed) Research methods in weed science. POP Enterprises, Atlanta, Georgia, pp 161–194

Cushing CE, Porter NS (1969) Radionuclide cycling by periphyton: an apparatus for continuous in situ measurements and initial data on zinc-65 cycling. (Symposium on radioecology) US Atomic Energy Commission, Oak Ridge, TID-4500 CONF-670503, Tennessee, pp 285–290

*Cushing CE, Thomas JM, Eberhardt LL (1975) Modeling mineral cycling by periphyton in a simulated stream system. Verh Int Ver Limnol 19:1593–1598

Cushing CE, Rose FL (1970) Cycling of zinc-65 by Columbia River periphyton in a closed lotic microcosm. Limnol Oceanogr 15(5):762–767

Davis CC (1966) Notes on the ecology and reproduction of *Trichcorixa reticulata* in a Jamaican salt-water pool. Ecology 5:850–851

*Day FP Jr (1983) Effects of flooding on leaf litter decomposition in microcosms. Oecologica 56:180–184

*Day FP Jr, Megonigal JP, Lee LC (1989) Cypress root decomposition in experimental wetland mesocosms. J Soc Wetland Sci 9(2):263–282

Day JW Jr (1970) Carbon metabolism of estuarine ponds receiving treated sewage wastes. (PhD Dissertation) University of North Carolina, Chapel Hill, North Carolina

Debano LF, Klopatek JM (1988) Phosphorus dynamics of pinyon–juniper soils following simulated burning. Soil Sci Soc Am J 52(1):271–277

*DeCatanzaro JB, Hutchinson TC (1985) Leaching and distribution of nitrogen and nickel in nickel-perturbed Jack pines forest microcosms. Water Air Soil Poll 26:281–292

*DeMott W, Kerfoot WC (1982) Competition among cladocerans: nature of the interactions between *Bosmina* and *Daphnia*. Ecology 63:1949–1966

*Dempster WF (1988) Biosphere II: design of a closed manned terrestrial ecosystem. (SAE Technical Paper Series #881096) 18th Intersociety Conference on Environmental Systems, SAE, Warrendale, PA

*Dempster WF (1989) Biosphere II: technical overview of a manned closed ecological system. (SAE Technical Paper Series #891599) 19th Intersociety Conference on Environmental Systems, SAE, Warrendale, PA

deNoyelles F Jr, Kettle WD (1985) Experimental ponds for evaluating bioassay predictions. In: Boyle TP (ed) Validation and predictability of laboratory methods for assessing the fate and effects of contaminants in aquatic ecosystems. (ASTM STP 865) American Society for Testing Materials, Philadelphia, pp 91–103

deNoyelles F Jr, Reinke D, Treanor D, Altenhofen C (1980) In situ continuous culturing of lake phytoplankton communities In: Giesy JP Jr (ed) Microcosms in ecological research. (DOE Symposium Series 52) US Department of Energy, Washington DC, pp 489–512

DePinto JV, Guminiak RJ, Howell RS, Edzwald JK (1980) Use of microcosms to evaluate acid lake recovery techniques. In: Giesy JP Jr (ed) Microcosms in ecological research. (DOE Symposium Series 52) US Department of Energy, Washington DC, pp 562–582

*Devik O (1976) Marine Greenhouse systems: the entrainment of large water masses. In: Devik O (ed) Harvesting polluted waters. Plenum Press, New York, pp 113–142

deWilde PA, Kuipers BR (1977) A large indoor tidal mud-flat ecosystem. Helgol Wiss Meeresunters 30:334–342

Dewitt TH, Ditsworth GR, Swartz RC (1988) Effects of natural sediment features on survival of the phoxocephalid amphipod *Rhepoxynius abronius*. Mar Environ Res 25(2):99–124

*Dickerson JE Jr, Robinson JV (1984) The assembly of microscopic communities: patterns of species importance. Trans Am Microsc Soc 103(2):164–171

*Dickerson JE Jr, Robinson JV (1985) Microcosms as islands: a test of the Macarthur–Wilson equilibrium theory. Ecology 66(3):966–980

*Dickerson JE Jr, Robinson JV (1986) The controlled assembly of microcosmic communities: the selective extinction hypothesis. Oecologia (Berlin) 71:12–17

Dickie LM, Kerr SR, Boudreau PR (1987) Size-dependent processes underlying regularities in ecosystems structure. Ecol Monogr 57(3):233–250

Dighton JED, Thomas PM (1987) Interactions between tree roots, *Mycorrhizas*, a saprotrophic fungus, and the decomposition of organic substrates in a microcosm. Biol Fertil Soils 4(3):145–150

Dillon RD, Bierle DA (1980) Microbiocoenoses in an antarctic pond. In: Giesy JP Jr (ed) Microcosms in ecological research. (DOE Symposium Series 52) US Department of Energy, Washington DC, pp 446–457

*Doering PH, Oviatt CA (1986) Application of filtration rate models to field population of bivalves: an assessment using experimental mesocosms. Mar Ecol Progr Ser 31:265–275

*Doering PH, Oviatt CA, Kelly JR (1986) The effects of the filter-feeding clam *Mercenaria mercenaria* on carbon cycling in experimental marine mesocosms. J Mar Res 44:839–861

*Doering PH, Kelly JR, Oviatt CA, Sowers F (1987) Effect of the hard clam *Mercenaria mercenaria* on benthic fluxes of inorganic nutrients and gases. Mar Biol 94:377–383

*Dollar AM, Taub FB, Hung R, Orsi J (1964) Research in fisheries. College of Fisheries, University of Washington, Seattle, Washington

*Donaghay PL (1985) An experimental test of the relative significance of food quality and past feeding history to limitation of egg production of the estuarine copepod *Acartia tonsa*. Arch Hydrobiol Beih Erg Limnol 21:235–245

*Donaghay PL, Klos E (1985) Physical, chemical, and biological responses to simulated wind and tidal mixing in experimental marine ecosystems. Mar Ecol Progr Ser 26:35–45

Downing AL (1958) Aeration in aquaria. Aquarist Pondkeeper 23:2–7

Downing AL, Truesdale GA (1956) Aeration in aquaria. Zoologica 41(4):129–143

*Draggan S (1976) The role of microcosms in ecological research (an introduction to this special issue on microcosms). Int J Environ Stud 10:1–2

Draggan S (1976) The microcosm as a tool for estimation of environmental transport of toxic materials. Int J Environ Stud 10:65–70

Draggan S (1976) The microprobe analysis of ^{60}Co uptake in sand microcosms. Xenobiotica 6:557–563

*Drake JF, Jost L, Fredrickson AG, Tsuchiya HM (1968) The food chain in bioregenerative systems. (NASA SP-165) US National Aeronautics and Space Administration, Washington DC, pp 87–95

Duggin JA, Voight GK, Bormann FH (1991) Autotrophic and heterotrophic nitrification in response to clear cutting northern hardwood forest. Bull Biol Biochem 23(8):779–788

*Duke MEL (1967) A production study of a thermal spring. (PhD Dissertation) University of Texas, Austin, Texas

*Durbin AG, Nixon SW, Oviatt CA (1979) Effects of the spawning migration of the alewife, *Alosa pseudoharengus*, on freshwater ecosystems. Ecology 60(1):8–17

*Dwyer RL, Perez KT (1983) An experimental examination of ecosystem linearization. Am Nat 121(3):305–323

Edmondson WT (1961) Factors affecting productivity in fertilized salt water. Mar Biol Oceanogr 3:451–464

*Edmondson WT, Edmondson YH (1947) Measurements of production in fertilized salt water. J Mar Res 6(3):228–246

*Ehrlich PR, Harte J, Harwell MA, Raven PH, Sagan C, Woodwell GM, Berry J, Ayensu ES, Ehrlich AH, Eisner T, Gould SJ, Grover HD, Herrera R, May RM, Mayr E, McKay CP, Mooney HA, Myers N, Pimentel D, Teal J (1983) Long-term biological consequences of nuclear war. Science 222:1293–1300

*Eichenberger E (1975) On the quantitative assessment of the effects of chemical factors on running water ecosystems. Schweiz Z Hydrol 37(1):21–34

Elliott ET, Coleman DC, Anderson RV, Cole CV, Hunt HW, Woods LE, Gould WD, McClellan JF (1980) Microbial trophic structure and habitable pore space in soil. In: Giesy JP Jr (ed) Microcosms in ecological research. (DOE Symposium Series 52) US Department of Energy, Washington DC, pp 1050–1070

Elmgren R, Vargo GA, Grassle JF, Grassle JP, Heinle DR, Langlois G, Vargo SL (1980) Trophic interactions in experimental marine ecosystems perturbed by oil. In: Giesy JP Jr (ed) Microcosms in ecological research. (DOE Symposium Series 52) US Department of Energy, Washington DC, pp 779–800

Elster HJ, Ohle HCW (1985) Food limitation and the structure of zooplankton communities. Adv Limnol 21:235–245

*Elwood JW, Newbold JD, O'Neill RV, Winkle WV (1981) Resource spiraling: an operational paradigm for analyzing lotic ecosystems. In: Fontaine TD, Bartell SM (eds) Dynamics of lotic ecosystems. Ann Arbor Science, Ann Arbor, Michigan, pp 3–28

Eppley RW, Dyer DL, Maciasr FM (1964) Growth and culture characteristics of certain marine algal flagellates for mass culture. (prepared under contract No AF 41(609)-1608) Northrop Space Laboratories, Hawthorne, California

Erb FJ, Dequidt J, Pommery P, Colein C, Gontier F (1980) Migration of some chlorine- or nitrogen-containing herbicides in a trophodynamic chain in fresh water. Environ Technol Lett 1(1):58–63

Erickson DC, Spaniel K, Loehr RC (1988) Abiotic loss of chemicals in soil. Hazard Waste Hazard Mater 5(2):121–128

Estrada MM, Alcaraz C, Marrase (1987) Effect of reversed light gradients on the phytoplankton composition in marine microcosms. Invest Pesq 51(3):443–458

Evans EW (1991) Intraspecific versus interspecific interactions of lady beetles (*Coleoptera*, *Coccinellidae*) attacking aphids. Oecologia (Heidelberg) 87(3): 401–408

*Everest JW, Davis DE (1979) Studies of phosphorus movement using salt marsh microecosystems. J Environ Qual 8(4):464–468

*Federle TW, McKinley VL, Vestal JR (1982) Physical determinants of microbial colonization and decomposition of plant litter in an arctic lake. Microb Ecol 8:127–138

Federle TW, Mckinley VL, Vestal JR (1982) Effects of nutrient enrichment on the colonization and decomposition of plant detritus by the microbiota of an arctic lake. Can J Microbiol 28(11):1199–1205

Federle TW, Livingston RJ, Metter DA, White C (1983) Modifications of estuarine sedimentary microbiota by exclusion of epibenthic predators. J Exp Mar Biol Ecol 73:81–94

Federle TW, Livingston RJ, Wolfe LE, White C (1986) A quantitative comparison of microbial community structure of estuarine sediments from microcosms and the field. Can J Microbiol 32:319–325

*Ferens MC (1974) The impact of mercuric ions on benthos and periphyton of artificial streams. (PhD Dissertation) University of Georgia, Athens, Georgia

*Ferens MC, Beyers RJ (1972) Studies of a simple laboratory microecosystem: effects of stress. Ecology 53:709–713

Fernandez IJ (1987) Vertical trends in the chemistry of forest soil microcosms following experimental acidification. Maine Agric Exp Stn Tech Bull (126):1–19

Fernandez IJ, Kosian PA (1986) Chemical response of soil leachate to alternative approaches to experimental acidification. Comm Soil Sci Plant Anal 17(9):953–973

Fewson CA (1988) Microbial metabolism of mandelate, a microcosm of diversity. FEMS Microbiol Rev 54(2):85–110

*Findlay RH, White DC (1983) The effects of feeding by the sand dollar *Mellita quinquiesperforata* on the benthic microbial community. J Exp Mar Biol Ecol 72(1):25–41

Findly RH, Trexler MB, Guckert JB, White DC (1990) Laboratory study of disturbance in marine sediments: response of a microbial community. Mar Ecol Progr Ser 62(1–2):121–134

Firth PL (1990) Controlled ecological life support system (CELSS) for long-duration space missions: potential human productivity benefits. Aviat Space Environ Med 61(5):504

Fish D, Carpenter SR (1982) Leaf litter and larval mosquito (*Aedes triseriatus*) dynamics in tree-hole ecosystems. Ecology 63(2):283–288

*Fisher JB, Matisoff G (1981) High resolution vertical profiles of pH in recent sediments. Hydrobiologia 79:277–284

Fisher SW (1985) Effects of pH upon the environmental fate of ^{14}C fenitrothion in an aquatic microcosm. Ecotoxicol Environ Safety 10:53–62

Fisher SW, Lohner TW (1986) Studies on the environmental fate of carbaryl as a function of pH. Arch Environ Contam Toxicol 15:661–667

Fisher SW, Lohner TW (1987) Changes in the aqueous behavior of parathion under varying conditions of pH. Arch Environ Contam Toxicol 16:79–84

Flemming CA, Trevors JT (1988) Copper retention and toxicity in a freshwater sediment. Water Air Soil Pollut 40(3–4):419–432

Flemming CA, Trevors JT (1988) Effect of copper on nitrous oxide reduction in freshwater sediment. Water Air Soil Pollut 40(3–4):391–398

Flum TF, Shannon LJ (1987) The effects of three related amides on microecosystem stability. Ecotoxicol Environ Safety 13(2):239–252

Folosm C (1985) Microbes. In: Snyder, TP (ed) The Biosphere Catalogue. Synergetic Press, London, pp 51–56

*Folsome C, Hanson JA (1986) The emergence of materially-closed-system ecology. In: Polunin N (ed) Ecosystem theory and application. John Wiley, New York, pp 269–288

*Fontaine TD, Bartell SM (1981) Dynamics of lotic ecosystems. Ann Arbor Science, Ann Arbor, Michigan

*Forbes SA (1887) The lake as a microcosm. Bull Sci Assoc Peoria: 77–87 reprinted (1925) Illinois Nat Hist Sur 15:537–550

*Forrester JW (1961) Industrial dynamics. MIT Press, Cambridge, Massachussetts

*Fraleigh PC (1971) Ecological succession in an aquatic microcosm and a thermal spring. (PhD Dissertation) University of Georgia, Athens, Georgia

*Fraleigh PC, Dibert PC (1980) Inorganic carbon limitation during ecological succession in aquatic microcosms. In: Giesy Jr JP (ed) Microcosms in ecological research. (DOE Symposium Series 52) US Department of Energy, Washington DC, pp 369–401

*Fraleigh PC, Wiegert RG (1975) A model explaining successional change in standing crop of thermal blue-green algae. Ecology 56(3):656–664

Francis BM, Metcalf RL (1984) Evaluation of mirex, photomirex, and chlordecone in the terrestrial aquatic laboratory model ecosystem. Environ Health Perspect 54:341–346

Francis BM, Magnus R, Metcalf RL (1981) Laboratory model ecosystem evaluations of twenty-six new pesticides. (IES Research Report No 9:1–98) University of Illinois, Umband-Champaign

Francis BM, Magnus R, Lampman L, Metcalf RL (1985) Model ecosystem studies of the environmental fate of five herbicides used in conservation tillage. Arch Environ Contam Toxicol 14:693–704

Friedman LB (1971) The effects of temperature of succession in aquatic microecosystems. (MS Thesis) Cleveland State University, Cleveland, Ohio

Frithsen JB (1984) Metal incorporation by benthic fauna: relationships to sediment inventory. Estuar Coastal Shelf Sci 19:523–539

*Frithsen JB, Doering PH (1986) Active enhancement of particle removal from the water column by tentaculate benthic polychaetes. Ophelia 25(3):169–182

*Frithsen JB, Elmgren R, Rudnick DT (1985) Responses of benthic meiofauna to long-term low-level additions of No. 2 fuel oil. Mar Ecol Progr Ser 23:1–14

*Frey RJ (1989) Preliminary report on the relationship of CO_2 level in closed ecological systems and the initial soil mass/plant biomass ratio. (unpublished report) Environmental Research Laboratory, University of Arizona, Tucson Arizona

*Fuhr F (1987) Non-extractable pesticide residues in soil. In: Greenhalgh R, Roberts TR (eds) Pesticide scicence biotechnology. Blackwell Scientific, Boston, pp 381–389

Fukuhara H, Sakamoto M (1987) Enhancement of inorganic nitrogen and phosphate release from lake sediment by tubificid worms and chironomid larvae, Oikos 48(3):312–320

*Gachter R (1979) Melimex: an experimental heavy-metal pollution study. Schweiz Z Hydrol 41:169–176

*Galat DL, McConnell WJ (1981) Effects of increasing total dissolved solids on the dynamics of Pyramid Lake microcosm communities. Colorado Cooperative Fishery Research Unit, Fort Collins, Colorado

*Galat DL, Robinson R (1983) Predicted effects of increasing salinity on the crustacean zooplankton community of Pyramid Lake, Nevada. Hydrobiologia 105:115–131

*Galat DL, Coleman M, Robinson R (1988) Experimental effects of elevated salinity on three benthic invertebrates in Pyramid Lake, Nevada. Hydrobiologia 158:133–144

*Gallagher JL, Daiber FC (1973) Diel rhythms in edaphic community metabolism in a Delaware salt marsh. Ecology 54(5):1160–1163

*Gambel JC, Davies JM, Steele JH (1977) Loch Ewe Bay experiment, 1974. Bull Mar Soc 27:146–175

*Ganning B, Wulff F (1969) The effects of bird droppings on chemical and biological dynamics in brackish water rockpools. Oikos 20:274–286

*Ganning B, Wulff F (1970) Measurement of community metabolism in some Baltic brackish water rockpools by means of diel oxygen curves. Oikos 21: 292–298

Gard TC (1981) Persistence for ecosystem microcosm models. Ecol Model 12(4):221–230

Gardner WS, Nalepa TF, Malczyk JM (1987) Nitrogen mineralization and denitrification in Lake Michigan, USA sediments. Limnol Oceanogr 32(6):1226–1238

Gauthier MJ, Munro PM, Mohajer S (1987) Influence of salts and sodium chloride on the recovery of *Escherichia coli* from seawater. Curr Microbiol 15:(1):5–10

Gauthier MJ, Munro PM, Breittmayer VA (1989) Influence of prior growth conditions on low nutrient response of *Escherichia coli* in seawater. Can J Microbiol 35(3):379–383
*Gearing JN (1989) The role of aquatic microcosms in ecotoxicologic research as illustrated by large marine systems. In: Levin SA et al. (eds) Ecotoxicology: problems and approaches. (Springer advanced texts in life sciences) Springer, Berlin Heidelberg New York, pp 411–472
*Gearing JN, Gearing PJ (1983) Suspended load and solubility affect sedimentation of petroleum hydrocarbons in controlled estuarine ecosystems. Can J Fish Aquat Sci 40(Suppl 2):54–62
*Gearing JN, Gearing PJ, Rudnick DT, Requejo AG, Hutchins MJ (1984) Isotopic variability of organic carbon in a phytoplankton-based temperate estuary. Geochim Cosmochim Acta 48(5):1089–1098
*Gearing PJ, Gearing JN (1982) Behavior of No 2 fuel oil in the water column of controlled ecosystems. Mar Environ Res 6:115–132
Gearing PJ, Gearing JN (1982) Transport of No 2 fuel oil between water column surface microlayer and atmosphere in controlled ecosystems. Mar Environ Res 6:133–143
*Gearing PJ, Gearing JN, Pruell RJ, Wade TL, Quinn JG (1980) Partitioning of No. 2 fuel oil in controlled estuarine ecosystem sediments and suspended particulate matter. Environ Sci Technol 14(9):1129–1136
Genthner BRS, Genthner FJ (1987) Influences of selective agents on the enumeration of pseudomonads for use in microcosm studies (abstract). Abstr Annu Meet Am Soc Microbiol 87:304
Giddings JM (1980) Types of aquatic microcosms and their research applications. In: Giesy JP Jr (ed) Microcosms in ecological research. (DOE Symposium Series 52) US Department of Energy, Washington DC, pp 248–266
Giddings JM (1982) Effects of the water-soluble fraction of a coal-derived oil on pond microcosms. Arch Environ Toxicol 11:735–747
Giddings JM (1986) A microcosm procedure for determining safe levels of chemical exposure in shallow-water. In: Community toxicity testing. (ASTM STP 920) American Society for Testing Materials, Philadelphia, pp 121–134
*Giddings J, Eddlemon GK (1987) Photosynthesis/respiration ratios in aquatic microcosms under arsenic stress. Water Air Soil Pollut 9:207–212
Giddings JM, Franco PJ, Bartell SM, Cushman RM, Herbes SE, Hook LA, Newhold JD, Southworth GR, Stewart AJ (1985) Effects of contaminants on aquatic ecosystems: Experiments with microcosms and outdoor ponds. Publication Number 2381. (ORNL/TM-9536) US Atomic Energy commission Oak Ridge, Tenneessee 104 pp
*Giesy JP Jr (1978) Cadmium inhibition of leaf decomposition in an aquatic microcosm. Chemosphere 6:467–475
*Giesy JP Jr (ed) (1980) Microcosms in ecological research. (DOE Symposium Series 52) US Department of Energy, Washington DC
*Giesy JP Jr, Geiger RA (1980) Relative mobilization of zinc, cerium, and americium from sediment in an aquatic microcosm. In: Giesy JP Jr (ed) Microcosms in ecological research. (DOE Symposium Series 52) US Department of Energy, Washington DC, pp 304–318
*Giesy JP Jr, Odum EP (1980) Microcosmology: introductory comments. In: Giesy JP Jr (ed) Microcosms in ecological research. (DOE Symposium Series 52) US Department of Energy, Washington DC, pp 1–13

Giesy JP Jr, Kania HJ, Bowling JW, Mashburn RL, Clarkin S (1979) Fate and biological effects of cadmium introduced into channel microcosms. (EPA-600/3-79-039) US Environmental Protection Agency, Washington DC

*Giesy JP Jr, Bowling JW, Kania HJ, Knight RL, Mashburn S (1981) Fates of cadmium introduced into channel microcosms. Environ Int 5:159–175

Gile JD (1982) Biological effects and interactions of pesticides in a soil–plant–water microcosm. (EPA Environmental Research Brief EPA-600/D-82-272) US Environmental Protection Agency, Washington DC

Gile JD (1983) 2,4-D: Its distribution and effects in a ryegrass ecosystem. J Environ Qual 12(3):406–412

*Gile JD (1983) Relative airborne losses of commercial 2,4-D formulations from a simulated wheat field. Arch Environ Contam Toxicol 12:465–469

Gile JD, Collins JC, Gillett JW (1979) The soil core microcosm — a potential screening tool. (EPA-600/3-79-089) Environmental Research Laboratory, US Environmental Protection Agency, Corvallis, Oregon

*Gile JD, Collins JC, Gillett JW (1980) Fate of selected herbicides in a terrestrial laboratory microcosm. Environ Sci Technol 14(9):1124–1128

*Gile JD, Collins JC, Gillett JW (1982) Fate and impact of wood preservatives in a terrestrial microcosm. J Agric Food Chem 30(2):295–301

Gillett JW (1989) The role of terrestrial microcosms and mesocosms in ecotoxicologic research. In: Levin SA, Harwell MA, Kelly JR, Kimball KD (eds) Ecotoxicology: problems and approaches. (Springer advanced texts in life sciences) Springer, Berlin Heidelberg New York, pp 367–410

Gillett JW, Gile JD (1976) Pesticide fate in terrestrial laboratory ecosystems. Int J Environ Stud 10:15–22

*Gillett JW, Witt JM (eds) (1978) Proceedings of the NSF Workshop: Terrestrial microcosms. National Science Foundations, Washington DC

*Gillett JW, Witt JM (1980) Chemical evaluation: projected application of microcosm technology. In: Giesy JP Jr (ed) Microcosms in ecological research. (DOE Symposium Series 52) Department of Energy, Washington DC, pp 1008–1049

Gillett JW, Hill J, Jasrvinen A, Schoor W (1974) A conceptual model for the movement of pesticides through the environment. (EPA 660/3-74-024) US Environmental Protection Agency Corvallis, Oregon

Gillett JW, Witt JM, Wyatt CJ (1978) Terrestrial microcosms. In: Gillett JW, Witt JM (eds) The proceedings of the NSF workshop on terrestrial microcosms. National Science Foundation, Washington DC

*Gillett JW, Gile JD, Russell LK (1983) Predator–prey (vole–cricket) interactions: the effects of wood preservatives. Environ Toxicol Chem 2:81–93

*Gilliland MW (1973) Man's impact on the phosphorous cycle in florida. (PhD Dissertation) Environmental Engineering Sciences, University of Florida, Gainesville, Florida

*Gilliland MW (1975) Systems models for phosphorous management in Florida. In: Howell FG, Gentry JB, Smith MH (eds) Mineral cycling in southeastern ecosystems. (DOE Symposium Series CONF-781101) US Department of Energy, Washington DC, pp 179–208

Gitelson II (1975) Experimental ecological managed systems (in Russian). Probl Space Biol Vol 28 Moscow "Nauka" 1–312

Gitelson II (1979) Man–higher plants closed system (in Russian). Novosibirak, "Nauka"

Gitelson II (1992) Biological life-support systems for Mars mission. Adv Space Res 12(5):167–192
*Gitelson II, Terskov IA, Kovrov BG (1973) Life support: system with internal control based on photosynthesis of higher and unicellular plants. LV Kirensky Institute of Physics, SD Academy of Sciences, USSR
Gitelson II, Terskov IA, Kovrov BG, Lisovky GM, Okladnikov Yu N, Sid'ko F Ya, Trubachev IN, Shilenko MP, Alekseev SS, Pan'kova IM, Tirranen LS (1989) Long-term experiments on man's stay in biological life-support system. Adv Space Res 9(8):65–71
*Gladyshev MI, Gribovskaya IV, Shchur LA (1990) Seasonal dynamics of the kinetics of natural phenol decontamination in the Sydinskii Bay of the Krasnoyarsk Reservoir, Russian SFSR, USSR. Dokl Akad Nauk USSR 313(6):1512–1514
Glaser D (1988) Simultaneous consumption of bacteria and dissolved organic matter by *Tetrahymena pyriformis*. Microb Ecol 15(2):189–202
*Glenn E, Clement C, Brannon P, Leigh L (1990) Sustainable food production for a complete diet. Hort Sci 25(12):1507–1512
Godsy EMD (1988) Grbic–Galic anaerobic degradation pathways for benzothiophene in aquifer-derived methanogenic microcosms (abstract). Abstr Annu Meet Am Soc Microbiol 88:301
*Goldman CR (1962) A method of studying nutrient limiting factors in situ in water columns isolated by polyethylene film. Limnol Oceanogr 7(1):99–101
*Golueke CG (1960) The ecology of a biotic community consisting of algae and bacteria. Ecology 41(1):65–73
*Golueke CG, Oswald WJ (1964) Role of plants in closed systems. Ann Rev Plant Physiol 15:387–407
*Golueke CG, Oswald WJ, McGauhey PH (1959) The biological control of enclosed environments. Sewage Ind Wastes 31(10):1125–1142
*Goodyear CP, Boyd CE, Beyers RJ (1972) Relationships between primary productivity and mosquitofish (*Gambusia affinis*) production in large microcosms. Limnol Oceanogr 17(3):445–450
*Gordon RW, Beyers RJ, Odum EP, Eagon RG (1969) Studies of a simple laboratory microecosystem: bacterial activities in a heterotrophic succession. Ecology 50(1):87–100
*Gorman OT (1988) An experimental study of habitat use in an assemblage of Ozark minnows. Ecology 69(4):1239–1250
*Grassle JF, Grassle JP, Brown-Leger LS, Petrecca RF, Copley NJ (1986) Subtidal macrobenthos of Narragansett Bay: field and mesocosm studies of the effects of eutrophication and organic input on benthic populations. In: Gray JS, Christiansen ME (eds) Marine biology of polar regions and effects of stress on marine organisms. John Wiley, New York, pp 421–434
*Gray JS (1987) Oil pollution studies of the Solbergstrand mesocosms. Phil Trans R Soc Lond B 316:641–654
*Grice GD, Reeve MR (eds) (1982) Marine mesocosms, biological and chemical research in experimental ecosystems. Springer, Berlin Heidelberg New York
Griffith PC, Cubit JD, Adey WH, Norris JN (1987) Computer-automated flow respirometry metabolism measurements on a Caribbean reef flat and in a microcosm. Limnol Oceanogr 32(2):442–451
Grill EV, Richards FA (1964) Nutrient regeneration from phytoplankton decomposing in seawater. J Mar Res 22(1):51–69

Grime JP, Mackey JML, Hillier SH, Read DJ (1987) Floristic diversity in a model system using experimental microcosms. Nature (London) 328(6129):420–422

Grime JP, Mackay JML, Hillier SH, Read DJ (1988) Floristic diversity in a model system using experimental microcosms. Proc R Soc Edinboro Sect B (Biol Sci) 94(1):173

Grossman GN (1988) Replicability of closed experimental microecosystems composed of microorganisms. Mikrobiol Zh (Kiev) 50(3):15–20

*Grossman GN (1990) Regulatory role of periodic illumination in the function of the microorganism community in a closed microecosystem. Mikrobiol Zh (Kiev) 52(3):3–10

Grossman GN (1990) Functioning of microbial communities in closed microecosystems. Izv Akad Nauk SSSR Ser Biol 0(2):261–268

Grossman G, Riis K (1990) Effect of mineral fertilizers on the microecosystem of soil microorganisms. Eesti Tead Akad Toim Biol 39(1):16–20

Guhl W (1987) Contribution to the biological evaluation of environmental chemicals using laboratory ecological models. I. Conception of the surface water model and substance evaluation. Z Angew Zool 74(4):385–410

Gunkel G (1983) Investigations of the ecotoxicological effect of a herbicide in an aquatic model ecosystem. I. sublethal and lethal effects. Arch Hydrobiol (Suppl) 65:235–267

*Guterstam B, Todd J (1990) Ecological engineering for wastewater treatment and its application in New England and Sweden. Ambio 19:173–175

*Hackney JM, Hensley P (1987) Use of an enzyme assay to detect glycolate in aquatic systems. Aquat Bot 27(4):395–402

*Haeckel E (1866) Generelle Morphologie der Organismem, Reimer, Berlin

Haefner JW, Gillett JW (1976) Aspects of mathematical models and microcosm research. In: Ott W (ed) Environmental modeling and simulation. (EPA-600/9-76-016) US Environmental Protection Agency, Washington DC, pp 624–628

Hagvar S (1988) Decomposition studies in an easily-constructed microcosm: effects of microarthropods and varying soil pH. Pedobiologia 31(5–6):293–303

*Hairston NG Sr (1989) Ecological experiments: purpose, design, and execution. Cambridge University Press, New York

*Hale ME (1983) The biology of lichens, 3rd ed. Edward Arnold, Baltimore, Maryland

*Hall CAS (1972) Migration and metabolism in a temperate stream ecosystem. Ecology 53(4):585–604

Hamala JA, Kollig HP (1985) The effects of atrazine on periphyton communities in controlled laboratory ecosystems. Chemosphere 14(9):1391–1408

Hamilton WE, Dindal DL (1989) Influence of earthworms and leaf litter on edaphic variables in sewage-sludge-treated soil microcosms. Biol Fertil Soils 7(2):129–133

Hamilton WE, Dindal DL, Parkinson CM, Mitchell MJ (1988) Interaction of earthworm species *Lumbricus terrestris* and *Eisenia foetida* in sewage sludge-amended soil microcosms. J Appl Ecol 25(3):847–852

*Hanson J (1982) Workshop on closed system ecology: summary report. (JPL Publ 82-64) Jet Propulsion Laboratory, California Institute of Technology, San Diego

Haque AH, Gruttke W, Kratz (1988) Environmental fate and distribution of sodium carbon-14 pentachlorophenate in a section of urban wasteland ecosystem. Sci Total Environ 68:127–140

*Hardy JT, Apts CW (1984) The sea-surface microlayer: phytoneuston productivity and effects of atmospheric particulate matter. Mar Biol 82:293–300

Hardy JT, Sullivan MF, Crecelius EA, Apts CW (1984) Transfer of cadmium in a phytoplankton–oyster–mouse food chain. Arch Environ Contam Toxicol 13:419–425

*Hardy JT, Apts CW, Crecelius EA, Fellingham GW (1985) The sea-surface microlayer: fate and residence times of atmospheric metals. Limnol Oceanogr 30(1):93–101

Hardy JT, Crecelius EA, Cowan CE, Wentsel S Sr (1988) Toxicity and metal speciation relationships for *Daphnia magna* exposed to brass powder. Arch Environ Contam Toxicol 17(5):575–582

*Harrass MC, Taub FB (1985) Comparison of laboratory microcosms and field responses to copper. In: Boyle TP (ed) Validation and predictability of laboratory methods for assessing the fate and effects of contaminants in aquatic ecosystems. (ASTM STP 865) American Society for Testing Materials, Philadelphia, pp 57–74

*Harris RP, Reeve MR, Grice GD, Evans GT, Gibson VR, Beers JR, Sullivan BK (1982) Trophic interactions and production processes in natural zooplankton communities in enclosed water columns. In: Grice GD, Reeve MR (eds) Marine mesocosms, biological and chemical research in experimental ecosystems. Springer, Berlin Heidelberg New York, pp 353–388

*Harrison PG (1978) Growth of *Ulva fenestrata* (Chlorophyta) in microcosms rich in *Zostera marina* (Anthophyta) detritus. J Phycol 14:100–103

Hart CM, Mulholland RJ (1979) Structural identifiability of compartmental systems based upon measurements of accumulated tracer in closed pools. Math Biosci 47:239–253

Hart CM, Mulholland RJ (1980) Structural identifiability of compartmental systems with respect to data obtained from constant-infusion tracer experiments. Math Biosci 50:195–205

*Harte J, Oldfather J (1987) Effects of prolonged darkness on a natural phytoplankton assemblage from a eutrophic lake. Curr Pract Environ Sci Eng 3:15–26

*Harte J, Levy D, Rees J, Saegebarth E (1980) Making microcosms an effective assessment tool. In: Giesy JP Jr (ed) Microcosms in ecological research. (DOE Symposium Series 52) US Department of Energy, Washington DC, pp 105–137

*Harte J, Levy D, Rees J (1983) Pelagic diatom populations in lentic freshwater microcosms. Inst Rev Ges Hydrobiol 68(2):255–267

Harvey RS (1973) A flowing stream laboratory for studying the effects of water temperature on the ecology of stream organisms. ASB Bull 20(1):3–7

Harwell MA, Harwell CC (1989) Environmental decision making in the presence of uncertainty. In: Levin SA, Harwell MA, Kelly JR, Kimball KD (eds) Ecotoxicology: problems and approaches. (Springer advanced texts in life sciences) Springer, Berlin Heidelberg New York, pp 517–540

Hatcher AI, Larkum AWD (1982) The effects of short-term exposure to bass strait crude oil and corexit 8667 on benthic community metabolism in *Posidonia australis* Hook dominated microcosms. Aquat Bot 12:219–228

Have A (1987) Experimental island biogeography: immigration and extinction of ciliates in microcosms. Oikos 50(2):218–224

Have A (1990) Microslides as microcosms for the study of ciliate communities. Trans Am Microsc Soc 109(2):129–140

*Hawes P, Nelson M, Augustine M (1990) Life systems for a lunar base. In: Mendell WW (ed) Lunar bases and space activities in the 21st century. (Proc Second Annual Conference Mendell WW (ed) Lunar and Planetary Institute, Houston
*Hawkes HA (1963) The ecology of waste water treatment. Pergamon Press, New York
Hay ME, Paul VJ, Lewis SM, (1988) Can tropical seaweeds reduce herbivory by growing at night? Diel patterns of growth, nitrogen content, herbivory, and chemical versus morphological defenses. Oecologia (Berlin) 75(2):233–245
Heath RT (1980) Are microcosms useful for ecosystem analysis? In: Giesy JP Jr (ed) Microcosms in ecological research. (DOE Symposium Series 52) US Department of Energy, Washington DC, pp 333–347
*Hedtke SF (1984) Structure and function of copper-stressed aquatic microcosms. Aquat Toxicol 5:227–244
Heitkamp MA, Johnson BT (1984) Impact of an oil field effluent on microbial activities in a Wyoming river. Can J Microbiol 30:786–792
*Heikamp MA, Cerniglia CE (1986) Microbial degradation of t-butylphenyl diphenyl phosphate: a comparative microcosm study among five diverse ecosystems. Toxic Assess 1(1):103–122
*Heitkamp MA, Cerniglia CE (1987) Effects of chemical structure and exposure on the microbial degradation of polycyclic aromatic hydrocarbons in freshwater and estuarine ecosystems. Environ Toxicol Chem 6:535–546
*Heitkamp MA, Huckins JN, Petty JD, Johnson JL (1984) Fate and metabolism of isopropylphenyl diphenyl phosphate in freshwater sediments. Environ Sci Technol 18(6):434–439
*Heitkamp MA, Freeman JP, Cerniglia CE (1986) Biodegradation of tert-butylphenyl diphenyl phosphate. Appl Environ Microbiol 51(2):316–322
*Heitkamp MA, Freeman JP, Cerniglia CE (1987) Naphthalene biodegradation in environmental microcosms: estimates of degradation rates and characterization of metabolites. Appl Environ Microbiol 53(1):129–136
Henderson RS, Smith SV (1980) Semitropical marine microcosms: facility design and an elevated-nutrient-effects experiment. In: Giesy JP Jr (ed) Microcosms in ecological research. (DOE Symposium Series 52) US Department of Energy, Washington DC, pp 869–910
Henrichs SM, Doyle AP (1986) Decomposition of ^{14}C-labeled organic substances in marine sediments. Limnol Oceanogr 31(4):765–778
*Herbert D (1956) The continuous culture of bacteria: a theoretical experimental study. J Gen Microbiol 14:601–662
*Herbert D (1964) Multi-stage continuous culture. In: Malek I, Beran K, Hospodka J (eds) Continuous cultivation of microorganisms. Publishing House Czech Academy of Science, Prague and Academic Press, New York, pp 23–44
*Herbes L, Hook A, Newbold JD, Southworth GR, Stewart AJ (1985) Effects of contaminants on aquatic ecosystems: experiments with microcosms and outdoor ponds. (ORNL/TM-9536 Publ No 2381) Martin Marietta Energy Systems Inc and Oak Ridge National Laboratory, Oak Ridge, Tennessee
*Hershey AE (1987) Tubes and foraging behavior in larval Chironomidae: implications for predator avoidance. Oecologica (Berlin) 73:236–241
*Hershey AE, Dodson SI (1985) Selective predation by a sculpin and a stonefly on two chironomids in laboratory feeding trials. Hydrobiologia 124:269–273
Hershey AE, Dodson S (1987) Predator avoidance by *Cricotopus*: cyclomorphosis and the importance of being big and hairy. Ecology 68(4):913–920

Hershey AE, Hiltner AL (1988) Effect of a caddisfly on black fly density: interspecific interactions limit black flies in an arctic river. North Am Benthol Soc 7(3):188–196

Hickman GT, Novak JT (1989) Relationship between subsurface biodegradation rates and microbial density. Environ Sci Technol 23(5):525–532

Hill JI, Porcella DB (1973) Component description and analysis of environmental systems: oxygen utilization in aquatic microcosms in Modeling the Eutrophication Process. Ann Arbor Science Publ, Ann Arbor, Michigan, pp 187–203

Hill JI, Wiegert RG (1980) Microcosms in ecological modeling. In: Giesy JP Jr (ed) Microcosms in ecological research. (DOE Symposium Series 52) US Department of Energy, Washington DC, pp 138–163

Hirata H, Yamasaki S (1980) Steady-state zooplankton community in a feedback culture system. In: Giesy JP Jr (ed) Microcosms in ecological research. (DOE Symposium Series 52) US Department of Energy, Washington DC, pp 402–415

*Hirsch P (1988) The Perfil'ev convectional flow technique for modeling stratified natural aquatic communities. In: Wimpenny JWT (ed) Handbook of laboratory model systems for microbial ecosystems, vol 2. CRC Press, Boca Raton, Florida, pp 99–108

Hockin DC (1981) Maintenance of a diverse harpacticoid copepod community in microcosm culture. Mar Biol (Berlin) 65:209–214

*Hockin DC (1982) The effects of sediment particle diameter upon the meiobenthic copepod community of an intertidal beach: a field and a laboratory experiment. J Anim Ecol 51:555–572

*Hodges C, Frye R (1990) Soil bed reactor work of the Environmental Research Lab of the University of Arizona in support of the Biosphere 2 project. In: Nelson M, Soffen GA (eds) Biological life support systems. (NASA CP-3094 33-40) Synergetic Press, Tucson, Arizona and US National Aeronautics and Space Administration, Washington DC

Hoffman RW, Horne AJ (1980) On-site flume studies for assessment of effluent impacts on stream aufwuchs communities. In: Giesy JP Jr (ed) Microcosms in ecological research. (DOE Symposium Series 52) US Department of Energy, Washington DC, pp 610–624

*Holling CS (1992) Cross-scale morphology, geometry, and dynamics of ecosystems. Ecol Monogr 62:447–502

*Hood DW (1978) Upwelled impoundments as a means of enhancing primary productivity. Rapp P-V Reun Cons Int Explor Mer 173:22–30

Hood MA, Macdonell MT (1987) Distribution of ultramicrobacteria in a Gulf Coast estuary and induction of ultramicrobacteria. Microb Ecol 14(2):113–128

*Hook LA, Franco PJ, Giddings JM (1986) Zooplankton community responses to synthetic oil exposure. In: Cairns J Jr (ed) Community toxicity testing. (ASTM STP 920) American Society for Testing Materials, Philadelphia, pp 291–321

Hornor SG (1984) Microbial leaching of zinc concentrate in fresh-water microcosms: comparison between aerobic and oxygen-limited conditions. Geomicrobiol J 3:359–371

*Huckins JN, Petty JD (1984) Modular containers for microcosm and process model studies on the fate and effects of aquatic contaminants. Chemosphere 13(12):1329–1341

*Huckins JN, Petty JD, England DC (1986) Distribution and impact of trifluralin, atrazine, and fonofos residues in microcosms simulating a northern prairie wetland. Chemosphere 15:563–588

*Hughes RM, Davis GE (1986) Production of coexisting juvenile coho salmon and steelhead trout in heated model stream communities. In: Cairns J (ed) Community toxicity testing. (ASTM STP 920) American Society for Testing Materials, Philadelphia, pp 322–337

Huhta V, Setala H, Haimi J (1988) Leaching of nitrogen and carbon from birch leaf litter and raw humus with special emphasis on the influence of soil fauna. Soil Biol Biochem 20(6):875–878

Humter M, Stephenson T, Kirk PWW, Perry R, Lester JN (1986) Effect of salinity gradients and heterotrophic microbial activity on biodegradation of nitrilotriacetic acid in laboratory simulations of the estuarine environment. Appl Environ Microbiol 51(5):919–925

*Hunt CD (1983) Incorporation and deposition of Mn and other trace metals by flocculent organic matter in a controlled marine ecosystem. Limnol Oceanogr 28(2):302–308

*Hunt CD (1983) Variability in the benthic Mn flux in coastal marine ecosystems resulting from temperature and primary production. Limnol Oceanogr 28(5):913–923

Hunt CD, Kelly JR (1988) Manganese cycling in coastal regions: response to eutrophication. Estuar Coastal Shelf Sci 26(5):527–558

Hunt CD, Smith DL (1980) Conversion of dissolved manganese to particulate manganese during diatom bloom: effects on the manganese cycle in the MERL microcosms. In: Giesy JP Jr (ed) Microcosms in ecological research. (DOE Symposium Series 52) US Department of Energy, Washington DC, pp 850–868

*Hunt CD, Smith DL (1983) Remobilization of metals from polluted marine sediments. Can J Fish Aquat Sci 40(Suppl 2):132–142

Huq A, West P, Small EB, Colwell RR (1984) Influence of water temperature salinity and pH on survival and growth of toxigenic *Vibrio*. Appl Environ Microbiol 48:420–424

*Hurlbert SH, Zedler J, Fairbanks D (1972a) Ecosystem alteration by mosquitofish (*Gambusia affinis*) predation. Science 175:639–641

*Hurlbert SH, Mulla MS, Wilson HR (1972b) Effects of an organophosphorus insecticide on the phytoplankton zooplankton and insect populations of freshwater ponds. Ecol Monogr 42(3):269–299

*Hutchinson GE (1964) The lacustrine microcosm reconsidered. Am Sci 52(3): 334–341

*Hutchinson GE (1978) An introduction to population ecology. Yale University Press, New Haven, Conn

Hutchinson GE (1961) The paradox of the plankton. Amer Nat 95:135–146

*Hyman LH (1940) The invertebrates: Protozoa through Ctenophora. McGraw-Hill, New York

Idso SB, Idos KE (1984) Conserving heat in a marine microcosm with a surface layer of fresh or brackish water: the "semi-solar pond." Solar Energy 33(2):149–154

Ineson P, Leonard MA, Anderson JM (1982) Effect of collembolan grazing upon nitrogen and cation leaching from decomposing leaf litter. Soil Biol Biochem 14:601–605

*Ingham RE, Coleman DC (1983) Effects of an ectoparasitic nematode on bacterial growth in gnotobiotic soil. Oikos 41:227–232

*Ingham RE, Trofymow JA, Ingham ER, Coleman DC (1985) Interactions of bacteria, fungi, and their nematode grazers: effects on nutrient cycling and plant growth. Ecol Monogr 55(1):119–140

Isensee AR (1976) Variability of aquatic model ecosystem-derived data. Int J Environ Stud 10(1):35–41

*Isensee AR, Kaufman DD, Jones GE (1982) Fate of 3,4-dichloroaniline in a rice (*Oryza sativa*) paddy microecosystem. Weed Sci 30(6):608–613

Isensee AR (1987) Persistence and movement of atrazine in a salt marsh sediment microecosystem. Bull Environ Contam Toxicol 39:516–523

*Ishimoto T (1956) The art of growing miniature trees plants and landscapes. New York Crown Publishers, New York

Jackson DR, Washburne CD, Ausmus (1977) Loss of Ca and NO_3–N from terrestrial microcosms as an indication of soil pollution. Water Air Soil Pollut 8:279–284

*Jagow RB (ed) (1967) Analysis of New Concepts. (Study of life support systems for space missions exceeding one year in duration, phase 1A, vol I) Lockheed Missiles and Space Company, Sunnyvale, California and NASA Environmental Control Research Branch, Ames Research Center, Moffett Field, California

Jagow RB (ed) (1967) Mission studies. (Study of life support systems for space missions exceeding one year in duration, phase 1A, vol II) Lockheed Missiles and Space Company, Sunnyvale, California and NASA Environmental Control Research Branch, Ames Research Center, Moffett Field, California

*Jagow RB, Thomas RB (1966) Study of life-support systems for space missions exceeding one year in duration. In: The closed life-support system. US National Aeronautics and Space Administration, Washington DC, pp 75–144

Jain RK, Sayler GS, Wilson JT, Houston L, Pacia D (1987) Maintenance and stability of introduced genotypes in groundwater aquifer material. Appl Environ Microbiol 53(5):996–1002

James DE (1972) Maintaining the marine aquarium. Carolina Tips 25(2):5–6

Jannasch HW (1960) Denitrification as influenced by photosynthetic oxygen production. J Gen Microbiol 23:55–63

*Jenkins DW (1968) Introduction. In: Bioregenerative systems. (NASA-SP-165) National Aeronautics and Space Administration, Washington DC, pp 1–6

Jensen K, Albrechtsen HJ, Nielsen J, Kruse B (1988) The use of ecocores to evaluate biodegradation in marine sediments. Water Air Soil Pollut 39(1–2): 89–100

*Johnson BT (1986) Potential impact of selected agricultural chemical contaminants on a northern prairie wetland: a microcosm evaluation. Environ Toxicol Chem 5:473–485

*Johnson BT, Heitkamp MA, Jones JR (1984) Environmental and chemical factors influencing the biodegradation of phthalic acid esters in freshwater sediments. Environ Pollut 8:101–118

Jullivan BK, Ritacco PJ (1985) Ammonia toxicity to larval copepods in eutrophic marine ecosystems: a comparison of results from bioassays and enclosed experimental ecosystems. Aquat Toxicol 7:205–217

Kanazawa JA, Isensee R, Kearney PC (1975) Distribution of carbaryl and 3,5-xylylmethyl carbamate in an aquatic model ecosystem. J Agric Food Chem 23:760–763

Kania HJ, Knight RL, Beyers RJ (1976) Fate and biological effects of mercury introduced into artificial streams. (EPA-600/3-76-060:129) US Environmental Protection Agency, Cincinnati, Ohio

*Kania HJ (1981) The fate of mercury input into artificial stream systems. (PhD Dissertation) University of Georgia, Athens, Georgia

Kapoor IP, Metcalf RL, Nystrom RF, Sangha K (1970) Comparative metabolism of methoxychlor methiochlor and DDT in mouse insects and in a model ecosystem. Agric Food Chem 18(6):1145–1152

*Karickhoff S, Morris KR (1985) Impact of tubificid oligochaets on pollutant transport in bottom sediments. Environ Sci Technol 19:51–56

*Kaushik NK, Solomon KR, Stephenson GL, Day KE (1986) Use of limnocorrals in evaluating the effects of pesticides on zooplankton communities. In: Cairns J Jr (ed) Community Toxicity Testing. (ASTM STP 920) American Society for Testing and Materials, Philadelphia, pp 269–290

*Kawabata Z, Kurihara Y (1978a) Effects of the consumer on the biomass and spatial heterogeneity in the aquatic microcosm. Sci Rep Tohoku Univ Ser IV 37:219–234

*Kawabata Z, Kurihara Y (1978b) Computer simulation study on the nature of the steady state of the aquatic microcosm. Sci Rep Tohoku Univ Ser IV 37:205–218

*Kawabata Z, Kurihara Y (1978c) Computer simulation study on the relationships between the total system and subsystems in the early stages of succession of the aquatic microcosm. Sci Rep Tohoku Univ, PR IV 37(3):179–204

Kazano H (1983) Studies on insecticidal characteristics of carbamate compounds and their metabolism in insects, soils, and model ecosystems. Bull Natl Inst Agric Sci (37):31–89

*Kearns EA (1983) Energetics of materially closed microbial ecosystems. (PhD Dissertation) University of Hawaii, Honolulu, Hawaii

*Kearns EA, Folsom CE (1981) Measurements of biological activity in materially closed microbial ecosystems. Biosystems 14:205–209

*Kearns EA, Folsom CE (1982) Closed microbial ecosystems and gas exchange units in CELSS. (SAE Technical Paper Series 820857)

Kearney PC, Oliver JE, Helling CS, Isensee AR, Kontson A (1977) Distribution, movement, persistence, and metabolism of N-nitrosoatrazine in soils and a model aquatic ecosystem. J Agric Food Chem 25(5):1177–1181

Keilty TJ, White DS, Landrum PF (1988) Sublethal responses to endrinin sediment by *Stylodrilus heringianus* Lumbriculidae as measured by a cesium-137 marker layer technique. Aquat Toxicol 13(3):251–270

*Kelly JR (1984) Microcosms for studies of sediment–water interactions. In: Parsone G, Jaspers E, Claus C (eds) Ecotoxicological testing for the marine environment, vol 2. State University of Gentand Institute of Marine Science Research Bredene, Belgium, pp 315–330

*Kelly JM, Strickland RC (1984) CO_2 efflux from deciduous forest litter and soil in response to simulated acid rain treatment. Water Air Soil Pollut 23:431–440

*Kelly JM, Strickland RC (1986) Throughfall and plant nutrient concentration response to simulated acid rain treatment. Water Air Soil Pollut 29:219–231

*Kelly JM, Strickland RC (1987) Soil nutrient leaching in response to simulated acid rain treatment. Water Air Soil Pollut 34(2):167–182

Kelly JR (1989) Ecotoxicology beyond sensitivity: a case study involving unreasonableness of environmental change. In: Levin SA, Harwell MA, Kelly JR,

Kimball KD (eds) Ecotoxicology: problems and approaches. (Springer advanced texts in life sciences) Springer, Berlin Heidelberg New York, pp 473–496

*Kelly JR, Duke TW, Harwell MA, Harwell CC (1987) An ecosystem perspective on potential impacts of drilling fluid discharges on seagrasses. Environ Manage 11(4):537–562

*Kelly JR, Berounsky VM, Nixon SW, Oviatt A (1985) Benthic–pelagic coupling and nutrient cycling across an experimental eutrophication gradient. Mar Ecol 26:207–219

Kelly JR, Nixon SW (1984) Experimental studies on the effect of organic deposition on the metabolism of a costal marine bottom community. Mar Ecol Progr Ser 17:157–169

*Kelly JR, Levin SA (1986) A comparison of aquatic and terrestrial nutrient cycling and production processes in natural ecosystems with reference to ecological concepts of relevance to some waste disposal. In: Kullenberg G (ed) The role of the oceans as a waste disposal option. D Reidel, pp 165–203

Kelly JR, Rudnick DT, Morton RD, Buttel LA, Levine SN, Carr KA (1990) Tributylin and invertebrates of a seagrass ecosystem: exposure and response of different species. Mar Environ Res 29(4):245–276

Kelly JR, Levine SN, Buttel LA, Carr KA, Rudnick DT, Morton RD (1990) The effect of tributylin within a *Thalassia* seagrass ecosystem. Estuaries 13(3):301–310

*Kelly RA (1970) The effects of fluctuating temperature on the metabolism of freshwater microcosms. (MS Thesis) Department of Zoology, University of North Carolina, Chapel Hill, North Carolina

*Kemp GP, Conner WH, Day JW Jr (1985) Effects of flooding on decomposition and nutrient cycling in a Louisiana swamp forest. Wetlands 5:35–51

Kemp GP, Day JW Jr (1981) Floodwater nutrient processing in a Louisiana swamp forest receiving agriculture runoff. (Louisiana Water Resources Research Institute Completion Report A-043-LA) US Department of the Interior, Washington DC

Kemp WM, Boynton WR, Cunningham JJ, Stevenson JC, Jones TW, Means JC (1985) Effects of artrazine and linuron on photosynthesis and growth of the macrophytes *Potamogeton perfoliatus* and *Myriophyllum spicatum* in an estuarine environment. Mar Environ Res 16(4):255–280

*Kemp WM, Lewis MR, Cunningham JJ, Stevenson JC, Boynton WR (1980) Microcosms, macrophytes, and hierarchies: environmental research in the Chesapeake Bay. In: Giesy JP Jr (ed) Microcosms in ecological research. (DOE Symposium Series 52) US Department of Energy, Washington DC, pp 911–936

*Kevern WR, Ball RC (1965) Primary productivity and energy relationships in artificial streams. Limnol Oceanogr 10:74–87

*Kerfoot WC, Demott WR (1980) Foundations for evaluating community interactions: the use of enclosures to investigate coexistence of *Daphnia* and *Bosmina*. In: Kerfoot WC (ed) Evolution and ecology of zooplankton communities. University Press of New England, Hanover, New Hampshire, pp 725–741

Kerger BD, Nichols PD, Sand W, Bock E, White DC (1987) Association of acid-producing thiobacilli with degradation of concrete. Analysis by signature fatty acids from the polar lipids and lipopolysaccharides. J Ind Microbiol 2(2):63–70

*Kersting K (1988) Normalized ecosystem strain in micro-ecosystems using different sets of state variables. Verh Int Ver Limnol 23:164–166

*Ketchum BH (1954) Relation between circulation and planktonic populations in estuaries. Ecology 35:191–200
*Kevern NR, Ball RC (1965) Primary productivity and energy relationships in artificial streams. Limnol Oceanogr 10(1):74–87
*Kevern NR, Wilhm JL, Van Dyne GM (1966) Use of artificial substrata to estimate the productivity of periphyton. Limnol Oceanogr 11(4):499–502
Kimmel BL, Lind OT (1970) Factors influencing orthophosphate concentration decline in the water of laboratory mud–water systems. Texas J Sci 21(4):439–445
King DL (1980) Some cautions in applying results from aquatic microcosms. In: Giesy JP Jr (ed) Microcosms in ecological research. (DOE Symposium Series 52) US Department of Energy, Washington DC, pp 164–191
*Kinne O (1978) Cultivation of marine organisms. In: Kinne O (ed) Water quality management and Technology. (Marine ecology, vol III) John Wiley, New York, pp 19–30
*Kirensky LV, Terskov IA, Gitelson II, Kovrov BG, Sidko FY, Lisovsky GM, Okladnikov JN (1967) Continuous cultivation of algae as a link of a closed ecosystem. In: Brown AH, Favorite FG (eds) Committee on space research life sciences and space research, VI. North Holland Publishing Co Amsterdam, pp 165–203
*Kirensky LV, Gitelson II, Terskov IA, Kovrov BG, Lisovsky GM, Okladnikov JN (1971) Theoretical and experimental decisions of the problem of creation of artificial ecosystems for human life support in space. Life Sciences and Space Research IX:75–80
Kirk PWW, Lester JN, Perry R (1983) Amenability of nitrilotriacetic acid to biodegradation in a marine simulation. Mar Pollut Bull 14(3):88–93
*Kitchens WM (1977) Salt marsh unit microecosystems for assessment of pollutant addition perturbation: the dynamics of an estuary as a natural system (EPA-600/3-77-016 Publ 67–71) US Environmental Protection Agency, Washington DC
*Kitchens WM, Edwards RT, Johnson WV (1979) Development of a "living" salt marsh ecosystem model: a microecosystem approach. In: Dame RF (ed) Marsh–estuarine systems simulation. (Belle W Baruch Library in Marine Science, No 8) University of South Carolina Press, Columbia, South Carolina, pp 107–116
Kitchens WM, Copeland BJ (1980) Succession in laboratory microecosystems subjected to thermal and nutrient-addition stresses. In: Giesy JP Jr (ed) Microcosms in ecological research. (DOE Symposium Series 52) US Department of Energy, Washington DC, pp 536–561
Kitching RL (1971) An ecological study of water-filled tree-holes and their position in the woodland ecosystem. J Anim Ecol 40:281–302
Kleiber M (1961) The fire of life: an introduction to animal energetics. John Wiley, New York
Klir GJ (1969) An approach to general systems theory. Van Norstrand Reinhold, New York
Klopatek CC, Debano LF, Klopatek JM (1988) Effects of simulated fire on vesicular–arbuscular mycorrhizae in pinyon–juniper woodland soil. Plant Soil 109(2):245–250
*Klopatek JM, Klopatek CC, Debano LF (1990) Potential variation of nitrogen transformations in pinyon-juniper ecosystems resulting from burning. Biol Fertil Soils 10(1):35–44

Kloskowski R, Fuhr F (1983) Die Buidung von gebundene Simazin-Ruckstanden in einer Parabraunerde. Chemosphere 12(11–12):1545–1556

Kloskowski R, Fuhr F (1983) Formation of bound residues of (ring-^{14}C)simazine in a parabraunerde (alfisol) and their availability to maize. (IAEA-TECDOC-306) International Atomic Energy Agency, Vienna, pp 133–147

Kloskowski R, Fuhr F (1983) Versuche zur Charakterisierung und Bioverfugbarkeit von nicht-extrahierbaren Simazin-Ruckstanden im Boden. Chemosphere 12: 1557–1574

Kloskowski R, Fuhr F, Mittelstaedt W (1986) Plant availability of bound anilazine residues in a degraded loess soil. J Environ Sci Health (Part B) (6):487–505

Kloskowski R, Fuhr F (1987) Aged and bound herbicide residues in soil and their bioavailability. I. Uptake of aged and non-extractable bound 3 carbon-14 metamitron residues by sugar beets. J Environ Sci Health (Part B) 22:509–536

Kloskowski R, Fuhr F, Mittelstaedt W (1987) The uptake of non-extractable soil-bound pesticide residues by roots — standardized experiments with four posticides. Pest Sci Biotechnol 3:405–410

Klump JV, Krezoski JR, Smith ME, Kaster JL (1987) Dual tracer studies of the assimilation of an organic contaminant from sediments by deposit-feeding oligochaetes. Can J Fish Aquat Sci 44(9):1574–1583

Knaus RM (1981) Double activation analysis: the simultaneous use of radiotracers and activable traces in a stream microcosm. Environ Sci Technol 15(7):809–819

*Knight RL (1980) Energy basis of control in aquatic ecosystems. (PhD Dissertation) Department of Environmental Engineering Sciences, University of Florida, Gainesville, Florida

*Knight RL (1981) A control hypothesis for ecosystems-energetics and quantification with the toxic metal cadmium. In: Mitsch WJ, Bosserman RW, Klopatek JM (eds) Energy and Ecological Modeling. Elsevier Scientific, Amsterdam, pp 601–615

*Knight RL (1983) Energy basis for ecosystem control at Silver Springs, Florida. In: Fontaine TD, Bartell SM (eds) Dynamics of lotic ecosystems. Ann Arbor Science, Ann Arbor, Mich., pp 161–180

Knudsen GR, Walter MV, Posteous LA, Prince VJ, Armstrong JL, Seidler RJ (1988) Predictive model of conjugative plasmid transfer in the rhizosphere and phyllosphere. Appl Environ Microbiol 54(2):343–347

*Kok B (1956) Photosynthesis in flashing light. Bioclem Biophys Acta 21:245–258

*Kolkwitz R, Marsson M (1908) Oekologe der Pflanzlichen Saprobian. Ber Dtsch Bot Ges 26a:505–519

*Kostitzin VA (1939) Mathematical biology (translated by Savory TH). George Harrap, London

*Kostitzin VA (1935) Evolution de l'atmosphere. Act Sci Indust Hermann, Paris

Kovrov BG, Terskov IA, Gitelson II, Lisovsky GM, Pankova IM, Trubachev IN, Shilenko MP, Alekseyev SS, Tirranen LS, Sidko F Ya, Okladnikov Yu N (1985) Artificial closed ecosystem: Man–higher plants with full regeneration of atmosphere, water, and ration. Vegetable part. (Preprint YAF-85-308 36) Congress of YAF Stockholm

*Krantzbert G, Stokes PM (1985) Benthic macroinvertebrates modify copper and zinc partitioning in freshwater–sediment microcosms. Can J Fish Aquat Sci 42:1465–1473

*Kremer JN (1979) An analysis of the stability characteristics of an estuarine ecosystem model. In: Dame RF (ed) Marsh–estuarine systems simulation. (Bell W Baruch Library in Marine Science, No 8) University of South Carolina Press, Columbia, South Carolina, pp 189–206
*Kremer JN, Nixon SW (1978) A coastal marine ecosystem: simulation and analysis. (Ecological studies #24) Springer, Berlin Heidelberg New York
Kristensen E, Blackburn TH (1987) The fate of organic carbon and nitrogen in experimental marine sediment systems: influence of bioturbation and anoxia. J Mar Res 45(1):231–257
*Kuhl H, Mann H (1951) Über die periodischen Anderungen im Chemismus. Deutsche Zoologische Gesellschaft Verhandlugen:378–381
*Kuhl H, Mann H (1955) Modellversuche zum Stickstoffhaushalt in Aquarien. Arch Hydrobiol Suppl 22(3–4):409–414
*Kuhl DH, Mann DH (1956a) Über den Stickstoffkreislauf im Aquarium. Aquarien Terrarien (2.4, 5):2–10
*Kuhl H, Mann H (1956b) Unperiodische Anderungen im Stoffhaushalt von Seewasseraquarien Hydrobiologia 8(1–2):66–77
Kuhl H, Mann H (1961) Vergleich des Stickstoffabbaus in Seen und Süsswasseraquarien. Vie Milieu 11(4):532–545
*Kuhl H, Mann H (1962) Modellversuche zum Stoffhaushalt in Aquarien bei verschiedenem Salzgehalt. Kieler Meeresforsch 18(3):89–92
Kuiper J (1981) Fate and effects of mercury in marine plankton communities in experimental enclosures. Ecotoxicol Environ Saf 5:106–134
*Kurihara Y (1954a) Studies on the changes of the community–structure of the benthonic protist community in bamboo containers. Saito Ho-on Kai Museum Res Bull 23:22–28
*Kurihara Y (1954b) Synecological study on the relationship between the benthonic microorganism community and dipterous insect larvae in the bamboo container. Sci Rep Tohoku Univ 20(2):130–138
*Kurihara Y (1957a) Synecological analysis of the biotic community in microcosm. I. The effect of mosquito larvae upon the succession of Protozoa in bamboo containers. Sci Rep Tohoku Univ 23(3–4):131–137
*Kurihara Y (1957b) Synecological analysis of the biotic community in microcosm. II. Studies on the relations of diptera larvae to Protozoa in bamboo containers. Sci Rep Tohoku Univ 23(3–4):139–142
*Kurihara Y (1958) Synecological analysis of the biotic community in microcosm. III. Studies on the relations of diptera larvae to Protozoa in the containers made of bamboos of different ages. Sci Rep Tohoku Univ 24(1):15–21
*Kurihara Y (1959a) Synecological analysis of the biotic community in microcosm. VIII. Studies on the limiting factor in determining the distribution of mosquito larvae in the polluted water of bamboo container with special reference to relation of larvae to bacteria. Jpn J Zool 12(3):391–400
*Kurihara Y (1959b) Synecological analysis of the biotic community in microcosm. IV. Studies of the relations of diptera larvae to pH in bamboo containers. Sci Rep Tohoku Univ 25(2):165–171
*Kurihara Y (1960a) Biological analysis of the structure of microcosm with special reference to the relations among biotic and abiotic factors. Sci Rep Tohoku Univ 26(4):269–296

*Kurihara Y (1960b) Synecological analysis of the biotic community in microcosm. VII. Studies on the nature of bacterial nutrient related to the appearance of the protozoan sequences. Sci Reports Tohoku Univ 26(2):219–226

*Kurihara Y (1960c) Synecological analysis of the biotic community in microcosm. VI. The biological significance of the starting concentration of bacterial nutrient to protozoan sequences in bamboo container. Sci Rep Tohoku Univ 26(2):213–218

*Kurihara Y (1978a) Studies of "succession" in a microcosm. Sci Rep Tohoku Univ 37(3):151–160

*Kurihara Y (1978b) Studies of interaction in a microcosm. Sci Rep Tohoku Univ 37(3):161–177

*Kurihara Y (1983) Study of domestic sewage waste treatment by the polychaetes *Neanthes japonica* and *Perineres nuntia* var. *vallata* on an artificial tidal flat. Int Rev Ges Hydrobiol 68(5):649–670

*Kurihara Y, Kato M (1952) Associative ecology of the mosquito larvae and their biotic environments in the stone basin. Sci Rep Tohoku Univ 19(4):339–347

*Kurihara Y, Shikano S, Toda M (1990) Trade-off between interspecific competitive ability and growth rate in bacteria. Ecology 71(2):645–650

*Lacaze JC (1972) Utilisation d'un dispositif experimental simple pour l'etude de la pollution des eaux in situ effects compares de trois agents emulsionnants anti-petrole. Extrait Tethys 3(4):705–716

Lacaze JC (1974) Ecotoxicology of crude oil and the use of experimental marine ecosystems. Mar Pollut Bull 5:153–158

Lack TJ, Lund JWG (1972) Observation and experiments on the phytoplankton of Blelham Tarn, English Lake District. Freshwater Biol 4:399–415

Lalli CM (ed) (1990) Enclosed experimental marine ecosystems: a review and recommendations. (Coastal and Estuarine Studies, 37) Springer, Berlin Heidelberg New York

*Lamberti GH, Gregory SV, Ashkenas LR, Steinman AD, McIntire CD (1989) Productive capacity of periphyton as a determinant of plant–herbivore interactions in streams. Ecology 70(6):1840–1856

*Lane PA, Blouin AC (1984) Plankton of an acid stressed lake, part 3. Community network systems. Ver Int Ver Limnol 22:395–400

Lane PA, Collins TM (1985) Food web models of a marine plankton community network: an experimental mesocosm approach. J Exp Mar Biol Ecol 94:41–70

Lang C, Lang-Dobler B (1979) Melimex and experimental heavy metal pollution study: oligochaetes and chironomid larvae in heavy metal-loaded and control limnocorrals. Schweiz Z Hydrol 41:217–276

Larkin RP, Kelly JM (1988) A short-term microcosm evaluation of carbon dioxide evolution from litter and soil and influenced by sulfur dioxide and sulfate additions. Water Air Soil Pollut 37(3–4):273–380

*Larsen DP, deNoyelles F, Stay F, Shiroyama T (1986) Comparisons of single-species microcosm and experimental pond responses to atrazine exposure. Environ Toxicol Chem 5:179–190

*Lassiter RR (1983) Prediction of ecological effects of toxic chemicals: overall strategy and theoretical basis for the ecosystem model. (EPA-600/S3-83-084 1–4) US Environmental Protection Agency, Washington DC

Lasserre P, Tournie T (1984) Use of microcalorimetry for the characterization of marine metabolic activity at the water–sediment interface. J Exp Mar Biol Ecol 74(2):124–140

Lasserre P (1984) The measurement of enthalpy in marine organisms. In: Fasham MTR (ed) Flows of energy and materials in marine ecosystems theory and practice. Plenum New York, pp 247–269

Lasserre P (1990) Marine microcosms: small-scale controlled ecosystems. In: Lalli CM (ed) Enclosed experimental marine ecosystems: a review and recommendations. Coastal and Estuarine Studies, 37, Springer, Berlin Heidelberg New York, pp 20–60

*Lauff GH, Cummins KW (1964) A model stream for studies in lotic ecology. Ecology 45(1):188–191

Lawrence GB, Mitchell MJ (1985) Use of ^{35}S to determine the influence of *Hexagenia* on sulfur cycling in lake sediments. Hydrobiologia 128:91–95

Lawrence GB, Mitchell MJ, Landers DH (1982) Effects of the burrowing mayfly, *Hexagenia*, on nitrogen and sulfur fractions in lake sediment microcosms. Hydrobiologia 87:273–283

Lawson TJ, Grice GD (1977) Zooplankton sampling variability: controlled ecosystem pollution experiment. Bull Mar Sci 27:80–84

Lay JP, Muller A, Peichl L, Lang R, Korte F (1987) Effects of δ-bhc lindane on zooplankton under outdoor conditions. Chemosphere 16(7):1527–1538

Lee K, Anderson JW (1977) Fate and effect of naphthalenes: controlled ecosystem pollution experiment. Bull Mar Sci 27:80–84

Lee RF (1983) Microbial and photochemical degradation of polycyclic aromatic hydrocarbons in estuarine waters and sediments. Can J Fish Aquat Sci 40:86–94

Lee RF, Takahahhi M (1977) The fate and effect of petroleum in controlled ecosystem enclosures. Rapp P-V Reun Cons Int Explor Mer 171:150–156

Leffler JW (1980) Microcosmology: theoretical applications of biological models. In: Giesy JP Jr (ed) Microcosms in ecological research. (DOE Symposium Series 52) US Department of Energy, Washington DC, pp 14–29

Leffler JW (1984) The use of self-selecter generic aquatic microcosms for pollution effects assessment. In: White HH (ed) Concepts in marine pollution measurements. (Maryland Sea Grant Publication) University of Maryland, College Park, Maryland, pp 139–157

Legrendre L, Demers S (1984) Towards dynamic biological oceanography and limnology. Can J Fish Aquatic Sci 41:2–19

Legrendre L, Demers S, Terriault JC, Boudreau CA (1985) Tidal variations in the photosynthesis of estuarine phytoplankton isolated in a tank. Mar Biol 88:301–309

*Lehman JT (1980) Nutrient recycling as an interface between algae and grazers in freshwater communities. In: Kerfoot WC (ed) Evolution and ecology of zooplankton communities. University Press of New England, Hanover, New Hampshire, pp 251–262

*Leigh L, Fitzsimmons K, Norem M, Stumpf D (1987) An introduction to the intensive agriculture biome of Biosphere II. In: Faughnan B, Maryniak G (eds) Space manufacturing. (nonterrestrial resources, biosciences, and space engineering vol 6) American Institute of Aeronautical Engineering, Washington DC, pp 76–81

Levin SA, Harwell MA, Kelly JR, Kimball KD (eds) (1989) Ecotoxicology, problems and approaches. Springer, Berlin Heidelberg New York

*Levy D, Lockett G, Oldfather J, Rees J, Saegebarth E, Schneider R, Harte J (1985) Realism and replicability of lentic freshwater microcosms. In: Boyle TP

(ed) Validation and predictability of laboratory methods for assessing the fate and effects of contaminants in aquatic ecosystems. (ASTM STP 865) American Society for Testing Materials, Philadelphia, pp 43–56

Lewis DL, Kellogg RB, Holm HW (1985) Comparison of microbial transformation rate coefficients of xenobiotic chemicals between field-collected and laboratory microcosm microbiota. In: Boyle TP (ed) Validation and predictability of laboratory methods for assessing the fate and effects of contaminants in aquatic ecosystems. (ASTM STP 865) American Society for Testing Materials, Philadelphia, pp 3–13

*Liang TT, Lichtenstein EP (1980) Effects of cover crops on the movement and fate of soil-applied carbon-14-labeled fonofos in a soil–plant–water microcosm. J Econ Entomol 73(2):204–210

Lichtenstein EP (1980) Fate and behavior of pesticides in a compartmentalized microcosm. In: Giesy JP Jr (ed) Microcosms in ecological research. (DOE Symposium Series 52) US Department of Energy, Washington DC, pp 954–970

*Lichtenstein EP, Liang TT (1987) Effects of simulated rain on the transport of fonofos and carbofuran from agricultural soils in a three-part environmental microcosm. J Agric Food Chem 35(2): 173–178

*Lichtenstein EP, Liang TT, Koeppe MK (1982) Effects of fertilizers captafol and atrazine on the fate and translocation of carbon-14-labeled fonofos and carbon-14-labeled parathion in a soil–plant microcosm. J Agric Food Chem 30(5):871–878

Lisovsky GM (ed) (1979) Closed system: man-higher plants. Nauka, Novosibirsk

*Lisovsky GM, Terskov IA, Gitelson II, Sidko FY, Kovrov BG, Shilenko MP, Polonskii VI, Ushakova SA (1981) Experimental estimation of the functional possibilities of higher plants as medium regenerators in life support systems. Preprint 32nd Congress International Astronautical Federation Rome, Italy: 1–4

*Liston J, Wiebe WJ, Lighthart B (1963) The activities of marine benthic bacteria. In: Contribution No 147, Research in Fisheries. Fisheries Research Institute, University of Washington, Seattle, Washington, pp 47–48

Lopez-Fugueroa FF, Niell X (1987) Feeding behavior of *Hydrobia ulvae* Pennant in microcosms. J Exp Mar Biol Ecol 114(2–3):153–168

*Lotka AJ (1922a) Contribution to the energetics of evolution. Proc Natl Acad Sci 8:191–200

*Lotka AJ (1922b) Natural selection as a physical principal. Proc Natl Acad Sci 8:151–154

*Lotka AJ (1925) Physical biology. Williams and Wilkins, Baltimore, Maryland

*Lovelock JE (1979) Gaia: a new look at life on earth. Oxford University Press, Oxford, England

Luckinbill LS (1979) Regulation stability and diversity in a model experimental microcosm. Ecology 6(6):1098–1102

*Luecke CW, O'Brien WJ (1981) Phototoxicity and fish predation: selective factors in color morphs in *Heterocope*. Limnol Oceanogr 26:454–460

*Lugo AE (1969) Energy, water, and carbon budgets of a granite outcrop community. (PhD Dissertation) University of North Carolina, Chapel Hill, North Carolina

*Lund JWG (1972) Preliminary observations on the use of large experimental tubes in lakes. Verh Int Ver Limnol 18:71–77

*Lund JWG (1975) The uses of large experimental tubes in lakes. In: The effects of storage on water quality. Medmenham Water Research Center, pp 291–312

*Lund JWG (1978) Experiments with lake phytoplankton in large enclosures. Ann Rep Freshwater Biol Assoc 46:32–39

Lund JWG, Reynolds CS (1982) The development and operation of large limnetic enclosures in Blelhan Tarn, English Lake District and their contribution to phytoplankton ecology. Prog Phycol Res 1:1–65

Lynch TR, Johnson HE, Adams WJ (1985) Impact of atrazine and hexachlorobiphenyl on the structure and function of model stream ecosystems. Environ Toxicol Chem 4(3):399–413

Lynch TR, Johnson HE, Adams WJ (1982) The fate of atrazine and a hexachlorobiphenyl isomer in naturally-derived model stream ecosystems. Environ Toxicol Chem 1(2):179–192

*MacArthur RJ, Wilson EO (1967) Theory of island biogeography. Princeton University Press, Princeton, New Jersey

*Maguire BJ (1971) Phytotelmata: biota and community structure determination in plant-held waters. Ann Rev Ecol Syst 2:439–463

*Maguire BJ (1980) Some patterns in post-closure ecosystem dynamics (failure). In: Giesy JP Jr (ed) Microcosms in ecological research. (DOE Symposium Series 52) US Department of Energy, Washington DC, pp 319–332

Maguire BJ, Slobodkin LB, Moroqitz JJ, Moore I, Botkin DB (1980) A new paradigm of the examination of closed ecosystems In: Giesy JP Jr (ed) Microcosms in ecological research. (DOE Symposium Series 52) US Department of Energy, Washington DC, pp 30–68

Major J, Pyott WT (1966) Buried viable seeds in two California bunchgrass sides and their bearing on the definition of a flora. Vegetation 13:253–282

Maki AW (1980) Evaluation of toxicant effects on structure and function of model stream communities: correlation with natural stream effects. In: Giesy JP Jr (ed) Microcosms in ecological research. (DOE Symposium Series 52) US Department of Energy, pp 583–609

*Malanchuk JL, Joyce K (1983) Effects of 2,4-D on nitrogen fixation and carbon dioxide evolution in a soil microcosm. Water Air Soil Pollut 20(2):181–190

Malanchuk JL, Kollig HP (1985) Effects of atrazine on aquatic ecosystems: a physical and mathematical modeling assessment. In: Boyle TP (ed) Validation and predictability of laboratory methods for assessing the fate and effects of contaminants in aquatic ecosystems. (ASTM STP 865) American Society for Testing Materials, Philadelphia, pp 212–224

*Malanchuk JL, Mueller CA, Pomerantz SM (1980) Microcosm evaluation of the agricultural potential of fly-ash-amended soils. In: Giesy JP Jr (ed) Microcosms in ecological research. (DOE Symposium Series 52) US Department of Energy, Washington DC, pp 1034–1049

Malone TC, Haines KC, Roals OA (1975) Nitrate uptake and growth of *Chaetocerus* sp. in large, outdoor, continuous cultures. Limnol Oceanogr 20:9–19

Mann H (1957) Untersuchungen über die Durchluftung in Subwasseraquarien. Aquarien Terrarien Z (Datz) 2:44–45

Manning BW, Federle TW, Cerniglia CE (1987) Use of a semicontinuous culture system as a model for determining the role of human intestinal microflora in the metabolism of xenobiotics. J Microbiol Meth 6:81–94

Manuel CY, Minshall GW (1980) Limitations on the use of microcosms for predicting algal response to nutrient enrichment in lotic systems. In: Giesy JP Jr (ed) Microcosms in ecological research. (DOE Symposium Series 52) US Department of Energy, Washington DC, pp 645–667
*Marais GVR (1966) New factors in the design, operation, and performance of waste-stabilization ponds. Bull WHO 34:737–763
*Margalef R (1958) Temporal succession and spatial heterogeneity in phytoplankton. In: Perspectives in marine biology. University of California Press, Berkeley, California, pp 323–349
*Margalef R (1963) Desarrollo experimental de picnoclinas en pequenos volumenes de agua. Invest Pesq 23:3–10
*Margalef R (1967) Laboratory analogues of estuarine plankton systems. In: Lauff G (ed) Estuaries. (Publ 83) American Association for the Advancement of Science, Washington DC, pp 515–521
*Margalef R (1969) Diversity and stability: a practical proposal and a model of interdependence. In: Diversity and stability in ecological systems. (Brookhaven Symposia in Biology, No 33, BNL-50175(C-56) Brookhaven, New York
*Margulis L, Sagan D (1986) Microcosmos. Summit Books, New York
Marshall JS, Mellinger DL (1980) Dynamics of cadmiun-stressed plankton communities. Can J Fish Aquat Sci 37:403–414
Martin GD, Mulholland RJ, Thornton KW (1976) Ecosystem approach to the simulation and control of an oil refinery waste treatment facility. J Dyn Syst Meas Control 98:20–29
Martinetz D, Wenzel KD, Wiessflog L (1989) Possibilities and limits of laboratory microecosystems. Z Chem 29(7):236–239
Matisoff G, Fisher JB, Matis S (1985) Effects of benthic macroinvertebrates on the exchange of solutes between sediments and freshwater. Hydrobiologia 122:19–33
*May RM (1974) Biological populations with non-overlaping generations, stable points, stable cycles, and chaos. Science 186:645–647
McAllister CD, Parsons TR, Stephens SK, Strickland JDH (1961) Measurements of primary production in coastal sea water using a large-volume plastic sphere. Limnol Oceanogr 6(3):237–258
McCall PL, Matisoff G, Tevesz MJS (1986) The effects of a unionid bivalve on the physical, chemical, and microbial properties of cohesive sediments from Lake Erie. Am J Sci 286:127–159
*McConnell WJ (1962) Productivity relations in carboy microcosms. Limnol Oceanogr 7(3):335–343
*McConnell WJ (1965) Relationship of herbivore growth to rate of gross photosynthesis in microcosms. Limnol Oceanogr 10(4):539–543
*McCormick PV, Cairns J Jr (1991) Limited verses unlimited membership in microbial communities: evaluation and experimental tests of some paradigms. Hydrobiologia 218(1):77–92
*McCormick JF, Platt RB (1962) Effects of ionizing radiation on a natural plant community. Radiat Bot 2:161–188
McCormick PV, Pratt JR, Cairns JJ (1986) Effects of 3-trifluoromethyl-4-nitrophenol on the structure and function of protozoan communities established on artificial substrates. In: Cairns J Jr (ed) Community Toxicity Testing. (ASTM STP 920) American Society for Testing Materials, Philadelphia, pp 224–240

McCormick PV, Stewart PM, Carins JJ (1987) Effect of distance from a source pool on protozoan colonization of isolated aquatic systems. J Freshwater Ecol 4(1):1–16

McCormick PV, Smith Cairns J Jr (1991) The relative importance of population versus community processes in microbial primary succession. Hydrobiologia 213(2):83–98

*McDonald RC (1981) Vascular plants for decontaminating radioactive water and soils. (NSTL MS38529 TM-X-72740), National Aeronautics and Space Administration, Washington DC

*McDonald RC, Wolverton BC (1980) Comparative study of wastewater lagoons with and without water hyacinths. Econ Bot 34(2):101–110

McElroy AE, Tripp BW, Farrington JW, Teal JM (1987) Biogeochemistry of benz-a-anthracene at the sediment–water interface. Chemosphere 16:2429–2440

McElroy AE, Farrington JW, Teal JM (1988) Metabolism of polycyclic aromatic hydrocarbons in benthic microcosms (abstract). Abstr Pap Chem Cong North Am 3(1)

*McIntire CD (1966) Some factors affecting respiration of periphyton communities in lotic environments. Ecology 47(6):918–929

*McIntire CD, Colby JA (1978) A hierarchical model for lotic ecosystems. Ecol Monogr 48:167–190

*McIntire CD, Phinney HK (1965) Laboratory studies of periphyton production and community metabolism in lotic environments. Ecol Monogr 35(3):237–257

*McIntire CD, Colby JA, Hall JD (1975) The dynamics of small lotic ecosystems: a modeling approach. Verh Int Ver Limnol 19:1599–1601

*McIntire CD (1973) Periphyton dynamics in laboratory streams: a simulation model and its implications. Ecol Monogr 43(3):399–420

*McIntire CD (1981) A conceptual framework for process studies in lotic ecosystems. In: Fontain T, Bartell SM (eds) Dynamics of lotic ecosystems. Ann Arbor Science Publ, Ann Arbor, Michigan, pp 43–68

McIntyre CD, Garrison RL, Phinney HK, Warren CE (1964) Primary production in laboratory streams. Limnol Oceanogr 9:92–102

*McKellar H, Hobro R (1976) Phytoplankton-zooplankton relationships in 100-liter plastic bags. Contr Asko Lab Univ Stockholm, Sweden 13:1–83

*McQueen DJ, Lean DRS (1983) Hypolimnetic aeration and dissolved gas concentration. Water Res 17:1781–1790

McVea C, Boyd CE (1975) Effects of water hyacinth cover on water chemistry phytoplankton and fish in ponds. J Environ Qual 4(3):375–378

Medine AJ, Porcella DB, Adams VD (1980) Heavy-metal and nutrient effects on sediment oxygen demand in three-phase aquatic microcosms. In: Giesy JP Jr (ed) Microcosms in ecological research. (DOE Symposium Series 52) US Department of Energy, Washington DC, pp 279–303

Meharg AA, Killham K (1988) A comparison of carbon flow from prelabelled and pulse-labelled plants. Plant Soil 112(2):225–232

Mendez DW (1977) Concept and design: controlled ecosystem pollution experiment. Bull Mar Sci 27:1–7

Mendez DW, Steele JH (1978) The application of plastic enclosures to the study of pelagic marine biota. Rapp P-V Reun Cons int Explor Mer 173:7–12

Menzel DW (1977) Summary of experimental results: controlled ecosystem pollution experiment. Bull Mar Sci 27:142–145

Menzel DW (1980) Applying results derived from experimental microcosms to the study of natural pelagic marine ecosystems. In: Giesy JP Jr (ed) Microcosms in ecological research. (DOE Symposium Series 52) US Department of Energy, Washington DC, pp 742–752

Merkens JC (1957) Controlled aqueous environments for bioassay. Lab Pract 6(8):456–459

*Metcalf RL (1977) Model ecosystem approach to insecticide degradation: a critique. Ann Rev Entomol 22:241–261

*Metcalf RL, Sanuha GK, Kapoor IP (1971) Model ecosystem for the evaluation of pesticide biodegradability and ecological magnification. Environ Sci Technol 5(8):709–713

*Metcalf RL, Cole LK, Wood SG, Mandel DJ, Milbrath ML (1979) Design and evaluation of a terrestrial model ecosystem for evaluation of substitute pesticide chemicals. (Corvallis Environmental Research Laboratory EPA-600/3-79-004) US Environment Protection Agency, Washington DC

Meyers MB, Powell EN, Fossing H (1988) Movement of oxybiotic and thiobiotic meiofauna in response to changes in pore-water oxygen and sulfide gradients around macro-infaunal tubes. Mar Biol (Berlin) 98(3):395–414

*Michaelis L, Menten ML (1913) Die Kinetik der Invertinwirkung. Biochem Z 49:333–369

*Mitchell D (1971) Eutrophication of lake water microcosms: phosphate versus nonphosphate detergents. Science 174:827–829

*Mitchell D, Asce AM, Buzzell JCJ, Asce M (1971) Estimating eutrophic potential of pollutants. J Sanit Eng Div 97:453–465

Mitchell MJ, Hornor SG, Abrams BI (1980) Use of microcosms in studying decomposition processes in sewage sludge. In: Giesy JP Jr (ed) Microcosms in ecological research. (DOE Symposium Series 52) US Department of Energy, Washington DC, pp 458–472

*Mitsch WJ, Jorgensen SE (eds) (1989) Ecological Engineering: an introduction to ecotechnology. John Wiley New York, pp 348

Moore JC, St John TV, Coleman DC (1985) Ingestion of vesicular arbuscular mycorrhizal hyphae and spores by soil microarthropods. Ecology 66(6):1979–1981

*Moore JC, Trofymow JA, Morley CR (1985) A technique to decontaminate soil microarthropods for introduction to gnotobiotic systems. Pedobiologia 28(3):185–190

*Moriaty F (1977) Prediction of ecological effects by pesticides. In: Perring FH, Mellanby K (eds) Ecological Effects of Pesticides Academic Press New York, pp 165–173

*Moriarty DJ, Pollard PC, Alongi DM, Wilkinson CR, Gray J (1985) Bacterial productivity and trophic relationships with consumers on a coral reef. (MECOR I) Proc Int Coral Reef Cong 3:457–462

Morrisey DJ (1987) Effect of population density and presence of a potential competitor on the growth rate of the mud snail *Hydrobia ulvae* Pennant. J Exp Mar Biol Ecol 108(3):275–295

Morrisey DJ (1988) Differences in effects of grazing by deposit-feeders *Hydrobia ulvae* Pennant, Gastropoda Prosobranchia, and *Corophium arenarium*, Cawrford Amphipoda, on sediment microalgal populations. II. Quantitative effects. J Exp Mar Biol Ecol 118(1):43–54

*Morton RD, Duke TW, Maculey JM, Clark JR, Price WA, Hendricks SJ,

Owsley-Montgomery SL, Plaia GR (1986) Impact of drilling fluids on seagrasses: an experimental community approach. In: Cairns J Jr (ed) Community toxicity testing. (ASTM STP 920) American Society for Testing Materials, Philadelphia, pp 199–112

Mulholland RJ (1974) Analysis of linear compartment models for ecosystems. J Theor Biol 44:105–116

Mulholland RJ, Gowdy CM (1977) Theory and application of the measurement of structure and determination of function for laboratory ecosystems. J Theor Biol 69:3211–344

Mulholland RJ, Weidner RJ (1980) Sampling strategies for the identification of compartment models. Int J Syst Sci 11(3):377–387

Mulholland RJ, Weidner RJ (1981) Stochastic properties of compartment models. Int J Syst Sci 12(8):927–936

Mulholland RJ, Crux JR, Hill J (1986) State-variable canonical forms for Prony's method. Int J Syst Sci 17(1):55–64

Muller P (1980) Effects of artificial acidification on the growth of periphyton. Can J Fish Aquat Sci 37:355–363

Mullin MM (1982) How can enclosing seawater liberate biological oceanographers? In: Grice GD, Reeve MR (eds) Marine mesocosms, biological and chemical research in experimental ecosystems. Spinger, Berlin Heidelberg New York, pp 399–410

Mullin MM, Evans PM (1974) The use of a deep tank in plankton ecology. 2. Efficiency of a planktonic food chain. Limnol Oceanogr 19:902–911

Munro PM, Laumond FM, Gauthier MJ (1987) A previous growth of enteric bacteria on a salted medium increases their survival in seawater. Lett Appl Microbiol 4(6):121–124

Munro PM, Gauthier MJ, Laumond FM (1987) Changes in *Escherichia coli* cells starved in seawater or grown in seawater–wastewater mixtures. Appl Environ Microbiol 53(7):1476–1481

*Murray RE, Hodson RE (1986) Influence of macrophyte decomposition of growth rate and community structure of Okefenokee Swamp baterioplankton. Appl Environ Microbiol 51(2):293–301

*Muschenheim DK, Grant J, Mills EL (1986) Flumes for benthic ecologists: theory, construction, and practice. Mar Ecol Progr Ser 28:185–196

*Myers J (1954) Basic remarks on the use of plants as biological gas exchangers in a closed system. J Aviat Med 25:407–411

Myers J (1958) Study of a photosynthetc gas exchanger as a method of providing for the respiratory requirements of the human in a sealed cabin. (Publ 58–117) Air University School of Aviation Medicine, USAF, Randolph Air Force Base, Texas

*Myers J (1960) The use of photosynthesis in a closed ecological system. In: Physics and Medicine of the Atmosphere & Space. John Wiley, New York, pp 388–396

Myers J (1963a) Introductory remarks — space biology: ecological aspects. Am Biol Teacher 25:409–411

*Myers J (1963b) Use of algae for support of the human in space. Life Sci Space Res 2:323–336

*Myers J, Clark LB (1944) Culture conditions and the development of the photosynthetic mechanism. II. An apparatus for the continuous culture of *Chlorella*. J Gen Physiol 28:103

Nadeau NJ, Roush TH (1973) A salt marsh microcosm: an experimental unit for marine pollution studies. Proceedings of the 1973 Joint Conference on Prevention and Control of Oil Spills, American Petroleum Institute, Washington DC, pp 671–683

Nagy E, Scott BF, Hart J (1984) The fate of oil and oil-dispersant mixtures in freshwater ponds. Sci Total Environ 35:115–133

*Nakamoto N (1989) Diurnal variations of dissolved oxygen and photosynthesis of the filamentous algae in a slow sand-filtration pond. 5th International Symposium on Microbial Ecology (ISME 6) 189.8.27-9.1 Kyoto Japan

Nash RG, Beall ML Jr, Harris WG (1977) Toxaphene and DDT losses from cotton in an agroecosystem chamber. J Agric Food Chem 25:336–342

National Research Council (1981) Testing for effects of chemicals on ecosystems (A report by the Committee to Review Methods for Ecotoxicology). National Academy Press, Washington DC

Neame PA, Goldman CR (1980) Oxygen uptake and production in sediment — water microcosms. In: Giesy JP Jr (ed) Microcosms in ecological research. (DOE Symposium Series 52) US Department of Energy, Washington DC, pp 267–278

Neill WE (1975) Resource partitioning by competing microcrustaceans in stable laboratory microecosystems. Verh Int Ver Limnol 19:2885–2890

*Neill WE (1981) Impact of *Chaoborus* predation upon the structure and dynamics of a crustacean zooplankton community. Oecologia (Berlin) 48:164–177

*Neill WE (1984) Regulation of rotifer densities by crustacean zooplankton in an oligotrophic montane lake in British Columbia. Oecologia (Berlin) 61:175–181

*Nelson M (1990) Biotehcnology of space biology. In: Asahima M, Malacinski GM (eds) Fundamentals of space biospheres. Japan Scientific Society Press, Tokyo and Springer, Berlin Heidelberg New York, pp 185–200

*Nelson M, Allen JP, Dempster W (1991a) Biosphere 2: prototype project for a permanent and evolving life system for a Mars base. COSPAR XXVIII Plenary Meeting. In: MacElroy RD, Averner MM, Tibbitts TW, Bugbee BB, Horneck G, Dunlop EH (eds) Advances in space research. 12(5) Life Sciences and Space Research XXIV (4) Natural and Artificial Ecosystems Pergamon Press, 211–218

*Nelson M, Leigh L, Alling A, MacCallum T, Allen J, Alvarez-Romo N (1991b) Biosphere 2 Test Module: ground based sunlight-driven prototype of a closed ecological system. COSPAR XXVIII Plenary Meeting. In: MacElroy RD, Averner MM, Tibbitts TW, Bugbee BB, Horneck G, Dunlop EH (eds) Advances in Space Research 12(5) Life Sciences and Space Research XXIV (4) Natural and Artificial Ecosystems Pergamon Press, New York, pp 151–158

*Nelson M, Burgess TL, Alling A, Alvarez-Romo N, Dempster WF, Walford RL, Allen JP (1993) Using a closed ecological system to study earth's biosphere. Bioscience 43(4):225–236

Nicholas WD, Abernathy AR (1987) A method to assess periodic responses of aquatic microecosystems to photosynthetic inhibition. Chemosphere 16(1):287–295

Nichols WD, Abernathy AR (1987) Inhibition of metabolic hydrogen ion flux resulting from cadmium stress in aquatic microecosystems. Water Sci Technol 19(11):85–94

*Nicolis G, Prigogine I (1977) Self organization in non-equilibrium systems. John Wiley, New York

*Niederlehner BR, Pratt JR, Buikema AL, Cairns J Jr (1985) Laboratory tests evaluating the effects of cadmium on freshwater protozoan communities. Environ Toxicol Chem 4:155–165

*Niederlehner BR, Pratt JR, Buikema AL, Carins J Jr (1986) Comparison of estimates of hazard derived at three levels of complexity. In: Cairns J Jr (ed) Community toxicity testing. (ASTM STP 920) American Society for Testing Materials, Philadelphia, pp 30–48

*Nixon SW (1969) A synthetic microcosm. Limnol Oceanogr 14(1):142–145

*Nixon SW (1970) Characteristics of some hypersaline ecosystems. (PhD Dissertation) Department of Botany, University of North Carolina, Chapel Hill, North Carolina

Nixon SW, Pilson MEQ (1983) Nitrogen in estuarine and coastal marine ecosystems. In: Carpenter E, Capone D (ed) Nitrogen in the marine environment Academic Press, New York, pp 565–648

*Nixon SW, Oviatt CA, Kremer JN, Perez K (1979) The use of numerical models and laboratory microcosms in estuarine ecosystem analysis — simulations of a winter phytoplankton bloom. In: Dame RF (ed) Marsh-estuarine systems simulation. (Bell W Baruch Library in Marine Science, No 8) University of South Carolina Press, Columbia, South Carolina, pp 165–188

*Nixon SW, Alonson D, Pilson MEQ, Buckley BA (1980) Turbulent mixing in aquatic microcosms Giesy JP Jr (ed) Microcosms in ecological research. (DOE Symposium Series 52) US Department of Energy, Washington DC, pp 818–849

Nixon SW, Pilson MEQ, Oviatt CA, Donaghy P, Sullivan B, Seit-Zinger S, Radnick D, Frithsev J (1984) Eutrophication of a coastal marine ecosystem — an experimental study using the MERL microcosms. In: Fasham M (ed) Flows of Energy and Materials in Marine Ecosystems. Plenum NY, pp 105–135

*Nixon SW, Oviatt CA, Frithsen J, Sullivan B (1986) Nutrients and the productivity of estuarine and coastal marine ecosystems. J Limnol Soc Southern Africa 12:43–71

Notini M, Nagell B, Hagstrom A, Grahn O (1977) An outdoor model simulating a Baltic Sea littoral ecosystem. Oikos 28:2–9

*Novara M, Cullingford HS (1988) Bio-isolation analysis of plants and humans in a piloted Mars sprint. SAE Engineering Soc Advancing Mobility Land Sea Air Space 18th Intersociety Conference on Environmental Systems San Francisco CA 1:1–11

*Novick A, Szilard L (1950) Description of the chemostat. Science 112:715–716

*Noyes G (1977) Energy analysis of space operations. In: Odum HT, Alexander J (eds) Energy analysis of models of the united states (Report to DOE). Department of Environmental Engineering Science, University of Florida, Gainesville Florida, pp 401–422

*Obenhuber D (1985) Carbon cycling as a measure of biological activity in closed ecological systems. (PhD Dissertation) University of Hawaii, Honolulu, Hawaii

*Obenhuber D, Folsome CE (1984) Procaryote/eucaryote ratios as an indicator of stability in closed ecosystems. Biosystems 16:291

*Obenhuber DC, Folsome C (1988) Carbon recycling in materially closed ecological life support systems. Biosystems 21:165–173

*O'Brien WJ, deNoyelles FJ (1974) Relationship between nutrient concentration, phytoplankton density, and zooplankton density in nutrient enriched experimental ponds. Hydrobiologia 44(1):105–188

*Odum EC, Odum HT (1984) System of ethanol production from sugarcane in Brazil. Ciencia Cultura 37(11):1849–1855
*Odum EP (1962) Relationship between structure and function in the ecosystem. Jap J Ecol 12:108–118
*Odum EP (1971) Fundamentals of ecology. Saunders, Philadelphia
*Odum HT (1950) The biogeochemistry of strontium. (PhD Dissertation) Yale University, New Haven, Connecticut
*Odum HT (1951) Stability of the world strontium cycle. Science 114:407–411
*Odum HT (1960) Ten classroom sessions in ecology. Am Biol Teacher 22:71–78
*Odum HT (1963) Limits of remote ecosystems containing man. Am Biol Teacher 26(6):429–443
*Odum HT (1964) Photoelectric ecosystem. Science 143:256–258
*Odum HT (1967) Biological circuits and the marine systems of Texas. In: Olson TA, Burgess FJ (eds) Pollution and marine ecology. Wiley Interscience, New York, pp 99–157
*Odum HT (1970) Summary: an emerging view of the ecological system at El Verde. In: Odum HT, Pigeon RF (eds) A tropical rain forest. TID-24270 (PRNC-138) US Atomic Energy Commission, Oak Ridge, Tennessee, pp I191–I289
*Odum HT (1971) Environment power and society. Wiley Interscience, New York
*Odum HT (1972) Chemical cycles with energy circuit models. (Nobel Symposium, vol 20) John Wiley, New York, pp 223–257
*Odum HT (1976) Energy quality and carrying capacity of the earth. Tropic Ecol 16(1):1–8
Odum HT, Arding JE (1991) EMERGY analysis of Shrimp Mariculture in Ecuador. Working Paper prepared for the Coastal Resources Center, University of Rhode Island, Narragansett, RI. 144 pp
*Odum HT (1982) Pulsing power and hierarchy. In: Mitsch WJ, Ragade RK, Bosserman RW, Dillon JA (eds) Energetics and systems. Ann Arbor Science, Ann Arbor, Michigan, pp 33–59
*Odum HT (1983a) Systems ecology: an introduction. John Wiley, New York
*Odum HT (1983b) Maximum power and efficiency: a rebuttal. Ecol Model 20:71–82
*Odum HT (1985) Self organization of ecosystems in marine ponds receiving treated sewage. (University of North Carolina Sea Grant Office Publ No UNC-SG-85-04) North Carolina State University, Raleigh, North Carolina
*Odum HT (1986) Emergy in ecosystems. In: Polunin N (ed) Ecosystem theory and application. John Wiley, New York, pp 337–369
*Odum HT (1987) Living with complexity. Crafoord Prize in the Biosciences Royal Swedish Acad Sci Stockholm Sweden 19–85
*Odum HT (1988) Self organization, transformity and information. Science 242:1132–1139
*Odum HT (1989) Experimental study of self-organization in estuarine ponds. In: Mitsch WJ, Jorgensen SE (eds) Ecological engineering: an introduction to ecotechnology. John Wiley, New York, pp 291–340
*Odum HT, Hoskin CM (1957) Metabolism of a laboratory stream microcosm. Publ Inst Mar Sci Texas 4:116–133
*Odum HT, Johnson JR (1955) Silver Springs and the balanced aquarium controversy. The science counselor (December): Duquesne University Press, Pittsburg, Pennsylvania, pp 1–4

*Odum HT, Lugo AE (1970) Metabolism of forest floor microcosms. In: Odum HT, Pigeon R (eds) A tropical rain forest. TID-24270 (PRNC-138) US Atomic Energy Commission, Oak Ridge, Tennessee, pp 35–54

Odum HT, Odum EC (eds) (1983) Energy analysis overview of nations. (Working paper WP-83-82) International Institute for Applied Systems Analysis, Laxenburg, Austria

*Odum HT, Odum EC (eds) (1987) Ecology and economy: emergy analysis and public policy in Texas. (LBJ School of Public Affairs and Texas Department of Agriculture Special Publication, No 78) Austin, Texas

*Odum HT, Odum EC (1991) Computer minimodels and simulation exercises available for Apple II, PC, or MacIntosh. (Center for Wetlands Publ) University of Florida, Gainesville, Florida

*Odum HT, Pinkerton RC (1955) Time's speed regulator: the optimum efficiency for maximum power output in physical and biological systems. Am Sci 43:321–343

*Odum HT, Siler WL, Beyers RJ, Armstrong N (1963b) Experiments with engineering of marine ecosystems. Publ Inst Mar Sci Texas 9:373–403

*Odum HT, Beyers RJ, Armstrong NE (1963a) Consequences of small storage capacity in nannoplankton pertinent to measurement of primary production in tropical waters. J Mar Res 21:191–198

*Odum HT, Nixon SW, Di Salvo LH (1971) Adaptations for photo-regenerative systems. In: Cairns J (ed) The structure and function of freshwater microbial communities. American Microbiology Society, Blacksburg, Virginia, pp 11–29

*Odum HT, Brown WT, Christianson RA (1986) Energy systems overview of the amazon basin. (Center for Wetlands Publication 86-1) University of Florida, Gainesville, Florida

*Odum HT, Odum EC, Brown MT, LaHart D, Bersok C, Sendzimir J (1988) Environmental systems and public policy. (Center for Wetlands Publ) University of Florida, Gainesville, Florida

Odum HT, Arding JE (1991) Energy analysis of shrimp mariculture in Ecuador. Working paper prepared for the Coastal Resources Center, University of Rhode Island, Narragansett, RI. 114 pp

*Odum HT, Odum EC, Brown MT, LaHart D, Bersok C, Sendzimir J (1988) Energy, environment and public policy: a guide to the analysis of systems. (UNEP Regional Seas Reports and Studies, No 95) US Environment Programs Nairobi Kenya

Office of Environmental Processes and Effects Research, US EPA (1980) Interlaboratory evaluation of microcosm research. (Proceedings of the Workshop ORNL/EIS-160) Oak Ridge National Laboratory, Oak Ridge, Tennessee

Oiestad V (1982) Application of enclosures to studies on the early life history of fishes. In: Grice D, Reeve MB (eds) Marine mesocosms, biological and chemical research in experimental ecosystems. Springer, Berlin Heidelberg New York, pp 49–62

Oil and Gas Working Group (1978) Studies of the effects of hydrocarbons on laboratory aquatic ecosystems. Institute for Environmental Studies, University of Toronto, Toronto

Oliver JD, Nilsson L, Kjelleberb S (1991) Formation of nonculturable *Vibrio vulnificus* cells and its relationship to the starvation state. Appl Environ Microbiol 57(9):2640–2644

O'Neill E, Hood M, Cripe C, Pritchard P (1988) Field calibration of aquatic microcosms (abstract). Abstr Annual Meet Am Soc Microbiol 88:297

*Oosting HJ, Humphreys ME (1940) Buried viable seeds in a successional series of old field and forest soils. Torrey Bot Club Bull 67:253–273

*Orvos DR, Lacy GH, Cairns J Jr (1990) Genetically engineered *Erwinia carotovora* survival, intraspecific competition, and effects upon selected bacterial genera. Appl Enviror Microbiol 56(6):1689–1694

Osunade MAA (1987) A viable method of land capability classification for small farmers. J Environ Manag 25(1):81–94

*Oswald WJ, Golueke CG (1964) Fundamental factors in waste utilization in isolated systems. Soc Ind Microbiol 5:196–206

*Oswald WJ, Gotaas HB, Golueke CG, Kellen WR (1957) Algae in waste treatment. Sewage Ind Wastes 29(4):437–457

*Oswald WJ, Asce M, Golueke CG, Horning DO (1965) Closed ecological systems. J Sanit Eng Div 91:23–46

*Overton WS (1977) A strategy of model construction. In: Hall CAS, Day JW Jr (eds) Ecosystem modeling in theory and practice. John Wiley, New York, pp 50–73

*Oviatt CA (1981) Effects of different mixing schedules on phytoplankton, zooplankton, and nutrients in marine microcosms. Mar Ecol Progr Ser 4:57–67

*Oviatt C, Perez KT, Nixon SW (1977) Multivariate analysis of experimental marine ecosystems. Helgol Wiss Meeresunters 30:30–46

*Oviatt CA, Nixon SW, Perez KT, Buckley B (1979) On the season and nature of perturbations in microcosm experiments. In: Dame RF (ed) Marsh–estuarine systems simulation. (Belle W Baruch Library in Marine Science, No 8) University of South Carolina Press, Columbia, South Carolina, pp 143–164

*Oviatt CA, Walker H, Pilson ME (1980) An exploratory analysis of microcosm and ecosystem behavior using multivariate techniques. Mar Ecol Progr Ser 2:179–191

*Oviatt C, Buckley B, Nixon S (1981a) Annual phytoplankton metabolism in Narragansett Bay calculated from survey field measurements and microcosm observations. Estuaries 4(3):167–175

*Oviatt CA, Hunt CD, Vargo GA, Kopchynski KW (1981b) Simulation of a storm event in marine microcosms. J Mar Res 39(4):605–626

*Oviatt C, Frithsen J, Gearing J, Gearing P (1982) Low, chronic additions of No. 2 fuel oil: chemical behavior, biological impact, and recovery in a simulated estuarine environment. Mar Ecol Progr Series 9:121–136

*Oviatt CA, Pilson ME, Nixon SW, Frithsa S, Rudnick J, Kelly D, Grassie JF, Grassle JR (1984) Recovery of a polluted estuarine system: a mesocosm experiment. Mar Biol Progr Ser 16:203–217

*Oviatt CA, Keller AA, Sampou PS, Beatty LL (1986a) Patterns of productivity during eutrophication: a mesocosm experiment. Mar Ecol Progr Ser 28:69–80

*Oviatt CA, Lane P, French F III, Donaghay P (1989) Phytoplankton species and abundance in response to eutrophication in coastal marine mesocosms. Plankton Research 11:1223–1224

*Oviatt CA, Rudnick D, Keller LA, Sampou PS, Almquist G (1986b) A comparison of system (O_2 and CO_2) and ^{14}C measurements of metabolism in estuarine mesocosms. Mar Ecol Progr Ser 28:57–67

Palka L, Couteaux MM (1986) Aptitude a la croissance de colpoda *Aspera* (protozoaire cillie) dans un sol humide prealablement sterilize a l'oxyde de propylene a lautoclave et aux rayons gamma. Rev Ecol Biol Sol 23(4):405–421

Pandya MK (1983) Impact of pesticides in a microcosm (on bacteria population dynamics). Pesticides 17(4):3-7

Parsons F, Lage G, Rice R (1983) Transformation of chlorinated organic solvents in ground water environments in southern Florida. Am Chem Soc 23(2):286-287

Parsons TR (1978) Controlled aquatic ecosystem experiments in ocean ecology research. Mar Pollut Bull 9:203-205

*Parsons TR (1978) Controlled ecosystem experiments. Rapp P-V Reun Cons Int Explor Mer 173:5-6

Parsons TR (1982) The future of controlled ecosystem enclosure experiments. In: Grice D, Reeve MB (eds) Marine mesocosms, biological and chemical research in experimental ecosystems. Springer, Berlin Heidelberg New York, pp 411-418

*Parsons F, Lage GB (1984) Chlorinated organics in simulated groundwater environments. Res Technol J AWWA (May):52-59

*Parsons F, Lage GB (1985) Changes in inorganic ions and evolution of gases during biotransformation of trichloroethene in aquifer microcosms (abstract). Abs Ann Meeting Am Soc Microbol 3-7

Parsons TR, Li WK, Waters R (1976) Some preliminary observations on the enhancement of phytoplankton growth by low levels of mineral hydrocarbons. Hydrobiol 51:85-89

Parsons TR, Thomas WH, Seibert D, Beers JR, Gillespie P, Bawden C (1977) The effect of nutrient enrichment on the plankton community in enclosed water columns. Int Rev Ges Hydrobiol 62:565-572

Parsons TR, Vonbrockel K, Koeller K, Takahashi M, Reeve MR, Holm-Hansen O (1977) The distribution of organic carbon in a marine planktonic food web following nutrient enrichment. J Exp Mar Biol Ecol 26:235-247

Parsons TR, Harrison PH, Waters R (1978) An experimental simulation of changes in diatom and flagellate blooms. J Exp Mar Biol Ecol 32:285-294

Parsons TR, Albright LJ, Whitnew F, Wong CS, Leb PJ (1981) The effect of glucose on the productivity of seawater: an experimental approach using controlled aquatic ecosystems. Mar Environ Res 4:229-242

*Parsons F, Lage GB, Rice R (1984) Biotransformation of chlorinated organic solvents in static microcosms. Environ Toxicol Chem 4:739-742

*Parsons F, Wood PR, DeMarco J (1984) Transformations of tetrachloroethene and trichloroethene in microcosms and groundwater. Res Technol J AWWA (February):56-59

*Paterek JR (1983) Ecology of methanogenesis in two hypersaline biocoenosis: Great Salt Lake and a San Francisco Bay saltern. (PhD Dissertation) University of Florida, Gainesville, Florida

Paterson DM, Wright SJL (1978) A continuous-flow model-ecosystem for studying effects of herbicides on aquatic plants. Weed Sci 35(2):704-710

*Patten BC, Witkamp M (1967) Systems analysis of cesium kinetics in terrestrial microcosms. Ecology 48(5):813-824

Paul AJ, Paul JM, Shoemaker PA (1979) Artificial upwelling and phytoplankton production in Alaska. Mar Sci Comm 5:79-89

Pedersen A (1987) Community metabolism on rocky shore assemblages in a mesocosm. A. Flucuations in production, respiration, chlorophyll a content and C:N ratios of grazed and non-grazed assemblages. Hydrobiologia 151/152: 267-275

*Perez KT, Morrison GM, Lackie NF, Oviatt CA, Nixon SW, Buckley BA, Heltshe JF (1977) The importance of physical and biotic scaling to the experi-

mental simulation of a coastal marine ecosystem. Helgol Wiss Meeresunters 30:144–162

*Perez KT, Morrison GE, Davey EW, Lackie NF, Soper AE, Blasco RJ, Winslow DL, Johnson RL, Murphy PG, Heltshe JF (1991) Influence of size on fate and ecological effects of kepone in physical models. Ecol Appl 1(3):237–248

Perez-Rosas N (1988) The in situ survival of *Vibrio cholerae* and *Escherichia coli* in tropical coral reefs. Appl Environ Microbiol 54(1):1–9

*Perring FH, Mellanby K (eds) (1976) Ecological effects of pesticides. (Linnean Society Symposium Series, #5) Academic Press, London

Perry JA, Troelstrup JNH (1988) Whole ecosystem manipulation: a productive avenue for test system research. Environ Toxicol Chem 7(11):941–951

*Peterson KM, Billings WD, Reynolds DN (1984) Influence of water table and atmospheric CO_2 concentration on the carbon balance of arctic tundra. Arct Alp Res 16(3):331–335

Petterson H, Gross F, Koczy FF (1939) Large scale plankton culture. Nature 144:332–333

Pfister CA, Hay ME (1988) Associational plant refuges: convergent patterns in marine and terrestrial communities result from differing mechanisms. Oecologia (Berlin) 77(1):118–129

Phaup JDJ (1967) Ecology of *Sphaerotilus* in an experimental outdoor channel. Water Res 1:523–541

*Phelps HL, Pearson WH, Hardy JT (1985) Clam burrowing behavior and mortality related to sediment copper. Mar Pollut Bull 16(8):309–313

*Phinney HK, McIntire CD (1965) Effect of temperature on metabolism of periphyton communities developed in laboratory streams. Limnol Oceanogr 10(3):341–344

Pignatello JJ (1987) Microbial degradation of 1,2-dibromoethane in shallow aquifer materials. J Environ Qual 16(4):307–312

Pilson MEQ (1985) Annual cycles of nutrients and chlorophyll in Narragansett Bay, Rhode Island. J Arct Res 43:849–873

*Pilson MEQ, Nixon SW (1980) Marine microcosms in ecological research. In: Giesy JP Jr (ed) Microcosms in ecological research. (DOE Symposium Series 52) US Department of Energy, Washington DC, pp 724–741

Pilson MEQ, Vargo GA, Gearing PA, Gearing JN (1977) Investigations of effects and fates of pollutants. Energy/environment II. (Proc 2nd Nat'l Conf Interagency R&D Program, EPA-600/9-77-012) US Environmental Protection Agency, Washington DC, pp 513–516

*Pilson MEQ, Oviatt CA, Vargo GA, Vargo SL (1979) Replicability of MERL microcosms: initial observations. (Advances in Marine Environmental Research, EPA-600/9-79-035) US Environmental Protection Agency, Washington DC, pp 359–381

*Pilson MEQ, Oviatt CA, Nixon SW (1980) Annual nutrient cycles in a marine microcosm. In: Giesy JP Jr (ed) Microcosms in ecological research. (DOE Symposium Series 52) US Department of Energy, Washington DC, pp 753–778

Pipes WO (1961) Waste-recovery processes for a closed ecological system. (National Research Council Publication 898:1–22) National Academy of Sciences, Washington DC

Pirie NW (1966) Towards the seleno-microcosm. New Scientist 2:574–576

*Platt RB, McCormick JF (1964) Manipulatable terrestrial ecosystems. Ecology 45(3):649–650

*Plessis Y (1956) Note sur le controle de la salinite en milieu marin artificiel. Bull Museum (Lab Peches Coloniales Museum) 28:583–589
*Plessis Y (1963) Realisation d'habitats sableux en milieu experimental. Bull Inst Oceanogr Monaco No. Special 1D (Congres International d'Aquariologie) 61–70
Plumley FG, Davis DE (1980) The effects of photosynthesis inhibitor atrazine on salt marsh edaphic algae in culture microecosystems and in the field. Estuaries 3(4):271–277
Porcella DB, Adams VD, Medine AJ, Cowan PA (1982) Using 3-phase aquatic microcosms to assess fates and impacts of chemicals in microbial communities. Water Res 16:489–496
Post FJ, Stube JC (1988) A microcosm study of nitrogen utilization in the Great Salt Lake, Utah, USA. Hydrobiologia 158:89–100
Pratt DM (1950) Experimental study of the phosphorus cycle in fertilized salt water. J Mar Res 9(1):29–54
*Pratt JR, Bowers NJ (1990) Effect of selenium on microbial communities in laboratory microcosms and outdoor streams. Toxic Assess 5(3):293–308
Pratt JR, Bowers NJ (1990) A microcosm procedure for estimating ecological effects of chemicals and mixtures. Toxic Assess 5(2):189–205
Pratt JR, Mccormick PV, Pontasch KW, Cairns JJ (1987) Evaluating potential groundwater contamination from contaminated soils. VPI State Univ Water Resour Res Center Bull (155):1–30
*Pratt JR, Niederlehner BR, Pratt NB, Cairns JJ (1987) Prediction of permissible concentrations of copper from microcosm toxicity tests. Toxic Assess 2(4):1–29
Pratt JR, McCormick PV, Pontasch KW, Cairns JJ (1988) Evaluating soluble toxicants in contaminated soils. Water Air Soil Pollut 37:293–308
Pratt JR, Bowers NJ, Niederlehner BR, Cairns JJ (1988) Effects of chlorine on microbial communities in naturally derived microcosms. Environ Toxicol Chem 7(9):679–688
*Price HJ (1989) Swimming behavior of krill in response to algal patches: a microcosm study. Limnol Oceanogr 34:649–659
Price DJ, Murphy BR, Smith LM (1989) Effects of tebuthiuron on characteristic green algae found in Playa Lakes. J Environ Qual 18(1):62–66
*Price WA, Macauley JM, Clark JR (1986) Effects of drilling fluids on *Thalassia testudinum* and its epiphytic algae. Environ Exp Bot 26(4):321–330
*Pritchard PH, Bourquin AW (1984) A perspective on the role of microcosms in environmental fate and effects assessments. In: White HH (ed) Concepts in marine pollution measurements. Maryland Sea Grant College, Park, Maryland, pp 117–138
Pritchard PH, Oneill EJ, Spain CM, Ahearn DG (1987) Physical and biological parameters that determine the fate of p-chlorophenol in laboratory test systems. Appl Environ Microbiol 53(8):1833–1838
*Ragsdale HL, Witherspoon JP, Nelson DJ (1968) The effects of biotic complexity and fast neutron radiation on cesium-137 and cobalt-60 kinetics in aquatic microcosms. (ORNL-4318) Oak Ridge National Laboratory, Oak Ridge, Tennessee
*Ramm AE, Bella DA (1974) Sulfide production in anaerobic microcosms. Limnol Oceanogr 19(1):110–118
*Ravera O (1989) Lake ecosystem degradation and recovery studied by the enclosure method. In: Ravera O (ed) Ecological assessment of environmental degradation, pollution, and recovery. Elsevier, Amsterdam New York, pp 217–243

*Ravera O, Annoni D (1980) Congresso Associaszione Italiana Oceanolotgia e Limnologia 417–421 (Cited by Ravera O (1989) Lake ecosystem degradation and recovery studied by the enclosure method. In: Ravera O (ed) Ecological assessment of environmental degradation, Pollution, and Recovery. Elsevier, Amsterdam 369pp)

*Raymont JEG, Adams MNE (1958) Studies on the mass culture of *Phaeodactylum* Limnol Oceanogr 3(3):119–136

*Raymont JEG, Miller RS (1962) Production of marine zooplankton with fertilization in an enclosed body of sea water. Int Rev Ges Hydrobiol 47(2):169–209

Reddy KR, Jessup RE, Rao PSC (1988) Nitrogen dynamics in a eutrophic lake sediment. Hydrobiologia 159(2):177–188

Reed C (1978) Species diversity in aquatic microecosystems. Ecology 59(3):481–488

*Rees JT (1979) Community development in freshwater microcosms. Hydrobiologia 63:113–128

Reeve MR, Gamble JC, Walter MA (1977) Experimental observations on the effects of copper on copepods and other zooplankton: controlled ecosystem pollution experiment. Bull Mar Sci 27:92–104

Reeve MR, Walter MA, Darcy K, Ikeda T (1977) Evaluation of potential indicators of sub-lethal toxic stress on marine zooplankton (feeding fecundity, respiration, and excretion): controlled ecosystem pollution experiment. Bull Mar Sci 27:105–113

*Reeve MR, Grice GD, Harris RP (1982) The CEPEX approach and its implications for future studies in plankton ecology. In: Grice GD, Reeve MR (eds) Marine mesocosms, biological and chemical research in experimental ecosystems. Springer, Berlin Heidelberg New York, pp 389–398

*Reichgott M, Stevenson LH (1978) Microbiological and physical properties of salt marsh and microecosystem sediments. Appl Environ Microbiol 36:662–667

Reinert KH, Rodgers JJH, Hinman ML, Leslie TJ (1985) Compartmentalization and persistence of endothall in experimental pools. Ecotoxicol Environ Safety 10:86–96

*Reynolds CS (1988) The concept of ecological succession applied to seasonal periodicity of freshwater phytoplankton. Verh Int Ver Limnol 23:683–691

Reynolds CS, Wiseman SW, Godfrey BM, Butterwick C (1983) Some effects of artificial mixing on the dynamics of phytoplankton populations in large limnetic enclosures. J Plankton Res 5:203–234

Reynolds CS, Graham GP, Harris P, Gouldney DN (1985) Comparison of carbon-specific growth rates and rates of cellular increase in phytoplankton in large limnetic enclosures. J Plankton Res 7:791–820

Rhyne C, Crump L, Jordan P (1985) Growth and protein production in selected laboratory cultures of blue-green algae grown in *Tillapia* wastewaters. (Mississippi-Alabama Sea Grant Consortium MASGP-84-025) Jackson, Mississippi

*Rich LGM (1963) Unit Process operations of sanitary engineering. John Wiley, New York

*Rich LGM, Ingram WM, Berger BF (1959) A balanced ecological system for space travel. J Sanit Eng Div 85 (SA 6):87–94

*Rich LGM, Ingram WM, Berger BF (1961) The use of vegetable cultures as the photosynthetic component of isolated ecological cycles for space travel. In: Advances in the astronautical sciences. (Proc 6th Ann Meeting of the American Astronautical Society) McMillan, New York 6:369–379

*Richardson AMN, Morton HP (1986) Terrestrial amphipods (Crustacea, Amphipoda, Talitridae) and soil respiration. Soil Biol Biochem 18(2):197–220
*Richardson JR, Odum HT (1981) Power and a pulsing production model. In: Mitsch WJ, Bosserman KW, Klopatek JM (eds) Energy and ecological modeling. Elsevier, New York, pp 641–648
*Richardson RE (1930) Notes on the simulation of natural aquatic conditions in fresh water by the use of small non-circulating aquaria. Ecology 11:102–109
*Richey JE (1970) Role of disordering energy in ecosystems. (MS Thesis) Department of Environmental Science and Engineering, University of North Carolina, Chapel Hill, North Carolina
Ringelberg J (1976) The possibilities of a new kind of micro-ecosystem in the aquatic ecosystem research. Hydrobiol Bull 10:17–18
*Ringelberg J (1977) Properties of an aquatic micro-ecosystem. II. Steady state phenomena in the autotrophic subsystems. Helgol Wiss Meeresunters 30:134–143
*Ringelberg J, Kersting K (1978) Properties of an aquatic micro-ecosystem. I. General introduction to the prototypes. Arch Hydrobiol 83:47–68
*Riznyk RZ, Hardy JT, Pearson W, Jabs L (1987) Short-term effects of polynuclear aromatic hydrocarbons on sea-surface microlayer phytoneuston. Bull Environ Contam Toxicol 38:1037–1043
*Robinson JV, Dickerson JE (1984) Testing the invulnerability of laboratory island communities to invasion. Oecologia (Berlin) 61:169–174
*Robinson JV, Dickerson JE (1987) Does invasion sequence affect community structure? Ecology 68(3):587–595
Robinson JV, Edgemon MA (1988) An experimental evaluation of the effect of invasion history on community structure. Ecology 69(5):1410–1417
*Robinson JV, Edgemon MA (1989) The effect of predation on the structure and invasibility of assembled communities. Oecologia 79:150–157
Robinson JV, Sandgren CD (1984) An experimental evaluation of diversity indices as environmental discriminators. Hydrobiologia 108(2):188–196
Rocchio PM, Malanchuk JL (1986) The effects of atrazine on dissolved oxygen and nitrate concentrations in aquatic systems. Environ Int 12(6):597–602
Rocha CE, Bjornberg MH (1987) Copepods of the Jureia Ecological Reserve, State of Sao Paulo Brazil. II. The genera *Hesperocyclops*, *Muscocyclops*, and *Ryocyclops* Cyclopoida Cyclopidae Hydrobiologia 153(2):97–108
Rochelle PA, Fry JC, Day MJ (1989) Plasmid transfer between *Pseudomonas* spp. within epilithic films in a rotating disc microcosm. FEMS Microbiol Ecol 62(2):127–136
Rodgers EB (1980) Effects of elevated temperatures on macroinvertebrate populations in the Brown's Ferry experimental ecosystems. Giesy JP Jr (ed) Microcosms in ecological research. (DOE Symposium Series 52) US Department of Energy, Washington DC, pp 684–702
Rodgers JHJ, Harvey RS (1976) The effect of current on periphytic productivity as determined using carbon-14. Water Res Bull 12:1109–1118
*Rodgers JHJ, Clark JR, Dickson KL, Cairns JJ (1980) Nontaxonomic analyses of structure and function of aufwuchs communities in lotic microcosms. In: Giesy JP Jr (ed) Microcosms in ecological research. (DOE Symposium Series 52 USA) US Department of Energy, Washington DC, pp 625–644
Rodgers JHJ, Dickson EL, Saleh FY, Staples CA (1983) Use of microcosms to study transport, transformation, and fate of organics in aquatic systems. Environ Toxicol Chem 2:155–167

Rogers DW, Watson TA, Langan JS, Wheaton TJ (1987) Effects of pH and feeding regime on methylmercury accumulation within aquatic microcosms. Environ Pollut 45(4):261–274

*Rodgers PW, DePinto JV (1981) Algae–bacteria interaction in a light–dark cycle. J Freshwater Ecol 1:71–80

Rojan YA, Hazen TC (1989) Survival of *Vibrio cholerae* in treated and untreated rum distillery effluents. Water Res 23(1):103–114

*Rose FL, McIntire CD (1970) Accumulation of dieldrin by benthic algae in laboratory streams Hydrobiologia 35:481–493

*Rose KA, Swartzman GL, Kindig AC, Taub FA (1988) Stepwise iterative calibration of a multi-species phytoplankton–zooplankton simulation model using laboratory data. Ecol Model 42(1):1–32

Roszak DB, Colwell RR (1987) Metabolic activity of bacterial cells enumerated by direct viable count. Appl Environ Microbiol 53(12):2889–2893

*Rowe DR, Gloyna EF (1964) Radioactivity transport in water — the transport of zinc-65 in an aqueous environment. (Technical Report to the Atomic Energy Commission, No 5) University of Texas, Austin

Rudd JWM (1972) The design and operation of a model aerated sewage lagoon. (Technical Report No 294(21)) Fisheries Research Board of Canada, Winnipeg Canada

Rudd JWM, Turner MA (1983) The English–Wabigoon River system. V. Mercury and selenium bioaccumulation as a function a aquatic primary productivity. Can J Fish Aquat Sci 40:2251–2259

Rudd JWM, Turner MA (1983) The English–Wabigoon River system. II. Suppression of mercury and selenium bioaccumulation by suspended and bottom sediments. Can J Fish Aquat Sci 40:2218–2227

*Rudnick DT, Oviatt CA (1986) Seasonal lags between organic carbon deposition and mineralization in marine sediments. J Mar Res 44:815–837

Rutledge RW, Basore BL, Mulholland RJ (1976) Ecological stability: an information theory viewpoint. J Theor Biol 57:355–371

*Saeki A (1957) Studies on fish culture in the aquarium of a closed-circulating system: its fundamental theory and standard plan. Bull Jpn Soc Sci Fish 23:684–695

Saks NM, Lee JJ, Muller WA, Tietjen JH (1974) Growth of salt-marsh microcosms subjected to thermal stress. In: Gibbons JW, Sharitz RR (eds) Thermal ecology. US Atomic Energy Commission, Oak Ridge, Tennessee, pp 391–398

*Salki A, Turner M, Patalas K, Rudd J, Findlay D (1985) The influence of fish–zooplankton–phytoplankton interactions on the results of selenium toxicity experiments within large enclosures. Can J Fish Aquat Sci 42:1132–1143

*Salt GW (1967) Predation in an experimental protozoan population (*Woodruffia–Paramecium*) Ecol Monog 37(2):113–144

*Sampou P, Oviatt CA (1991) Seasonal patterns of sedimentary carbon and anaerobic respiration along a simulated eutrophication gradient. Mar Ecol Progr Ser 72(3):271–282

*Samsel GL Jr, Parker BC (1972) Nutrient factors limiting primary productivity in simulated and field antarctic microecosystems. Virginia J Sci 23(2):64–71

Sanborn JR, Yu CC (1973) The fate of dieldrin in a model ecosystem. Bull Environ Contam Toxicol 10:340–346

Sanders JG, Gibik SJ, D'Elia CF, Boynton WR (1987) Nutrient enrichment studies in a coastal plain estuary: changes in phytoplankton species composition. Can J Fish Aquat Sci 44(1):83–90

Sanders JG, Cibik SJ (1985) Adaptive behavior of euryhaline phytoplankton communities to arsenic stress. Mar Ecol 22:199–205

*Sanders JG, Cibik SJ (1985) Reduction of growth rate and resting spore formation in a marine diatom exposed to low levels of cadmium. Mar Environ Res 16:165–180

*Sanders JG, Osman RW (1985) Arsenic incorporation in a salt marsh ecosystem. Estuar Coastal Shelf Sci 20:387–392

Santschi PH (1982) Application of enclosures to the study of ocean chemistry. In: Grice D, Reeve MB (eds) Marine mesocosms, biological and chemical research in experimental ecosystems. Springer, Berlin Heidelberg New York, pp 63–80

Santschi PH, Adler D, Amdurer M, Li YH, Bell J (1980) Thorium isotopes as analogues for "particle-reactive" pollutants in coastal marine environments. Earth Planet Sci Lett 47:327–335

Santschi PH, Carson S, Li YH (1982) Natural radionuclides as tracers for geochemical processes in MERL mesocosms and Narragansett Bay. In: Grice D, Reeve MB (eds) Marine mesocosms, biological and chemical research in experimental ecosystems. Springer, Berlin Heidelberg New York, pp 97–109

Santschi PH, Li YH, Bell J, Adler D, Amdurer M, Nyfeller HP (1983) The relative mobility of natural (Th, Pb, Po) and fallout (Pu, Cs, Am) radionuclides in the coastal marine environment: results from model ecosystems (MERL) and Narragansett Bay studies. Geochim Cosmochim Acta 47:201–210

*Santschi PH, Adler DM, Amdurer M (1983) The fate of particles and particle-reactive trace metals in coastal waters: radioisotope studies in microcosms. In: Wong CS, Boyle E, Breland KW, Burton JD, Goldberg ED (eds) Trace metals in sea water. Plenum Press, New York, pp 331–349

*Saunders JG, Cibik SJ (1985) Adaptive behavior of euryhaline phytoplankton communities to arsenic stress. 22:199–205

Saward D, Stirling A, Topping G (1975) Experimental studies on the effects of copper on a marine food chain. Mar Biol 29:351–361

Scanferlato VS, Cairns J Jr (1990) Effect of sediment-associated copper on ecological structure and function of aquatic microcosms. Aquat Toxicol (Amsterdam) 18:23–34

Schindler DW, Hesslein RH, Wagemann R, Broecker WS (1980) Effects of acidification on mobilization of heavy metals and radionuclides from the sediments of a freshwater lake. Can J Fish Aquat Sci 37:373–377

*Schneider SH, Boston PJ (ed) (1991) Scientists on Gaia. MIT Press, Cambridge, Massachusetts 433pp

Schneidereit H, Schmidt FRJ (1990) The use of a *Sesbania rostrata* microcosm for studying gene transfer among microorganisms. In: Fry JC, Day MJ (eds) Bacterial genetics in natural environments. Chapman and Hall, London, pp 182–187

Schnurer JS, Clarholm M, Rosswall T (1985) Microbial biomass and activity in an agricultural soil with different organic matter contents. Soil Biol Biochem 17(5):611–618

Schreiner SP (1980) Use of microcosms in the study of habitat changes caused by the growth and perturbation of a water hyacinth community. In: Giesy JP

Jr (ed) Microcosms in ecological research (DOE Symposium Series 52) US Department of Energy, Washington DC, pp 473–488

Schroeder PB (1980) Trace metal cycling in tropical–subtropical estuaries dominated by the seagrass *Thalassia testudinum*. Am J Bot 67(7):1075–1088

Schulz S (1990) Baltic sea eutrophication: a case study using experimental ecosystems. In: Lalli CM (ed) Enclosed experimental marine ecosystems: a review and recommendations. Coastal and estuarine studies, 37, Springer, Berlin Heidelberg New York, pp 169–187

*Schwartzkopf SH (1992) Design of a controlled ecological life support system. Bioscience 42(7):526–535

*Sciandra A (1986) Study and modeling of a simple planktonic system reconstituted in an experimental microcosm. Ecol Model 34:61–82

*Scienceman D (1987) Energy and emergy. In: Pillet G, Murota (eds) Environmental economics. Roland Leimgruber, Geneva, Switzerland, pp 257–276

Scott BF, Nagy E, Dutka BJ, Sherry JP, Hart J, Jaylor WD, Glooschenko V, Wade PJ (1984) The fate and impact of oil and oil-dispersant mixtures in freshwater pond ecosystems: introduction. Sci Total Environ 35:105–113

Scott BF, Wade PJ, Taylor WD (1984) Impact of oil and oil-dispersant mixtures on the fauna of freshwater ponds. Sci Total Environ 35:191–206

Scott BF, Glooschenko V (1984) Impact of oil and oil-dispersant mixtures on flora and water chemistry parameters in freshwater ponds. Sci Total Environ 35:169–190

Seitzinger SP, Nixon SW, Pilson MEQ (1984) Denitrification and nitrous oxide production in a coastal marine ecosystem. Limnol Oceanogr 29:73–83

Seki H, Whitney FA, Wong CS (1983) Copper effect on dynamics of organic materials in marine controlled ecosystems. Arch Hydrobiol 96:176–189

Setala H, Haimi J, Huhta V (1987) A microcosm study on the respiration and weight loss in birch litter and raw humus as influenced by soil fauna. Biol Fertil Soils 5(4):282–287

Shannon CE, Weaver W (1949) Mathematical theory of communications, University of Illinois Press, Urbana

Shannon LJ, Harass MC, Yount JD, Walbridge CT (1986) A comparison of mixed flask culture and standardized laboratory model ecosystems for toxicity testing. In: Cairns J Jr (ed) Community Toxicity Testing. (ASTM STP 920) American Society for Testing Materials, Philadelphia, pp 135–157

*Shapiro J (1973) Blue-green algae. Why they became dominant. Science 179:382–384

Shaw C, Pawluk S (1986) The development of soil structure by *Octolasion tyrtaeum*, *Aporrectodea turgida*, and *Lumbricus terrestris* in parent materials belonging to different textural classes. Pedobiologia 29(5):327–339

Sheenan PJ, Axler RP (1986) NRC evaluation of simple generic aquatic ecosystem tests to screen the ecological impacts of pesticides. In: Cairns J Jr (ed) Community Toxicity Testing. (ASTM STP 920) American Society for Testing Materials, Philadelphia, pp 158–179

*Sheldon SP (1987) The effects of herbivorous snails on submerged macrophyte communities in Minnesota lakes. Ecology 68:1920–1931

*Shelford VE (1913) Animal communities in temperate America. Bull Geog Soc Chicago 5:1–368

*Shelford VE (1929) Labortory and field ecology. Williams and Wilkins, Baltimore

*Shepelev YY (1972) Biological life support systems. In: Foundations of space biology and medicine. Academy of Sciences USSR/NASA Joint Publication, Moscow/Washington DC

Shevyrnogov AP, Molvinskikh SI, Cbepilov VV, Gitelson II (1988) Determination of photosynthetic pigments in aqueous ecosystems Mitt Geol Paleont Inst Univ Hamburg SCOPE/UNEPS Sonderbannd Heft 66:331–340

Shikano S, Kurihara Y (1985) Community responses to organic loading in a microcosm. Jpn J Ecol 35:297–305

Shikano S, Kurihara Y (1988) Analysis of factors controlling responses of an aquatic microcosm to organic loading. Hydrobiologia 169(2):251–257

Shimp RJ (1989) Adaptation to a quaternary ammonium surfactant in aquatic sediment microcosms. Environ Toxicol Chem 8(3):201–208

Shimp RJ, Pfaender FK (1987) Effect of adaptation to phenol on biodegradation of monosubstituted phenols by aquatic microbial communities. Appl Environ Microbiol 53(7):1496–1499

Shugart HH (1989) The role of ecological models in long-term ecological studies. Likens GE (ed) Long-term studies in ecology: approaches and alternatives. (2nd Cary Conference, Millbrook, New York, XVI + 214P) Springer, Berlin Heidelberg New York, pp 90–109

Sigmon CF, Kania HF, Beyers RJ (1977) Reductions in biomass and diversity resulting from exposure to mercury in artificial streams. J Fish Res Board Can 34:493–500

*Simberloff PS, Wilson EO (1970) Experimental zoogeography of islands: a two-year record of colonization. Ecology 51:934–937

*Simpson DA (1946) Small marine aquaria for the home: a digest and guide for the hobbyist. Steinhart Aquarium, San Fransisco

Sjkoldal HR, Dale T, Haldorsen H, Pengerud B, Thingstad TF, Tjessem K, Anberg A (1982) Oil pollution and plankton dynamics. I. Controlled ecosystem experiment during the 1980 spring bloom in Lindaspollene, Norway. Neth J Sea Res 16:511–523

Sjkoldal HR, Johannessen P, Klinken J, Haldorsen SH (1983) Controlled ecosystem experiment in Lindaspollene, western Norway, June 1979: comparisons between the natural and two enclosed water columns. Sarsia 68:47–64

Smetacek V, vonBodungen B, vonBrockel K, Zeitzschel B (1976) The plankton tower. II. Release of nutrients from sediments due to changes in the density of bottom water. Mar Biol 34:373–378

Smetacek V, vonBodungen B, Knoppers B, Pollehne F, Zeitzschel B (1982) The plankton tower. IV. Interactions between water column and sediment in enclosure experiments in Kiel Bight. In: Grice D, Reeve MB (eds) Marine mesocosms, biological and chemical research in experimental ecosystems. Springer, Berlin Heidelberg New York, pp 205–216

Smith JA, Novak JT (1987) Biodegradation of chlorinated phenols in subsurface soils. Water Air Soil Pollut 33(1–2):29–42

Smith TP (1980) Application of the simple charge–discharge metabolic model to transient behavior of experimental marine microcosms. Ecol Model 10(1):13–30

*Smith TP (1980) Responses of a benthic marine microcosm subjected to changes in energy flow. In: Giesy JP Jr (ed) Microcosms in ecological research. (DOE Symposium Series 52) US Department of Energy, Washington DC, pp 801–817

Smith W, Gibson VR, Brown-Leger LS, Grassle JF (1979) Diversity as an indicator of pollution: cautionary results from microcosm experiments. In: Hall CAS, Day JW Jr (eds) Ecosystem modeling in theory and practice. John Wiley, New York, pp 269–277

*Smith W, Gibson VR, Grassle JF (1982) Replication in controlled marine systems: presenting the evidence. In: Grice GD, Reeve MR (eds) Marine mesocosms, biological and chemical research in experimental ecosystems, Springer, Berlin Heidelberg New York, pp 217–225

Smyly WJP (1976) Some effects of enclosure on the zooplankton in a small lake. Freshwater Biol 6:241–251

Snow NB, Scott BF (1975) The effect and fate of crude oil spilt on two arctic lakes. (Conference on Prevention and Control of Oil Spills) American Petroleum Institute, Washington DC 527–534

Solemdal P (1981) Overview — enclosure studies. Rapp P-V Reun Cons Int Explor Mer 178:117–120

*Sollins P (1970) Measurements and simulation of oxygen flows and storages in a laboratory blue-green algal mat ecosystem. (MS Thesis) University of North Carolina, Chapel Hill, North Carolina

Solomon KR, Smith K, Guest G, Yoo JY, Kaushik NK (1980) Use of limnocorrals in studying the effects of pesticide in the aquatic ecosystem. (Technical Report No 975) Fisheries Research Board of Canada, Winnipeg, Canada

Solomon KR, Yoo JY, Lean D, Kaushik NK, Day KE, Stephenson GL (1985) Dissipation of permethrin in limnocorrals. Can J Fish Aquat Sci 42:70–76

*Sommer U (1988) Phytoplankton succession in microcosm experiments under simultaneous grazing pressure and resource limitation. Limnol Oceanogr 33(5): 1037–1054

Sonntag NC (1979) TRP mixing and enclosed 1,300 m^3 water column: effects on the planktonic food web. J Plankton Res 1:85–102

Southeworth GR, Watson KW, Keller JL (1987) Comparison of models that describe the transport of organic compounds in macroporous soil. Environ Toxicol Chem 6:251–257

*Spencer R (1990) Solar aquatic treatment of sewage. Biocycle 31(5):66–70

*Sperling JA, Grunewald R (1969) Batch culturing of thermophilic benthic algae and phosphorus uptake in a laboratory stream model. Limnol Oceanogr 14(6): 944–949

Spies A, Parsons TR (1985) Estuarine microplankton: an experimental approach in combination with field studies. J Exp Mar Biol Ecol 92:63–81

Spotte SH (1970) Fish and invertebrate culture: water management in closed systems. Wiley Interscience, New York

Staley CS, Case MJ (1987) An evaluation of in situ microcosms for validating aquatic fate and transport models. Environ Monit Assess 8(2):103–112

Staples CA, Dickson KL, Saleh FY, Rogers JHJ (1983) A microcosm study of lindane and naphthalene for model validation. In: Aquatic Toxicology and Hazard Assessment: Sixth Symposium (ASTM STP 802) American Society for Testing Materials, Philadelphia, pp 26–41

Stay FS, Larsen DR, Katko A, Rohm CM (1985) Effects of atrazine on community-level responses in Taub microcosms. In: Boyle TP (ed) Validation and predictability of laboratory methods for assessing the fate and effects of contaminants in aquatic ecosystems. (ASTM STP 865) American Society for Testing Materials, Philadelphia, pp 75–90

Stay FS, Katko A, Rohm CM, Fix MF, Larsen DP (1988) Effects of fluorene on microcosms developed from four natural communities. Environ Toxicol Chem 7(8):635–644

Steber J, Gode P, Guhl W (1988) Fatty alcohol sulfates: the ecological evaluation of a group of important detergent surfactants. Fett Wiss Technol 90(1):32–38

Steele J (1977) Spatial pattern in plankton communities. Plenum Press, New York 465 pp

Steele JH (1979) The uses of experimental ecosystems. Phil Trans R Soc Lond B 286:583–595

*Steele JH (1985) A comparison of terrestrial and marine ecological systems. Nature 313(6001):355–359

Steele JH, Baird IE (1968) The ecology of O-group plaice and common dabs at Loch Ewe. I. Population and food. J Exp Mar Biol Ecol 2:215–238

*Steele JH, Gamble JC (1982) Predator control in enclosures. In: Grice GD, Reeve MR (eds) Marine mesocosms, biological and chemical research in experimental ecosystems. Springer, Berlin Heidelberg New York, pp 228–237

*Steele JH, Farmer DM, Henderson EW (1977) Circulation and temperature structure in large marine enclosures. J Fish Res Board Can 34:1095–1104

Steele JH (1991) Marine ecosystem dynamics: comparison of scales. Ecological research 6:175–183

*Steinmann AD, McIntire CD (1987) Effects of irradiance on the community structure and biomass of algal assemblages in laboratory streams. 44:1640–1648

*Steinmann AD, McIntire CD, Lowry RR (1988) Effects of irradiance and age on chemical constituents of algal assemblages in laboratory streams. Arch Hydrobiol 114:45–61

*Steinmann AD, McIntire CD, Gregory SV, Lamberti GA (1989) Effects of irradiance and grazing on lotic algal assemblages. J Phycol 25:478–485

Stephenson GL, Hamilton P, Kaushik NK, Robinson JB, Solomon KR (1984) Spacial distribution of plankton in enclosures of three sizes. Can J Fish Aquat Sci 41:1048–1054

Stevenson BG, Dindal DL (1987) Insect effects on decomposition of cow dung in microcosms. Pedobiologia 30(2):81–92

*Stewart PM, Sappington KG, Carins J Jr (1986) Community and population response of natural diatom communities to zinc stress in laboratory mesocosms. Curr Pract Environ Sci Eng 2:177–184

Stout RJ, Cooper WE (1983) Effect of p-cresol on leaf decomposition and invertebrate colonization in experimental outdoor streams. Can J Fish Aquat Sci 40:1647–1657

*Streeter HW, Phelps EB (1925) A study of the pollution and natural purification of the Ohio River. (US Public Health Service Bulletin, No 146) US Public Health Service, Washington DC

Streit B (1979) Uptake accumulation and release of organic pesticides by benthic invertebrates. III. Distribution of carbon-14-labelled lindane in an experimental three-step food chain microcosm. Arch Hydrobiol 55(3-4):373–400

*Strickland JDH, Terhune LDB (1961) The study of in situ marine photosynthesis using large plastic bag. Limnol Oceanogr 6:93–96

Strickland JDH, Holm-Hansen O, Eppley RW, Linn RJ (1969) The use of a deep tank in plankton ecology. I. Studies of the growth and composition of phytoplankton crops at low nutrient levels. Limnol Oceanogr 14:23–34

Stroganov NS, Filenko OF, Gusev MV (1982) Principles of toxicological testing in experimental aquatic ecosystems. Toxic substances in the aquatic environment: an international aspect:29–37

Sugiura K, Sato S, Goto M, Kurihara Y (1976a) Toxicity assessment using an aquatic microcosm. Chemosphere 5(2):113–118

*Sugiura K, Sato S, Goto M, Kurihara Y (1976b) Effects of beta-BHC on aquatic microcosm. Chemosphere 1:39–44

*Sullivan BK, Ritacco PJ (1985) Ammonia toxicity to larval copepods in eutrophic marine ecosystems: a comparison of results from bioassays and enclosed experimental ecosystems. Aquat Toxicol 7:205–217

Sullivan BK, Ritacco PJ (1985) The response of dominant copepod species to food limitation in a coastal marine ecosystem. Arch Hydrobiol Beih Ergeb Limnol 21:407–418

*Sun T, Liu Z, Lin S The function of the root microecosystem in the process of dyeing wastewater treatment by the hyacinth. Chin J Environ Sci (Beijing) 11(3):24–27, 95–96

Sundelin B (1988) Effects of sulfate pulp mill effluents on soft bottom organisms: a microcosm study. Water Sci Technol 20(2):175–178

Suorsa KE, Fisher SW (1986) Effects of pH on the environmental fate of ^{14}C-aldicarb in an aquatic microcosm. Ecotoxicol Environ Safety 11(1):81–90

*Swartzman GL, Rose KA (1984) Simulating the biological effects of toxicants in aquatic microcosm systems. Ecol Model 22:123–134

*Swartzman GL, Kaluzny SP (1987) Ecological simulation primer. McMillan, New York

*Swartzman GL, Rose KA, Kindig A, Taub F (1989) Modeling the direct and indirect effects of streptomycin in aquatic microcosms. Aquat Toxicol 14(2):109–130

*Swingle HS, Smith EV (1939) Fertilizers for increasing the natural food for fish in ponds. Trans Am Fish Soc 68:126–135

Swenson R (1989a) Emergent evolution and the global attractor: the evolutionary epistemology of entropy production maximization. Preprints Ann Meet Int Soc Syst Sci 3:46–53

*Swenson R (1989b) Emergent attractors and the laws of maximum entropy production: foundation to a theory of general evolution. Systems Res 6(3):187–197

Tagatz ME, Ivey JM, Gregory NR, Oglesby JL (1982) Effects of pentachlorophenol on field- and laboratory-derived estuarine benthic communities. Bull Environ Contam Toxicol 26:137–143

Tagatz ME, Deans CH, Moore JC, Plaia GR (1983) Alterations in composition of field- and laboratory-derived estuarine benthic communities exposed to di-n-butyl phthalate. Aquat Toxicol 3:239–248

Takahashi M, Whitney FA (1977) Physical features of controlled experimental ecosystems (CEE) with special reference to their temperature, salinity, and light penetration structures. Bull Mar Sci 27:8–16

Takahashi M, Thomas WH, Seibert DLR, Beers J, Koeller P, Parsons TR (1975) The replication of biological events in enclosed water columns. Arch Hydrobiol 76:5–23

Takahashi M, Wallace GT, Whitney FA, Menzel DW (1977) Controlled ecosystem pollution experiment: effect of mercury on enclosed water columns. I. Manipulation of experimental enclosures. Mar Sci Comm 3:313–329

Takahashi M, Koike I, Iseki K, Bienfang PK, Hattori A (1982) Phytoplankton species' responses to nutrient changes in experimental enclosures and coastal waters. In: Grice D, Reeve MB (eds) Marine mesocosms, biological and chemical research in experimental ecosystems. Springer, Berlin Heidelberg New York, pp 333–348

Takahashi M (1990) Pelagic mesocosms. I. Food chain analysis. In: Lalli CM (ed) Enclosed experimental marine ecosystems: a review and recommendations. Coastal and estuarine studies, 37, Springer, Berlin Heidelberg New York, pp 61–80.

Takano C (1984) ATP measurement of biomass in materially closed microbial ecosystems. M.S. thesis, University of Hawaii

*Takano C, Folsome CE, Karl DM (1983) ATP as a biomass indicator for closed ecosystems. Biosystems 16:75–78

Tangley L (1985) And live from the East Coast — a miniature Maine ecosystem. Bioscience 35:618–619

Tate RL III, Parmelee RW, Ehrenfeld JG, O'Reilly L (1991) Enzymatic and microbial interactions in response to pitch pine root growth. Soil Sci Soc Am J 55(4):998–1004

Tatrai I (1982) Oxygen consumption and ammonia excretion of herbivorous chironomid larvae in Lake Balaton. Hydrobiologia 96:129–135

*Tatrai I (1982) The influence of herbivorous chironomids on the exchange of nutrients between sediment and water. BFB-Bericht 43:231–248

Tatrai I (1986) Rates of ammonia release from sediments by chironomid larvae. Freshwater Biol 16:61–66

Tatrai I (1987) Nitrogen and phosphorus excretion by chironomid larvae. Gidrobiol Zh 23(4):59–63

*Taub FB (1963) Some ecological aspects of space biology. Am Biol Teacher 25(6):412–421

*Taub FB (1964) From wastes to resources. Univ Washington Coll Fish Activ Report 17(1):72–78

*Taub FB (1969a) A biological model of a freshwater community: a gnotobiotic ecosystem. Limnol Oceanogr 14(1):136–142

*Taub FB (1969b) Gnotobiotic models of freshwater communities. Verh Int Ver Limnol 17:485–496

*Taub FB (1969c) A continuous gnotobiotic (species defined) ecosystem. In: Cairns J Jr (ed) The structure and function of freshwater microbial communities. (Research Division Monograph) Virginia Polytechnic Institute and State University, Blacksburg, Virginia, pp 101–120

Taub FB (1984) Introduction: laboratory microcosms. In: White HH (ed) Concepts in marine pollution measurements. (Maryland Sea Grant Publication) University of Maryland, College Park, Maryland, pp 113–116

*Taub FB (1984) Measurement of pollution in standardized aquatic microcosms. In: White HH (ed) Concepts in marine pollution measurements. (Maryland Sea Grant Publication) University of Maryland, College Park, Maryland, pp 159–192

*Taub FB (1985) Toward interlaboratory (round-robin) testing of a standardized aquatic microcosm. In: Cairns J Jr (ed) Multispecies toxicity testing. Pergamon Press, United Kingdom, pp 165–168

*Taub FB (1989) Standardized aquatic microcosm — development and testing. In: Boudou A, Ribeyre F (eds) Aquatic ecotoxicology: fundamental concepts and methodologies, vol II. CRC Press, Boca Raton, Florida, pp 47–91

Taub FB, Crow ME (1980) Synthesizing aquatic microcosms. In: Giesy JP Jr (ed) Microcosms in ecological research. (DOE Symposium Series 52) US Department of Energy, Washington DC, pp 69–104

*Taub FB, Dollar AM (1964) A *Chlorella-Daphnia* food-chain study: the design of a compatible chemically defined culture medium. Limnol Oceanogr 9(1):61–74

Taub FB, Dollar AM (1968) Improvement of a continuous-culture apparatus for long-term use. Appl Microbiol 16(2):232–235

Taub FB, Dollar AM (1968) The nutritional inadequacy of *Chlorella* and *Chlamydomonas* as food for *Daphnia pulex*. Limnol Oceanogr 13(4):607–617

Taub FB, Kindig AC (1984) Potential use of microcosms to assess survival efficacy and environmental safety of genetically engineered microorganisms. In: Omenn GS, Hollaender A (eds) Genetic control of environmental pollutants. Plenum Press, New York, pp 383

*Taub FB, McKenzie D (1973) Continuous cultures of an alga and its grazer. Bull Ecol Res Comm (Stockholm) 17:371–377

Taub FB, Hung R, Tomlinson R (1968) A biological model of a freshwater community: a gnotobiotic ecosystem. (1967 Research in Fisheries, Contribution 280) College of Fisheries, University of Washington, Seattle, Washington, pp 136–142

Taub FB, Crow ME, Hartmann HJ (1980) Responses of aquatic microcosms to acute mortality. In: Giesy JP Jr (ed) Microcosms in ecological research. (DOE Symposium Series 52) US Department of Energy, Washington, pp 513–535

*Taub FB, Harras MC, Hartmann HJ, Kindig AC, Read PL (1981) Effects of initial algal density on community development in aquatic microcosms. Verh Int Ver Limnol 21:197–204

*Taub FB, Read PL, Kindig AC, Harass MC, Hartmann HJ, Conquest KK, Hardy FJ, Muro PT (1983) Demonstration of the ecological effects of streptomycin and malathion on synthetic aquatic microcosms. In: Aquatic Toxicology and Hazard Assessment: 6th Symposium (ASTM STP 802) American Society for Testing Materials, Philadelphia, pp 5–25

Taub FB, Kindig AC, Conquest LL (1986) Preliminary results of interlaboratory testing of a standardized aquatic microcosm. In: Cairns J Jr (ed) Community toxicity testing. (ASTM STP 920) American Society for Testing Materials, Philadelphia, pp 93–120

*Taub FB, Kindig AC, Conquest LL (1987) Interlaboratory testing of a standardized aquatic microcosm protocol. In: Adams WJ, Chapman GA, Landis WG (eds) Aquatic toxicology and environmental fate. (Vol 10 ASTM STP 971) American Society for Testing Materials, Philadelphia, pp 384–405

*Taub FB, Kindig AC, Conquest LL, Meador JP (1989) Results of interlaboratory testing of the standardized aquatic microcosm protocol. In: Sutter GW II, Lewis MA (eds) Aquatic toxicology and environmental fate. (Vol 11 ASTM STP 1007) American Society for Testing Materials, Philadelphia, pp 368–394

*Taub FB, Kindig AC, Meador JP, Swartzman GL (1991) Effects of "seasonal succession" and grazing on copper toxicity in aquatic microcosms. Verh Int Ver Limnol 24(4):2205–2214

Taylor BR, Parkinson D (1988) A new microcosm approach to litter decomposition studies. Can J Bot 66(10):1933–1939

Taylor BR, Parkinson D (1988) Annual differences in quality of leaf litter of aspen Populus tremuloides affecting rates of decomposition. Can J Bot 66(10): 1940–1947

Taylor BR, Parkinson D (1988) Aspen and pine leaf litter decomposition in laboratory microcosms. I. Linear versus exponential models of decay. Can J Bot 66(10):1960–1965

Taylor BR, Parkinson D (1988) Aspen and pine leaf litter decomposition in laboratory microcosms. II. Interactions of temperature and moisture level. Can J Bot 66(10):1966–1973

Taylor BR, Parkinson D (1988) Does repeated freezing and thawing accelerate decay of leaf litter? Soil Biol Biochem 20(5):657–656

Taylor BR, Parkinson D (1988) Does repeated wetting and drying accelerate decay of leaf litter? Soil Biol Biochem 20(5):647–656

Taylor BR, Parkinson D (1988) Respiration and mass loss rates of aspen and pine leaf litter decomposing in laboratory microcosms. Can J Bot 66:1948–1959

Taylor BR, Parkinson D, Parsons WJF (1989) Nitrogen and lignin content as predictors of litter decay rates: a microcosm test. Ecology 70(1):97–104

*Terami H, Watanabe M (1989) Excessive transitory migration of guppy populations. Zool Sci (Tokyo) 6(5):975–982

*Terskov IA, Gitelson BG (1968) Continuous culture of microalgae as a link of a closed ecological system. In: Continuous cultivation of microorganisms. (Proc 4th Symposium) Czechoslovak Academy of Sciences, Prague, pp 357–368

*Terskov IA, Gitelson BG, Kovrov BG (1979) Closed system: Man–higher plants (four month experiment). (Translation from Nauka Press, Siberian Branch, Novocibirsk, NASA Publication TM-76452) US National Aeronautics and Space Administration, Washington DC

Thomas WH, Seibert DLR (1977) Effects of copper on the dominance and the diversity of algae: controlled ecosystem pollution experiment. Bull Mar Sci 27:23–33

Thomas WH, Seibert DLR, Takahashi (1977) Controlled ecosystem pollution experiment: effect of mercury on enclosed water columns. III. Phytoplankton population dynamics and production. Mar Sci Comm 3:331–354

Thomas WH, Holm-Hansen O, Seibert DLR, Azam F, Hodson R, Takahashi M (1977) Effects of copper on phytoplankton standing crop and productivity: controlled ecosystem pollution experiment. Bull Mar Sci 27:34–43

*Thorp JH, Cothran ML (1982) Floating field microcosms for studying benthic communities. Freshwater Invert Biol 1(1):44–49

Throndsen J (1982) Oil pollution and plankton dynamics. III. Effects on flagellate communities in controlled ecosystem experiments in Lindaspollene, Norway, June 1980 and 1981. Sarsia 67:163–170

Tison DL, Wilde EW, Pope DH, Fliermans CB (1982) Productivity and species composition of algal mat communities exposed to a fluctuating thermal regime. Microb Ecol 7:151–166

Tjessem K, Pedersen D, Aaberg A (1984) On the environmental fate of a dispersed Ecofisk crude oil in sea-immersed plastic columns. Water Res 18:1129–1136

*Tolle DA, Arthur MF (1983) Microcosm–field comparison of trace element uptake in crops grown in fly ash-amended soil. Sci Total Environ 31:243–261

*Tolle DA, Arthur MF, Chesson J (1985) Comparison of pots versus microcosms for predicting agroecosystem effects due to waste amendment. Environ Toxicol Chem 4:501–509

Topping G, Windom H (1977) Biological transport of copper at Loch Ewe and Saanich Inlet: controlled ecosystem pollution experiment. Bull Mar Sci 27:134–141

Topping G, Davies IM, Pirie JM (1982) Processes affecting the movement and speciation of mercury in the marine environment. In: Grice D, Reeve MB (eds) Marine mesocosms, biological and chemical research in experimental ecosystems. Springer, Berlin Heidelberg New York, pp 416–479

Tolle DA, Arthur MF, Duke KM, Chesson J (1990) Ecological effects evaluation of two phosphorus smoke-producing compounds using terrestrial microcosms. In: Wang W, Gorsuch JW, Lower WR (eds) Plants for toxicity assessment. American Society for Testing Materials, Philadelphia, pp 127–142

Tournie TP (1984) Microcalorimetric characterization of seasonal metabolic trends in marine microcosms. J Exp Mar Biol Ecol 74(2):111–122

Trevors JT (1988) Use of microcosms to study genetic interactions between microorganisms. Microbiol Sci 5(5):132–136

*Tribbey BA (1965) A field and laboratory study of ecological succession in temporary ponds. (PhD Dissertation) University of Texas, Austin

*Triska FJ, Kenedy VC, Avanzino RJ, Reilly BN (1981) Effect of simulated canopy cover on regulation of nitrate uptake and primary production by natural periphyton assemblages. In: Fontaine T, Baretll SM (eds) Dynamics of Lotic Ecosystems. Ann Arbor Publishing, Ann Arbor, Michigan, pp 129–160

*Tsuchiya HM, Drake JF, Jost JL, Fredrickson AG (1972) Predator–prey interactions of *Dictyostelum* and *Escherichia coli* in continuous culture. J Bacteriol 110:1147–1153

*Turco RP, Toon OB, Ackerman TP, Pollack JB, Sagan C (1983) Nuclear winter: global consequences of multiple nuclear explosions. Science 222:1283–1292

*Tuxen SL (1944) The hot springs of Iceland. Einal Musksgaard, Copenhagen

*Uehlinger U, Bossard P, Bloesch J, Burge HR, Buhrer H (1984) Ecological experiments in limnocorrals: methodological problems and quantification of the epilimnetic phosphorus and carbon cycles. Verh Int Ver Limnol 22:163–171

Uhlmann D (1967) Limnology of saprotrophic shallow ponds. Arch Hydrobiol 63(1):1–85

Uhlmann D (1969) Primary production and decomposition in microecosystems with different proportions of illuminated and dark layers. Arch Hydrobiol 66(2):113–138

Uhlmann D (1971) Abstufungender Fioaktivität Hinitereinander-geschalteten Gewassermodellen. Linmologica (Berlin) 8(2):421–452

Uhlmann D (1971) Influence of dilution, sinking, and grazing rate on phytoplankton populations of hyperfertilized ponds and micro-ecosystems. Mitt Int Ver Limnol 19:100–124

Uhlmann D (1979) BOD Removal rates of waste stabilization ponds as a function of loading, retention time, temperature, and hydraulic flow pattern. Water Res 13:193–200

*Uhlmann D, Cramer F (1975) Equilibrium and stability in laboratory models of sewage ponds and polluted rivers. Int Rev Ges Hydrobiol 60:1–16

*Uhlmann D (1985) Scaling of microcosms and the dimensional analysis of lakes. Int Rev Ges Hydrobiol 70:47–62

Uhlmann D, Mihan H, Kuchta M (1980) Experimentation with plankton ecosystems on the basis of semicontinuous or continuous cultivation methods. (Proc 7th Sym Contin Cultiv Microorganisms) 383–394

Vaccaro R, Azam FF, Hodson RE (1977) Response of natural marine bacterial populations to copper: controlled ecosystem pollution experiment. Bull Mar Sci 27:17–22

*Vanni MJ, Findlay DL (1990) Trophic cascades and phytoplankton community structure. Ecology 71(3):921–937

*Van Voris P, O'Neill RV, Shugart HH, Emanuel WR (1978) Functional complexity and ecosystem stability: an experimental approach. (ORNL/TM-6199 Publ 1123) Oak Ridge National Laboratory, Oak Ridge, Tennessee

*Van Voris P, O'Neill RV, Emanuel WR, Sugart Jr HH (1980) Complexity and ecosystem stability. Ecology 61(6):1352–1360

*Van Voris P, Arthur MF, Tolle DA (1982) Evaluation of terrestrial microcosms for assessing ecological effects of utility wastes. (Research Project Report 1224–5) Electric Power Institute, Palo Alto, California

*Van Voris P, Arthur MF, Tolle DA, Morris JP, Larson M (1983) Use of microcosms for monitoring nutrient-cycling processes in agroecosystems. Univ Georgia Coll Agric Exp Stn Spec Pub 23:171–182

*Van Voris R, Tolle DA, Arthur MF (1985) A method for measuring the potential ecological effects, fate, and transport of chemicals in terrestrial ecosystems. (PNL-5450) Battelle Columbus Laboratories, Columbus, Ohio

Vargo G, Hutchins AM, Almquist G (1982) The effect of low chronic levels of No. 2 fuel oil on natural phytoplankton assemblages in microcosms. I. Species composition and seasonal succession. Mar Environ Res 6:245–264

Vargo SL (1981) The effects of chronic low concentrations of No. 2 fuel oil on the physiology of a temperate estuarine zooplankton community in the MERL microcosms. In: Vernberg FJ, Calabrese A, Thurberg FP, Vernberg WB (eds) Biological monitoring of marine pollutants. Academic press, New York, pp 295–322

*Verduin J (1969) Critique of research methods involving plastic bags in aquatic environments. Trans Am Fish Soc 98:335–336

*Vernadsky V (1929) The biosphere. (Abridged version reprinted 1986) Synergetic Press, London and Oracle, Arizona

*Vernadsky WI (1944) Problems of biogeochemistry. II. The fundamental matter–energy difference between the living and inert natural bodies of the biosphere. Trans Conn Acad Arts Sci 85:488–517

Volva VG, Gitelson II, Sidko FYa, Trubachev IN (1991) Chemolithotrophic hydrogen-oxidizing bacteria and their possible functions in closed ecological life-support systems. Proc 4th European Symposium on Space Environmental and Control Systems, pp 811–816

*Von Brand R, Rakestraw NW (1941) Decomposition and regeneration of nitrogenous organic matter in sea water. Biol Bull Woods Hole 81:63–72

Wade TL, Quinn JG (1980) Incorporation, distribution, and fate of saturated petroleum hydrocarbons in sediments from a controlled marine ecosystem. Mar Environ Res 3:15–33

Waide JB, Schindler JE, Waldron MC, Hains JJ, Schreiner SP, Freedman ML, Benz SL, Pettigrew DR, Schissel LA, Clark PJ (1980) A microcosm approach to the study of biogeochemical systems. 2. Responses of aquatic laboratory microcosms to physical, chemical, and biological perturbations. In: Giesy JP Jr (ed) Microcosms in ecological research. (DOE Symposium Series 52) US Department of Energy, Washington DC, pp 204–223

Wainright SC (1978) Stimulation of heterotrophic microplankton production by resuspended marine sediments. Science 238(4834):1710–1712

Wakeham SG, Davis AC, Goodwin JT (1982) Biogeochemistry of volatile organic compounds in marine experimental ecosystems and the estuarine environment — initial results. In: Grice D, Reeve MB (eds) Marine mesocosms, biological and chemical research in experimental ecosystems. Springer, Berlin Heidelberg New York, pp 137–151

*Wakeham SG, Davis AC, Karas JL (1983) Mesocosm experiments to determine the fate and persistence of volatile organic compounds in coastal seawater. Environ Sci Technol 17:511–617

Walker ED, Poirier SJ, Veldman WT (1987) Effects of *Ascogregarina barretti* Eugregarinida Lecudinidae infection on emergence, success, development time, and size of *Aedes triseriatus* Diptera Culicidae in microcosms and tires. J Med Entomol 24(3):303–309

*Walker JCG (1991) Feedback processes in the biogeochemical cycles of carbon. In: Schneider SH, Boston PJ (eds) Scientists and Gaia. MIT Press, Cambridge, Massachussetts, pp 183–190

Wallace GTJ, Siebert DL, Holzknecht SM, Thomas WH (1982) The biogeochemical fate and toxicity of mercury in controlled experimental ecosystems. Estuar Coastal Shelf Sci 15:151–182

Walsh SM (1988) Culturability and repair of chlorine-injured *Escherichia coli* in an aquatic microcosm (abstract). Abstr Annu Meet Am Soc Microbiol 88:298

Walter DE (1987) Life history, trophic behavior and description of *Gamasellodes vermivorax* new species Mesostigmata Ascidae, a predator of nematodes and arthropods in semiarid grassland soils. Can J Zool 65(7):1689–1695

Walter DE, Kethley J, Moore JC (1987) A heptane flotation method for recovering microarthropods from semiarid soils with comparison to the Merchant–Crossley high-gradient extraction method and estimates of microarthropod biomass. Pedobiologia 30(4):221–232

Walter MV, Porteous VA, Seidler RJ (1987) Formation of false positive transconjugants following in vitro conjugation experiments (abstract). Abstr Annu Meeting Am Soc Microbiol 87:302

Walter MV, Porteous VA, Prince VJ, Ganio L, Seidler RJ (1991) A microcosm for measuring survival and conjugation of genetically engineered bacteria in rhizosphere environments. Curr Microbiol 22(2):117–122

Walter-Echols G, Lichtenstein EP (1977) Microbial reduction of phorate sulfoxide in a soil–lake mud–water microcosm. J Econ Entomol 70:505–509

Walter-Echols G, Lichtenstein EP (1978) Movement and metabolism of ^{14}C-phorate in a flooded soil system. J Agric Food Chem 26:559–604

Walton WE, Tietze NS, Mulla MS (1990) Ecology of *Culex tarsalis* (Diptera, Culicidae): factors influencing larval abundance in mesocosms in southern California, USA. J Med Entomol 27(1):57–67

*Ward CH, Wilks SS, Craft HL (1963) Use of algae and other plants in the development of life support systems. Am Biol Teacher 25:512–521
*Warington R (1857) On the aquarium. Notices Proc Royal Inst 2:403–408
Warren CE (1963) Ecological studies of an experimental stream. (Progress Report USPHS) US Public Health Service, Washington DC
*Warren CE, Davis GE (1971) Laboratory stream research: objectives, possibilities, and constraints. Ann Rev Ecol Syst 2:111–144
*Warren CE, Phinney HK, Davis GE, McIntire DC (1962) Studies of the trophic dynamics of simplified communities in artificial streams. (Progress Report to NSF) Department of Fish and Game and Department of Botany, Oregon State University, Corvallis, Oregon
Warren CE, Wales JH, Davis GE, Doudoroff P (1964) Trout production in an experimental stream enriched with sucrose. J Wildlife Manage 28:617–660
Watanabe M, Kohata K, Kunugi M (1988) Phosphate accumulation and metabolism by *Heterosigma akashiwo* Raphidophyceae during diel vertical migration in a stratified microcosm. J Phycol 24(1):22–28
*Watts JR, Harvey RS (1963) Uptake and retention of cesium-137 by a blue-green algae in continuous flow and batch culture systems. Limnol Oceanogr 8:45–49
*Weaver PL (1970) Species diversity and ecology of tidepool fishes in three Pacific coastal areas of Costa Rica. Rev Biol Trop 17(2):165–185
Weinberg JR (1983) Enhanced growth of a filter-feeding bivalve by means of nutrient regeneration. J Mar Res 41:557–570
Weinberger P, Greenhalgh R, Moody RP, Boulton B (1982) Fate of fenitrothion in aquatic microcosms and the role of aquatic plants. Environ Sci Technol 16:470–473
Werner W, Ellenberg H, Stickan W (1983) Effects and distribution of pollutants in a terrestrial-ecosystem model. Verh Ges Oekol 11:425–436
White DS, Klahr PC, Robbins JA (1987) Effects of temperature and density on sediment reworking by *Stylodrilus heringianus* Oligochaeta Lumbriculidae. J Great Lakes Res 13(2):147–156
*Whitford LA (1960) The current effect and growth of freshwater algae. Trans Am Microscop Soc 79:302–309
*Whitford LA, Schumacher GJ (1961) Effect of current on mineral uptake and respiration by a fresh-water alga. Limnol Oceanogr 6:423–425
*Whitford LA, Dillard GE, Schumacher GJ (1964) An artificial stream apparatus for the study of lotic organisms. Limnol Oceaongr 9:598–600
*Whittaker RH (1961) Experiments with radiophosphorus tracer in aquarium microcosms. Ecol Monogr 31(2):157–188
*Wiegert RG, Fraleigh PC (1972) Ecology of Yellowstone thermal effluent systems: net primary production and species diversity of a successional blue-green algal mat. Limnol Oceanogr 17:215–228
*Wiegert RG, Mitchell R (1973) Ecology of Yellowstone thermal effluent systems: intersects of blue-green algae, grazing flies (*Paracoenia*, Ephydridae), and water mites (*Partnuniella* Hydrachnellae). Hydrobiologia 41:251–271
*Wiener N (1948) Cybernetics. Wiley, New York
Wilhm JL (1970) Transfer of radioisotopes between detritus and benthic macroinvertebrates in laboratory microecosystems. Health Physics 18:277–284

Wilhm JL, Long J (1969) Succession in algal mat communities at three nutrient levels. Ecology 50(4):645–652
*Wilks SS (1962) Preliminary report on the photosynthetic gas exchange potentialities of the family Lemnaceae (Duckweed). (Biologists for Space Systems Symposium, USAF Report AMRL-TDR-62-116) Wright Patterson Air Force Base, Ohio, pp 265–278
Williams DR, Giesy JP Jr (1978) Relative importance of food and water sources to cadmium uptake by *Gambusia affinis* (Poeciliidae). Environ Res 16:326–332
Williams IP, Gibson VR, Smith WK (1977) Horizontal distribution of pumped zooplankton during a controlled ecosystem pollution experiment: implications for sampling strategy in large volume, enclosed water columns. Mar Sci Comm 3:239–253
*Williams LL, Mount DI (1965) Influence of zinc on periphytic communities. Am J Bot 52(1):26–34
*Williams SL, Adey WH (1983) *Thalassia testudinum* seedling success in a coral reef microcosm. Aquat Bot 16(2):181–188
*Wilson RF (1963) Studies of organic matter in aquatic ecosystems. (PhD Dissertation) University of Texas, Austin
*Wilson WH Jr (1984) An experimental analysis of spatial competition in a dense infaunal community: the importance of relative effects. Estuar Coastal Shelf Sci 18:673–684
*Wilson MV, Botkin D (1990) Models of simple microcosms: emergent properties and effect of complexity on stability. Amer Naturalist 135:414–434
*Wimpenny JWT (1988) Introduction. In: Wimpenny JWT (ed) CRC handbook of laboratory model systems for microbial ecosystems. CRC Press, Boca Raton, Florida, pp 1–18
*Winkler LW (1888) Die Bestimmung des in Wasser geloston Saurstoffs. Ber Deutsch Chem Ges 21:2843–2845
*Winogradsky S (1949) Microbiologie du sol: problemes et methodes. Masson, Paris
*Witkamp M (1972) Transfer of ^{137}Cs from detritus to primary producer. (Isotopes and Radiation in Soil-Plant Relationships Including Forestry, Publication No 461) Oak Ridge National Laboratory, Oak Ridge, Tennessee, pp 341–348
*Witkamp M, Frank ML (1967) Cesium-137 kinetics in terrestrial microcosms. (Symposium on Radioecology) US Atomic Energy Commission, Oak Ridge, Tennessee TID-4500 (CONF-670503), pp 635–643
*Witkamp M, Frank ML (1970) Effects of temperature, rainfall, and fauna on transfer of ^{137}CS, K, Mg, and mass in consumer–decomposer microcosms. Ecology 51(3):465–474
Witherspoon JP, Bondietti EA, Draggan S, Taub FP, Pearson N, Trabalka JR (1976) State-of-the-art and proposed testing for environmental transport of toxic substances. (ORNL/EPA-1) Oak Ridge National Laboratory, Oak Ridge, Tennessee
*Wolfe J (1981) A computer simulation model of the solar–algae pond ecosystem. New Alchemy Institute Falmouth, Massachusetts
*Wolfe J, Zweig RD, Engstrom DG (1986) A computer simulation model of the solar–algae pond ecosystem. (Report to Appropriate Technology Program of the NSF Project ISP 8016577) New Alchemy Institute, Falmouth, Massachusetts

*Wolverton BC (1975) Aquatic plants for removal of mevinphos from the aquatic environment. (TM-X-72720 NSTL) NASA National Space Technology Labs, Bay St. Louis, Mississippi

*Wolverton BC (1979) Engineering design data for small vascular aquatic plant wastewater treatment systems. In: Aquaculture Systems for Wastewater Treatment. US Environmental Protection Agency, Washington DC, pp 179–189

*Wolverton BC (1980) Higher plants for recycling human waste into food, potable water, and revitalized air in a closed life support system. (ERL Report No 192 NSTL) NASA National Space Technology Labs, Bay St. Louis, Mississippi

*Wolverton BC (1982) Hybrid wastewater treatment system using anaerobic microorganisms and reed (*Phragmites communis*). Econ Bot 36(4):373–380

*Wolverton BC (1984) A review of aquaculture wastewater treatment systems. In: Carlile BL (ed) Workshop on low cost wastewater treatment systems for communities and municipalities. Texas Agricultural Extension Service, Texas A & M University College Station, Texas

*Wolverton BC (1987a) Artificial marshes for wastewater treatment. In: Reddy KA, Smith WH (eds) Aquatic plants for water treatment and resource recovery. Magnolia Publications, Orlando, Florida, pp 141–152

*Wolverton BC (1989) Aquatic plant–microbial filters for treating septic tank effluent. In: Hammar DA (ed) Constructed wetlands for waste water treatment. Lewis Publishers, Chelsea, Michigan, pp 173–178

*Wolverton BC, Barlow RM, McDonald RC (1976) Application of vascular aquatic plants for pollution removal, energy, and food production in a biological system. In: Tourbier J, Person RW (eds) Biological control of water pollution. University of Pennsylvania Press, University of Pennsylvania

*Wolverton BC, McCaleb RC (1987) Pennywort and duckweed marsh system for upgrading wastewater effluent from a mechanical package plant. In: aquatic plants for waste and resource recovery. Magnolia Publications Orlando, Florida, pp 289–294

*Wolverton BC, McCaleb RC, Douglas WL (1989) Bioregenerative space and terrestrial habitat. (9th Biennial Princeton Conference on Space Manufacturing) Space Studies Institute, Princeton, New Jersey

*Wolverton BC, McDonald RC (1975) Water hyacinths for upgrading sewage lagoons to meet advanced wastewater treatment standards, Part I. (TM-X-72729 NSTL 39529) NASA National Space Technology Labs, Bay St. Louis, Mississippi

*Wolverton BC, McDonald RC (1976) Water hyacinths for upgrading sewage lagoons to meet advanced wastewater treatment standards, Part II. (TM-X-72730 NSTL 39529) NASA National Space Technology Labs, Bay St. Louis, Mississippi

*Wolverton BC, McDonald RC (1979a) Upgrading facultative wastewater lagoons with vascular aquatic plants. J Water Pollut Control Fed 51(2):305–313

*Wolverton BC, McDonald RC (1979b) Water hyacinth (*Eichornia crassipes*) productivity and harvesting studies. Econ Bot 33(1):1–10

*Wolverton BC, McDonald RC (1979c) The water hyacinth: from prolific pest to potential provider. Ambio 8(1):2–9

*Wolverton BC, McDonald RC (1981a) Natural processes for treatment of organic chemical waste. Environ Profess 3:99–104

*Wolverton BC, McDonald RC (1981b) Energy from vascular plant wastewater treatment systems. Econ Bot 35(2):224–232
*Wolverton BC, McDonald RC (1982a) Basic engineering criteria and cost estimations for hybrid microbial filter-reed (*Phragmites communis*) waste-water treatment concept. (TM-84669 NSTL 39529) NASA National Space Technology Labs, Bay St. Louis, Mississippi
*Wolverton BC, McDonald RC (1982b) Foliage plants for removing formaldehyde from contaminated air inside energy-efficient homes and future space stations. (TM-84674 NSTL 39529) NASA National Space Technology Labs, Bay St. Louis, Mississippi
*Wolverton BC, McDonald RC (1982c) The role of vascular aquatic plants in wastewater treatments. Herbalist 4:24–29
*Wolverton BC, McDonald RC, Duffer WR (1983) Microorganisms and higher plants for waste water treatment. J Environ Qual 12(2):236–242
*Wolverton BC, McDonald RC, Marble (1984a) Removal of benzene and its derivatives from polluted water using the reed/microbial filter technique. J Miss Acad Sci 29:119–127
*Wolverton BC, McDonald RC, Watkins EA (1984b) Foliage plants for removing indoor air pollutants from energy-efficient homes. Econ Bot 38(2):224–228
*Wolverton BC, McDonald-McCaleb RC (1986) Biotransformation of priority pollutants using biofilms and vascular plants. J Miss Acad Sci 31:79–88
*Wolverton BC, McKown (1976) Water hyacinths for removal of phenols from polluted waters. Aquat Bot 2:191–201
*Wolverton BC, Myrick CC, Johnson KM (1984c) Upgrading septic tanks using microbial/plant filters. J Miss Acad Sci 29:19–25
*Wolverton BC, Richter R, Lefstad SM (1971) Orthophosphorus insecticide decontaminant. In: Technology utilization ideas for the 70s and beyond, vol 26. Science and Technology Am Astronautical Society, Tarzana, California, pp 7–13
*Woodruff LL (1912) Observations on the origin and sequence of the protozoan fauna of hay infusions. J Exp Zool 1:205–264
Word JQ, Hardy JT, Crecelius EA, Kiesser SL (1987) A laboratory study of the accumulation and toxicity of contaminants at the sea surface from sediments proposed for dredging. Mar Environ Res 23(4):325–338
Wrenn WB, Grannemann KL (1980) Effects of temperature on bluegill reproduction and young-of-the-year standing stocks in experimental ecosystems. In: Giesy JP Jr (ed) Microcosms in ecological research. (DOE Symposium Series 52) US Department of Energy, Washington DC, pp 703–714
*Wulff BL, McIntire CD (1972) Laboratory studies of assemblages of attached estuarine diatoms. Limnol Oceanogr 17(2):200–214
*Wunder CC (1966) Life into space. FA Davis, Philadelphia
*Yensen NP (1988) Ecosystems in glass. Carolina Tips 51(4):13–15
Yochida K, Sigeoka T, Yamauchi F (1988) Evaluation of predictability of an aquatic environmental fate model by using an aquatic microcosm. Chemosphere 17(10):2063–2072
*Yount JL (1956) Factors that control species numbers in Silver Springs, Florida. Limnol Oceanogr 1:286–295
Yount JR, Shannon LJ (1987) Effects of aniline and three derivatives on laboratory microecosystems. Environ Toxicol Chem 6:463–468

Yousef YA, Padden TJ, Gloyna EF (1975) Diurnal changes in radionuclides uptake by phytoplankton in small scale ecosystems. Water Res 9:181–187

*Zedler PH (1987) The ecology of southern California vernal pools: a community profile. (Biol Rept 7.11) US Fish and Wildlife Service, Washington DC

*Zeitzschel B, Davies JM (1978) Benthic growth chambers. Rapp P-V Reun Cons Int Explor Mer 173:31–42

*Zieris FJ, Feind D, Huber W (1988) The split pond: one means of getting comparable model ecosystems. Z Wasser Abwasser Forsch 21(1):7–10

Zimmerman TL, Felder DL (1991) Reproductive ecology of an intertidal Brachyuran crab *Sesarma* sp NE. *Sesarma reticulatum* from the Gulf of Mexico. Biol Bull Woods Hole 181(3):387–401

*Zweig RD, Wolfe JR, Todd JH, Engstrom DG, Doolittle AM (1981) Solar aquaculture: an ecological approach to human food production Bioengineering Symposium for Fish Culture. (FCS Publ 1) 1:210–226

*Zweig RD, Engstrom DG (1982) Solar aquaculture: perspectives in renewable resource-based fish production. New Alchemy Institute, Falmouth, MA 151 pp

Index

Acid effects on terrestrial microcosms, 254–256
Activated sludge ecosystem, 379, 382, 383
Aerobic soils, 234
Aggregation, linear, of small-scale components, 70–71
Agriculture, ecological engineering and, 357–358
Air, microcosms open to, 23–24
Algal culture, 363
Algal dominated aquaria, 201–202
Algal mats, hypersaline, 348, 350–352
Algal photosynthesis, sewage ponds with, 384–387
Algal turf scrubber, 300–301
Anaerobic soils, 260
and sediment microcosms, 311–313
Animal activity, soil function and, 259
Aquaculture, 363–369
solar heated fish-pool, 369–375
Aquaria, 191–208
algal dominated, 201–202
balanced, 178–179
filtered, 202–203
phosphorus movement in, 193–195
strategies for maintaining, 191–193
Aquatic leaf litter microcosms, 281
Aquatic microcosm, standardized (SAM), 134–140
Aquatic model, with oxygen, 21–23

Aquatic plants, larger, waste microcosms with, 387–389
Aquatron marine mesocosm, 324
Arctic tundra microcosms, carbon dioxide and, 244, 245
Arrested succession, 213
Arsenic, 171
Artificial chaos, 126
Artificial marshes, 266, 267
Artificial reefs, 300
Artificial substrates for water quality studies, 300
Assemblages, synthetic versus natural, 7–8
Attractor shifts, 126, 128
Aufwuchs, 209
Autocatalytic consumption and production, 31–34
Autocatalytic plant growth, 31
Autotrophic succession, 56

Bacterial beds of trickling filters, 377–379, 380, 381
Balanced aquaria, 178–179
Balcones model river
diagram of, 210
tracer studies in, 214–216
Beach sand microcosms, 314–315
Benthic pore space pH, tubificid control of, 311

Benthic sediment microcosms, 306–321
Benthic systems, herbivory in, 197, 199–201
Bernard Cell, 62–64
Beyers' standardized successional-eutrophic microecosystem, 58–61
Bibliography, 481–545
Bifurcations, 126
Biochemical oxygen demand (BOD) bottles, 14
Biogeochemical cycling, 10, 62
Biogeochemistry, unusual, 344–345
Biomass, diversity and, 152
Biomass diversity model, 149, 151
Biosphere 2, 414–419
Biosphere 2 test module
 carbon dioxide in, 475, 477–479
 description of, 463–467
 ethyl benzene in, 475
 experimental on closed ecological system in, 463–479
 floor plan, 468
 human closure experiments, 467
 life systems, 466–467
 methane in, 472
 monitoring system, 465–466
 nitrate levels in irrigation water in, 474
 nitrate levels in waste recycling system in, 473
 nitrogenous gases in, 471
 ozone in, 471
 phosphate levels in irrigation water in, 474
 phosphate levels in waste recycling system in, 473
 relative humidity in, 470
 results, 467–479
 solar insulation in, 469
 sulfur dioxide in, 472
 technogenic outgassing, 479
 temperature in, 470
 tetrahydrofuran in, 476
 toluene in, 476
 water systems, 466
Biostats, 138
BOD (biochemical oxygen demand) bottles, 14
Bog, miniature, succession in, 260
Bottom-up models, 112
Boundaries, ecosystem, 4
Boundary conditions, 6

Brine microcosms, *see* Thermal and brine microcosms

Cadmium, 163, 173
 in glass tube microcosms, 230–233
 in terrestrial microcosms, 256
Carbon dioxide
 Arctic tundra microcosms and, 244, 245
 in Biosphere 2 test module, 475, 477–479
 metabolism, 238–240
  determined from pH changes, 455–459
Carboy microcosms, marine, 195, 196, 197
Carrier flow effect on tracer flow, 79
Carrying capacity, 427–431
Catastrophic stresses, 58
Causal minimodel, mineral cycle overlay for, 66–67
CELSS (Controlled Ecological Life Support System) program, 409, 412
Chaos, 121
 artificial, 126
Chaotic models, 126
Chaotic oscillations, 121–128
Charge-up curves, 92
Chemical cycling, 62
 ecosystem organization and, 244, 246
 element ratios in, 80–84
 steady-state, 70
Chemical effects, in synthetic standardized microcosms, 134–140
Chemicals, dangerous, microcosm apparatus for working with, 206
Chemostats, 138, 141, 142
 compound, 143–147
 flushing rate and diversity, 152
Chlorophyll quantities, 226–227
Classroom microcosms
 directions for constructing, 445–451
 directions for monitoring, 452–462
Climate-controlled chambers, 399
Climax
 classical concept of, 42–43
 patches of, 104
Climax eutrophy, 101–102
Closed cycle homeostasis, 67–68
Closed terrestrial ecosystems with

humans, simulation model for, 422–427
Closed-to-matter microcosms, 64
Closed-to-nutrient systems, 101
Compartmental diagrams, 64, 65
Compartments, within dynamic models, 68
Competition, 116
Complex self-organized mesoecosystem, 419–431
Compound microcosms and chemostats, 143–147
Connected unit processes, 406–412
Conservation of material, 68–69
Consumer reef microcosms, 295–299
Consumers, succession of, 47
Consumption
  accelerated by external energy source, 35–37
  autocatalytic, 31–34
  linear, production models with, 15–17
  pulses of, 104
  showing detail in, 38–40
Consumption rates, phase plane graphs of, 18, 19
Controlled Ecological Life Support System (CELSS) program, 409, 412
Controversy over human life support, 420
Copper, 175
Coral reef ecosystems, hierarchical control of, 302–305
Coral reef microcosms, ecological engineering of, 300–301
Crop microcosms, 358–363
Cross boundary influences, 12
Cross seeding, 9
Current velocity, self-organization and, 217–218

Denitrification, 260
Destructive pulses, 148
Diagram symbols, energy, 12–14
Dieldren, 167
Directions
  for classroom microcosms, 445–451
  for monitoring microcosms, 452–462
Disruption, 154
Dissected microcosms, 146
Dissemules, 8

Dissolved oxygen measurements, 452–454
Diurnal metabolism, 25
Diurnal oxygen curve, 182, 184
Diurnal rates of photosynthesis and respiration, 19–21
Diversity, 88, 90
  as balance of information inflow and extinction, 92–96
  biomass and, 152
  biomass diversity model, 149, 151
  chemostat flushing rate and, 152
  instability and, 106–107
  low, physiological adaptation and, 341–342
  minimodel of, 176, 177
  model of energy for maintaining, 96–98
  nutrients and longitudinal succession and, 101–104
  self-reinforcing, 91
  Shannon-Weaver index of, 102
  stress as generator of, 153–154
  toxic stress and reduction of, 166–167
  uniformly distributed stress and, 154
Dormant organisms, 49
Duke stream microcosms, 218–219
Dynamic models, 69–70
  compartments within, 68

Ecological education, perspectives on, 437
Ecological engineering
  agriculture and, 357–358
  of coral reef microcosms, 300–301
Ecological hierarchies, transformity and position in, 165–166
Ecological microcosms, *see* Microcosms
Ecological web, 39–40
Ecology, microcosms for teaching, 10
Ecosystem boundaries, 4
Ecosystem immunology, 112
Ecosystem organization, chemical cycling and, 244, 246
Ecosystems, 4–6, 88
  gnotobiotic, 8
  influences on, 104–106
  microbial decomposition, 235–240
  sustainable, perspectives on, 440–442
Ecotechnic mesocosms, 412–414

Education, ecological, perspectives on, 437
Electro-microbial system, 412, 413
Element ratios, in chemical cycles, 80–84
Emergy, 90
  energy and stress and, 155
  per unit energy, 106
  term, 57
Emergy requirements for human life support, 430
Emjoules, 12, 57
Empower, 103
Enclosure, transitions following, 51–55
Energy
  Emergy and stress and, 155
  model of, for maintaining diversity, 96–98
Energy diagram symbols, 12–14
Energy regimes, 56–58
Energy source, external, consumption accelerated by, 35–37
Estuaries, transitions in, 53–54
Estuarine bottom communities microcosm with salinity gradient, 310
Estuarine microcosms, treated with sewage, 197, 199
Estuarine ponds receiving treated domestic sewage, 287–293
Eutrophication experiments, in ponds, 272–273
Eutrophication in microcosms, 50–51
Eutrophy, climax, 101–102
Everglades mesocosm, Smithsonian, 270, 271
Experiments on closed ecological system in Biosphere 2 test module, 463–479
External stress, 148–152
Extinction, 101
  diversity as balance of information inflow and, 92–96

Filtered aquaria, 202–203
Filters, trickling, bacterial beds of, 377–379, 380, 381
Fish consumers, uncontrolled, over-consumption by, 273
Fish growth, pool microcosms relating nutrient concentration to, 273, 275–276

Fish hatchery ponds, 368–369
Fish-pool aquaculture, solar heated, 369–375
Floating field microcosms for benthic communities, 309–310
Flowing saltwater microcosms, 233
Flowing waters, 209
Flows, valve-controlled, 143
Flushing rate, chemostat, diversity and, 152
Fly-ash microcosms, 392–393, 394
Food chains, 118–119
Food microcosms, 357–375
  algal culture, 363
  aquaculture, 363–369
  ecological engineering and agriculture, 357–358
  technology mesocosms, 369–375
Food production, 357
Fouling surfaces, 297
Freshwater succession with multiple seeding, 202
Frictional devices, 56

Gaseous exchange, terrestrial microcosms with, 27–29
Genetic information, 156
Genetically altered organisms, 442
Glass tube microcosms, cadmium in, 230–233
Global homeostasis, perspectives on, 442–444
Global policy, perspectives on, 438–440
Gnotobiotic ecosystems, 8
Great Salt Lake degradation microcosms, 351–353
Ground-water decomposition microcosm, 393, 395–396

Heavy metal elements
  macroinvertebrates and partitioning of, 318
  organism response to, 171–177
  self-organization with, in Savannah River mesocosms, 229–230
  toxic, 168
  in wetland microcosms, 267
Herbivory
  in algal dominated aquaria, 201–202
  in benthic systems, 197, 199–201

Heterotrophic succession, 56
Hierarchical control
    of coral reef ecosystems, 302–305
    transformity and, 233
Hierarchical networks, 108
Hierarchy
    ecological, transformity and position in, 165–166
    models and, 112–113
    in space and time, 108–112
High metabolism microcosms, 382–383
Homeostasis, 11
    closed cycle, 67–68
    global, perspectives on, 442–444
    of terrestrial microcosms, 242–243
Human closure experiments, 467
Human life support
    controversy over, 420
    emergy requirements for, 430
Human microcosms, 397–432
    Biosphere 2, 414–419
    climate-controlled chambers, 399
    complex self-organized mesoecosystem, 419–431
    connected unit processes, 406–412
    ecotechnic mesocosms, 412–414
    electro-microbial system, 412, 413
    simple symbiosis, 401–406
    stored supplies from earth, 399–401
Hutchinsonian "paradox of the plankton" concept, 153
Hypersaline algal mats, 348, 350–352
Hyperspace, 87
    state, 87

Iceland hot streams, 345–346
Inferences, larger, scaling microcosms for, 7
Inflows, dominant, microcosms with, 31
Influences on ecosystems, 104–106
Information
    defined, 88
    genetic, 156
Information inflow, diversity as balance of extinction and, 92–96
Inhibition, in microcosms, 129–134
Initial conditions, 6
Initial storages, large, effect of, 29–31
Instability, diversity and, 106–107
Interactions, productive, 72
Ionizing radiation, 156–160

Isotope processing, small differences in, 80

Lake simulators, microcosms as, 203–206
Lakes, new, role of sediment in, 316–318
Landfill microcosm, 392
Lateral organization, exchange of excess material resources in, 85–87
Leaf litter microcosms, aquatic, 281
Light intensity, varying, self-organization and, 226–227
Limiting factors, to production processes, 72, 73, 74
Linear aggregation, of small-scale components, 70–71
Linear materials model, for non-linear systems minimodel, 65, 66
Linear mathematics, 70
Linear models, 69–70
Logistic model, 123, 126
Longitudinal succession, 103
    diversity and nutrients and, 101–104
    in stream microcosms, 212, 213
Lotic environments, 209
Lotka theory, 67–68

Macroinvertebrates
    heavy metal partitioning and, 318
    pollutant transport and, 318–320
Manganese, 169
Marine carboy microcosms, 195, 196, 197
Marine coastal environments, simulation of, 283, 284, 285
Marine mesocosm, Aquatron, 324
Marshes, artificial, 266, 267
Material, conservation of, 68–69
Material cycles, in self-organizing systems, 62–64
Material resources, exchange of excess, in lateral organization, 85–87
Materials model, linear, for non-linear systems minimodel, 65, 66
Mathematics, linear, 70
Maximum power, self-organization and, 84–85
Maximum power principle, 7
Maximum power theory, 91

Mercury, 167
MERL (Microcosm Estuarine Research Laboratory) microcosm, 323, 328–329, 330–337
Mesocosm, 6
  plankton column, *see* Plankton columns
  rocky shore, 305–306
  Smithsonian Everglades, 270, 271
Metabolic equation, 9
Metabolism
  carbon dioxide, 238–240
    determined from pH changes, 455–459
  community, 343
    in pools, 283–286
  diurnal, 25
  of microcosm, 460, 461–462
  microcosm oxygen, 454, 455
  of salt marsh tidal microcosm, 264–266
  of transplanted salt marsh, 261, 263
Metals, heavy, *see* Heavy metal elements
Michaelis-Menten model, 17
Michaelis-Menten algebra, 133
Michigan State laboratory streams, 224–225
Microbial decomposition ecosystem, 235–240
Microbial degradation of xenobiotics, 320–321
Microcosm Estuarine Research Laboratory (MERL) microcosm, 323, 328–329, 330–337
Microcosm oxygen metabolism, 454, 455
Microcosmic theories, 179–181
Microcosms
  aquatic, *see* Aquaria
  aquatic leaf litter, 281
  Arctic tundra, carbon dioxide and, 244, 245
  benthic sediment, 306–321
  brine, *see* Thermal and brine microcosms
  classroom, *see* Classroom microcosms
  closed-to-matter, 64
  compound, 143–147
  concepts in philosophy, 179–181
  concepts and uses, 1
  consumer reef, 295–299
  crop, 358–363
  defined, 3
  dissected, 146
  with dominant inflows and outflows, 31
  Duke stream, 218–219
  estuarine, treated with sewage, 197, 199
  eutrophication in, 50–51
  flowing saltwater, 233
  fly-ash, 392–393, 394
  food, *see* Food microcosms
  glass tube, cadmium in, 230–233
  ground-water decomposition, 393, 395–396
  high metabolism, 382–383
  history of, 178–187
  human, *see* Human microcosms
  inhibition in, 129–134
  introduction to, 3–10
  kinds of, 189
  as lake simulators, 203–206
  landfill, 392
  marine carboy, 195, 196, 197
  metabolism of, 460, 461–462
  natural, 8
  Neuston, 207, 208
  nutrient enrichment, 195–196, 198
  ocean floor microbial, 308, 309
  open to air, 23–24
  open-to-matter, 64
  Oregon stream, 211, 220–224
  perspectives, 433–444
  perspectives on ecological education, 437
  perspectives on global homeostasis, 442–444
  perspectives on global policy, 438–440
  perspectives on principles of self-organization, 434–435
  perspectives on scaling results from test microcosms, 436–437
  perspectives on science and microcosm uses, 435–436
  perspectives on sustainable ecosystems, 440–442
  pond type, 445–448
  practical uses of, 10
  rate graphs for representing performance of, 17–19
  relaxation, 14–15
  rock outcrop, 246–252

salt pan, 353, 354
scaling, for larger inferences, 7
science and, 181–186
sensitivity of organisms versus, 161–162
sewage, 377–389
for society, 355–356
soil, *see* Terrestrial microcosms
soil depression, 249
solid waste, 390, 392
split-level, 143–147
standardized, 58
standardized aquatic (SAM), 134–140
stream *see* Stream microcosms
teaching ecology with, 10
temperature role in, 37–38
terrestrial, *see* Terrestrial microcosms
test, scaling results from, 436–437
thermal, *see* Thermal and brine microcosms
toxicity tests with organisms versus those with, 161–162
turf, 358–363
variability in, 6–7
variety of, 5
waste, *see* Waste microcosms
water surface, 202
water treatment, 395, 396
wetland, *see* Wetland microcosms
Mimbres hot springs, 347–348, 349
Mineral cycle overlay for causal minimodel, 66–67
Mineral cycles, added to successional minimodel, 84
Mineral mobilization from sediments, 205–206
Models, 11–12
  bottom-up, 112
  chaotic, 126
  diagrams and equations for representing, 12–14
  dynamic, *see* Dynamic models
  of energy for maintaining diversity, 96–98
  hierarchy and, 112–113
  linear, 69–70
  production, with linear consumption, 15–17
  stream hierarchy and, 227–228
  successional, *see* Successional minimodel
  top-down, 113
Monads, 181
Multiple seeding, *see* Seeding, multiple

Natural assemblages, 7–8
Natural microcosms, 8
Net production, 101, 103
Network analysis, 91
Network mutualism, 434
Networks, hierarchical, 108
Neuston microcosms, 207, 208
Nickel, in terrestrial microcosms, 256
Nitrate reduction, 260
Noise events, 149
Non-linear systems minimodel, linear materials model for, 65, 66
Nutrient concentration, pool microcosms relating fish growth to, 273, 275–276
Nutrient control, 50
Nutrient cycles and limits, 17
Nutrient enrichment microcosms, 195–196, 198
Nutrient flows, of salt marsh tidal microcosm, 264–266
Nutrient spiralling, turnover length and, 213–214
Nutrients
  diversity and longitudinal succession and, 101–104
  effect of excess of, 74–75

Ocean floor microbial microcosm, 308, 309
Oil-drilling mud and turtle grass benthic core microcosms, 307
Open-to-matter microcosms, 64
Oregon stream microcosm, 211, 220–224
Organic compounds
  pond mesocosms treated with, 276–281
  processing of volatile, 338
  in terrestrial microcosms, 256–258
Organic storage, labile and dead, 25–26
Organic toxicity, 167
Organisms
  genetically altered, 442
  heavy metal response of, 171–177
  sensitivity of microcosms versus, 161–162
  size of, 105–106

toxicity tests with microcosms versus
those with, 161–162
Oscillations
  chaotic, 121–128
  population, 113–121
  prey-predator, 116
Outflows, dominant, microcosms with, 31
Outgassing, technogenic, 479
Outside sources, limitations on, 14
Over-consumption by uncontrolled fish consumers, 273
Overgrazing, 117–118
Oxygen, aquatic model with, 21–23
Oxygen curve, diurnal, 182, 184
Oxygen measurements, dissolved, 452–454
Oxygen metabolism, microcosm, 454, 455
Oxygen sag curves, 216
Oxygen saturation, 454

Patches of succession and climax, 104
Periphyton, 209
Pesticide transport, 360–363
Pesticides, in wetland microcosms, 267–269
pH changes, carbon dioxide metabolism determined from, 455–459
pH conditions, 254–256
Phase plane graphs, 18, 19
Philosophy, microcosm concepts in, 179–181
Phosphate mining wastewaters in estuaries, microcosms simulating, 389
Phosphorus movement in aquaria, 193–195
Photoelectric ecosystem, 350
Photorespiration, 35–37
Photosynthesis
  algal, sewage ponds with, 384–387
  diurnal rates of, 19–21
  radiocarbon uptake measurements of, 79–80
Photosynthetic photon flux density (PPFD), 478
Plankton columns, 322–340
  bags suspended in waters, 325, 326–328
  control tanks, 329, 332–333
  effects of varying physical
  turbulence, 333–334
  eutrophication series, 338–339
  fuel oil experiment, 334, 336
  long rigid cylinders, 328
  predation, 339
  processing of volatile organic compounds, 338
  radioactive tracer studies of chemical cycles, 338
  simulated storm, 338
  stratification experiments, 340
  time scales and sizes, 323–325
Plankton ecosystem in fertilized ponds, 273, 274
Plant growth, autocatalytic, 31
Pollutant transport, macroinvertebrates and, 318–320
Pond mesocosms treated with organic compounds, 276–281
Pond type microcosm, 445–448
Ponds, 272
  estuarine, receiving treated domestic sewage, 287–293
  eutrophication experiments in, 272–273
  fertilized, plankton ecosystem in, 273, 274
  fish hatchery, 368–369
  sewage, with algal photosynthesis, 384–387
  shrimp, 363–368
  waste stabilization, 384
Pool microcosms, fish growth and nutrient concentration and, 273, 275–276
Pools
  community metabolism in, 283–286
  natural vernal, succession in, 281–283
  as teaching tools, 286–287
Population effects, in synthetic standardized microcosms, 134–140
Population oscillations, 113–121
Population surges, 119
PPFD (photosynthetic photon flux density), 478
Prey-predator coupling, 141
Prey-predator oscillations, 116
Producers, succession of, 47
Product index, summation index versus, 87
Production
  autocatalytic, 31–34

food, 357
net, 101, 103
showing detail in, 38–40
Production-consumption models, 15–17
for terrestrial microcosms, 26–29
Production-consumption pattern, generalized, 19
Production models, with linear consumption, 15–17
Production processes, limiting factors to, 72, 73, 74
Production rates, phase plane graphs of, 18, 19
Productive interactions, 72
Productivity, 101
tight-cycle, 101
Propagules, 8
Pulses
of consumption, 104
destructive, 148

Quadratic denominators, 133

Radiation, ionizing, 156–160
Radioactive tracers, see Tracer entries
Radiocarbon uptake measurements of photosynthesis, 79–80
Rate graphs for representing microcosm performance, 17–19
Recapitulation theory, 180–181
Recirculating sea water through sand beds, 315–316
Recovery, measured, 170
Reefs, 294–306
artificial, 300
Refinery waste systems, 389–390, 391
Relaxation microcosms, 14–15
Respiration, diurnal rates of, 19–21
Rock outcrop microcosms, 246–252
Rocky shore mesocosm, 305–306

Salinity gradient, estuarine bottom communities microcosm with, 310
Salt marsh, transplanted, metabolism of, 261, 263
Salt marsh tidal microcosm, nutrient flows and metabolism of, 264–266

Salt pan microcosms, 353, 354
Saltwater microcosms, flowing, 233
SAM (standardized aquatic microcosm), 134–140
Sand beds, recirculating sea water through, 315–316
Saprobe series, 212
Savannah River mesocosms, self-organization with heavy metals in, 229–230
Scaling factors, 109
Scaling results from test microcosms, 436–437
Science
microcosms and, 181–186
perspectives on, and microcosm uses, 435–436
Sediment
mineral mobilization from, 205–206
role of, in new lakes, 316–318
Sediment control, of water chemistry, 307
Seeding, multiple, 7
freshwater succession with, 202
Self-organization, 7
current velocity and, 217–218
different temperatures and, 225–226
with heavy metals in Savannah River mesocosms, 229–230
maximum power and, 84–85
mechanisms of, 43–50
perspectives on principles of, 434–435
process of, 41
of stream microcosms with toxic substances, 228–229
of terrestrial microcosms, 240–242
varying light intensity and, 226–227
Self-organized life support, complex, principles for, 421–422
Self-organized mesoecosystem, complex, 419–431
Self-organizing systems, material cycles in, 62–64
Self-reinforcing diversity, 91
Seral stage or sere, 42
Sewage
estuarine microcosms treated with, 197, 199
treated domestic, estuarine ponds receiving, 287–293
Sewage microcosms, 377–389
Sewage ponds with algal photosynthesis, 384–387

Shannon-Weaver diversity index, 102, 120
Shannon-Wiener-Weaver equation, 90
Shrimp ponds, 363–368
Simulation models, 11–12
Size of organisms, 105–106
Small-scale components, linear aggregation of, 70–71
Smithsonian Everglades mesocosm, 270, 271
Smithsonian tropical reef microcosms, 301–303
Soil column studies, 252–254
Soil cores, transplanted, terrestrial microcosms containing, 252–254
Soil depression microcosms, 249
Soil function, animal activity and, 259
Soil microcosms, *see* Terrestrial microcosms
Soils
  aerobic, 234
  anaerobic, *see* Anaerobic soils
Solar emjoules, 12
Solar energy, waste microcosms and, 383
Solar heated fish-pool aquaculture, 369–375
Solar transformity, 12, 428
Solid waste microcosms, 390, 392
Space and time, hierarchy in, 108–112
Spaceship earth, 397, 398
Spatial organization, 110
Species, surplus, 106
Split-level microcosms, 143–147
Standardized aquatic microcosm (SAM), 134–140
Standardized microcosms, 58
State hyperspace, 87
State variables, 64
Steady-state chemical cycles, 70
Steady-state tracer and carrier distribution, 80
Storages, initial large, effect of, 29–31
Stored supplies from earth, 399–401
Stream consumer model, 228
Stream hierarchy, models and, 227–228
Stream microcosms, 209–233
  directions for preparing, 448–449
  self-organization of, with toxic substances, 228–229
Stress, 153–154
  catastrophic, 58
  as diversity generator, 153–154
  energy and Emergy and, 155
  external, 148–152
  toxic, 160–177
  transformity and, 155–156
  uniformly distributed, species diversity and, 154
Stress removal, 149
Succession, 41
  arrested, 213
  autotrophic, 56
  classical concept of, 41–42
  defined, 42–43
  freshwater, with multiple seeding, 202
  heterotrophic, 56
  of irradiate systems, 159
  longitudinal, *see* Longitudinal succession
  mechanisms of, 43–50
  in miniature bog, 260
  in natural vernal pools, 281–283
  patches of, 104
  of producers and consumers, 47
Successional-eutrophic microsystem, Beyers' standardized, 58–61
Successional minimodel, 42
  combining mechanisms for, 98–101
  mineral cycles added to, 84
Sulfate reduction, 260
Summation index, product index versus, 87
Surplus species, 106
Sustainable ecosystems, perspectives on, 440–442
Symbiosis, simple, 401–406
Symbiotic interactions, 155
Symbols, energy diagram, 12–14
Synthetic assemblages, 7–8

Taub microcosms, 279
Technogenic outgassing, 479
Temperatures
  different, self-organization at, 225–226
  role of, in microcosms, 37–38
Terraria, *see* Terrestrial microcosms
Terrestrial ecosystems with humans, closed, simulation model for, 422–427
Terrestrial microcosms, 26–29
  containing transplanted soil cores, 252–254
  directions for preparing, 449–451

effects of acid on, 254–256
for global perspectives, 244
homeostasis of, 242–243
minimodels of, 234–235
need for adaptive studies of, 258–259
organic chemicals in, 256–258
scale and hierarchy, 240
self-organization and succession and adaptation, 240–242
toxic elements in, 256
Test microcosms, scaling results from, 436–437
Thermal and brine microcosms, 341–354
  adaptation to drying, 343–344
  biogeochemistry, 344–345
  community metabolism, 343
  examples of, 345–354
  physiological adaptation and low diversity, 341–342
  taxonomic similarities, 342–343
Tidal sandflat waste microcosms, 389, 390
Tight-cycle productivity, 101
Time and space, hierarchy in, 108–112
Top-down models, 113
Toxic dose, effects of, 168–170
Toxic elements in terrestrial microcosms, 256
Toxic heavy metal elements, 168
Toxic stress, 160–177
Toxic substances, self-organization of stream microcosms with, 228–229
Toxicity, organic, 167
Toxicity tests, with organisms and microcosms, 161–162
Tracer, radioactive, 76–80
Tracer flow, carrier flow effect on, 79
Tracer studies in Balcones model river, 214–216
Transformity, 106, 110
  in ecological hierarchies, 165–166
  hierarchical control and, 233
  solar, 12, 428
  stress and, 155–156
Transitions
  in estuaries, 53–54
  following enclosure, 51–55
Tropical reef microcosms, 301–303

Tubificid control of benthic pore space pH, 311
Turbulence, physical, effects of varying, 333–334
Turbulence indices, 110
Turf microcosms, 358–363
Turnover length, nutrient spiralling and, 213–214
Turnover rate, 64
Turtle grass-coral microcosms, 306

Valve-controlled flows, 143
Variables, state, 64
Vernal pools, natural, succession in, 281–283

Warington, Robert, 178–179
Waste consumption units, 376
Waste microcosms, 376–396
  with larger aquatic plants, 387–389
  solar energy and, 383
  tidal sandflat, 389, 390
Waste stabilization ponds, 384
Water chemistry, sediment control of, 307
Water exchange rate, 6
Water quality studies, artificial substrates for, 300
Water surface microcosms, 202
Water treatment microcosm, 395, 396
Waters, flowing, 209
Wetland decomposition, 261, 262
Wetland microcosms, 260–271
  heavy metals in, 267
  pesticides in, 267–269
Wetlands, 260
"White mouse" microecosystem, 58–61
Winogradsky columns, 314

Xenobiotics, microbial degradation of, 320–321

Yellowstone hot springs, 346, 348

Zinc, 175, 229
  in terrestrial microcosms, 256

DATE DUE	
APR 1 7 1994	
MAR 1 3 1996	
JAN 0 3 2005	

DEMCO, INC. 38-2971